AF324917

WET GRANULAR MATTER

A Truly Complex Fluid

Second Edition

SERIES IN SOFT CONDENSED MATTER   ISSN: 1793-737X

Founding Advisor:
Pierre-Gilles de Gennes
(1932–2007)
Nobel Prize in Physics 1991
Collège de France
Paris, France

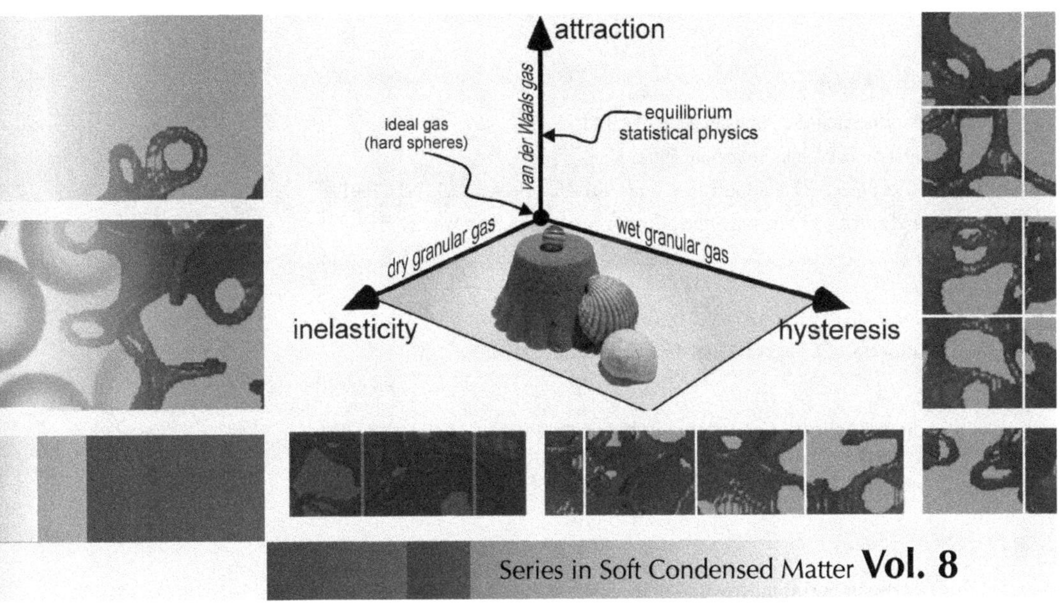

WET GRANULAR MATTER

A Truly Complex Fluid

Second Edition

Stephan Herminghaus

Max Planck Institute for Dynamics and Self-Organisation, Germany

World Scientific

NEW JERSEY · LONDON · SINGAPORE · BEIJING · SHANGHAI · HONG KONG · TAIPEI · CHENNAI

Published by

World Scientific Publishing Co. Pte. Ltd.

5 Toh Tuck Link, Singapore 596224

USA office: 27 Warren Street, Suite 401-402, Hackensack, NJ 07601

UK office: 57 Shelton Street, Covent Garden, London WC2H 9HE

Library of Congress Control Number: 2023946655

British Library Cataloguing-in-Publication Data
A catalogue record for this book is available from the British Library.

Series in Soft Condensed Matter — Vol. 8
WET GRANULAR MATTER
A Truly Complex Fluid
Second Edition

ISBN 978-981-12-8225-6 (hardcover)
ISBN 978-981-12-8226-3 (ebook for institutions)
ISBN 978-981-12-8227-0 (ebook for individuals)

For any available supplementary material, please visit
https://www.worldscientific.com/worldscibooks/10.1142/13567#t=suppl

Desk Editor: Nur Syarfeena Binte Mohd Fauzi

Typeset by Stallion Press
Email: enquiries@stallionpress.com

Dedicated to the
Nama
*indigenous people living in the Gamsberg region in
central Namibia*

in memoriam
Pierre-Gilles De Gennes and Isaac Goldhirsch

Preface to Second Edition

As the focus of interest usually shifts rather quickly in the scientific community, it is not often that a specialized textbook as the present one goes through more than one edition. Therefore, I was all the more pleased to be asked to prepare a second edition. Reading the book again, checking whether the presentation still represents the state of the art, it became clear that clarifications were necessary only in a few places.

Aside from correcting a number of typos and smoothing out clumsy wording in some places, I was able to include some new results which had been obtained in the meantime. Chapter 2 was extended by results from field trips to the Great Gamsberg, Namibia, demonstrating that mathematical results on granular flows can significantly contribute to the clarification of the geological history of an outcrop. In Chapter 3, recent results on the wetting phase diagram of rough surfaces could be incorporated, which confirm and illustrate the theory developed in that chapter. Finally, our remarks in Chapter 5 on the existence of general principles of structure formation in systems far from equilibrium could again be sharpened, to the effect that the quest for such principles must be regarded as ill-posed.

Göttingen
February 2023

Preface to First Edition

There is no need to stress the paramount importance of granular matter. The ground we dwell on, which sometimes scares us with devastating land slides, is the granular top layer formed by weathering and erosion of the earth's crust. Most of the raw materials which make our economical world go round are solids and hence come in granular form. Their skilful handling in process technology, including the minimization of energy consumption, requires a solid grasp of the mechanical properties of granulates. A considerable body of literature on granular matter has thus been produced over time, chiefly in the engineering sciences. Phenomenological descriptions of various aspects of granular systems have been set up and developed to considerable maturity. This allows us to construct dams and dikes, and design industrial plants dealing with granular matter, with some reasonable success.

As physicists, however, we are eager to penetrate the phenomenological facade and unravel the deeper principles and mechanisms hidden behind, driven by both curiosity and the hope for finding a complete picture. In 'granular physics', this endeavor was hampered for a long time by the immense complexity of the systems under study. An illustrative example (out of many) is the problem of random packings, which still challenges physicists and mathematicians alike, and whose possible relation to glass forming liquids is a matter of ongoing debate. Similarly, attempts to find a physically satisfactory description of *wet* granular matter have been scarce: the presence of a twisty liquid–gas interface spanning the interstitial space of the already challengingly complex granular pile seems to render the wet system forbiddingly complicated. This is particularly unsatisfactory, since moisture is almost omnipresent, and perfectly dry granulates merely represent a special case with limited relevance.

Two recent developments have ameliorated this situation. First, it has been recognized that the energy scale provided by the surface tension of the liquid allows to treat the system as quasi-thermal, enabling the use of a number of concepts from the physics of phase transitions and critical phenomena. Second, the advent of powerful computers has enormously sharpened our view into simulated versions of many-body systems far from thermal equilibrium. As a consequence, a lot of new insight into wet granular systems has been gained since the turn of the millennium. This was strongly supported by the finding that a number of their properties may be understood on the basis of surprisingly simple microscopic concepts, such that many apparent degrees of complexity actually turned out to be largely irrelevant. The field of physical research in wet granular matter meanwhile spans aspects as different as the rheology of dense granular paste, collective behavior far from thermal equilibrium, universal aspects of liquid interface morphologies within wet granular packings, and many more.

The present book tries to give a state-of-the-art overview on the physics of wet granular matter. It is written for those who have a solid background in physical sciences and want to understand wet granulates from a fundamental point of view. Microscopic models are presented which allow to understand the basic mechanisms, and how they synergize to yield the characteristic properties of wet granular matter. The book should be well suited, e.g., to serve as a text book for a one-semester course in graduate school.

If you have to construct a building on muddy ground, or design a technical solution for handling a particular granulate, you may refer to the vast engineering literature. If, however, you are interested in basic mechanisms and fundamental implications, or even strive to further our understanding of granular matter in general, you may find the present book useful.

I am grateful to my colleagues and friends, who supported me with numerous hints and inspiring discussions. Some read through chapter preprints and returned to me with invaluable comments. I feel particularly indebted to my wife and family, for their patience with 'the book' lurking through our lives for about a year. It does no more!

Göttingen
April 2013

Acknowledgments

I am grateful to my colleagues and friends who supported me with numerous hints and inspiring discussions. I am particularly indebted to Reinhold Wittig and Hubertus Porada for introducing me to the Great Gamsberg during several excursions in Namibia, and for teaching me geology.

Contents

Chapter 1

Introduction

Fig. 1.1 A sand sculpture[1] at the 'Sandsations' festival in Berlin, 2010. The sculpture is approximately 3 m high.

From our childhood, we remember the pasty consistency of wet sand, out of which we baked our first 'cakes' in the sand box; soft enough to be squished and formed by our little hands, but sufficiently stiff to stand bold against gravity, and hence maintain the shapes we imposed on it. We were more than satisfied with hours and hours of the sheer material experience, the ductile substance on our palms, breathing the humid mineral smell. As grown-ups, in contrast, we like to embark on larger scale projects, exploiting the peculiar properties of wet sand to decorate beaches with sophisticated sculptures. Every summer there are international sand sculpture competitions, where awesome artwork is being created. The photo displayed in Fig. 1.1, for instance, was taken on the 'Sandsations' festival in Berlin in 2010, on the banks of the Spree river.

Clearly, aside from the color, wet sand has few in common with its dry counterpart, as found abundantly in deserts. Dry sand will easily trickle through our fingers. No stable shape can be imposed upon it. With some resemblance of a liquid, it forms wave-like structures in the wind, as one can see from the dunes as well as the ripples in the foreground shown in Fig. 1.2. When slightly agitated, it tends to make a horizontal boundary

---

[1] Entitled *Climate change is an inconvenient truth*, by Sudarsan Pattnaik.

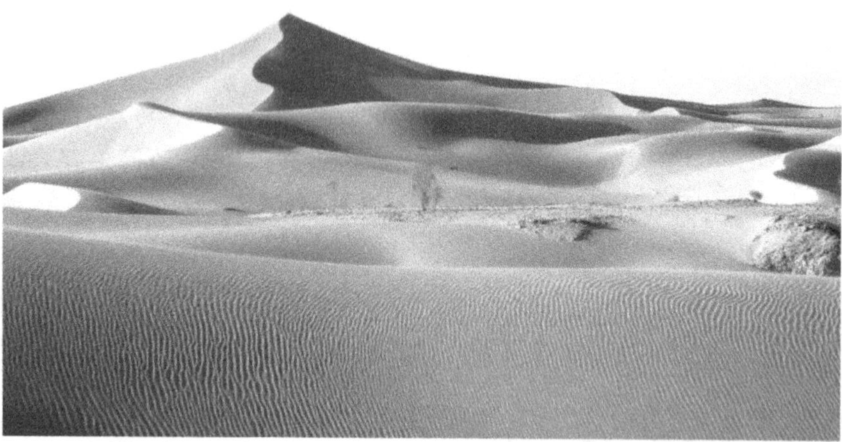

Fig. 1.2 Sand dunes in a desert, the paradigmatic example of granular matter. The slopes never exceed a certain characteristic angle: steeper slopes would be unstable and flatten spontaneously through avalanches. Dunes are formed by wind, which is the dominant force shaping this landscape. The same is true for the ripples in the foreground (image taken with permission from www.picture-newsletter.com).

with the space above, just as an ordinary liquid, like water. The addition of just a minute amount of the latter, however, transforms it into the ductile, moldable material of the sand box. Obviously, wet granular matter qualifies as a member of the 'soft matter family'. The dramatic change in properties between the dry and the wet state of granular matter is fascinating, and will be the main focus of this book.

1.1 The significance of wet granular matter

The practical significance of wet granular matter goes of course well beyond the construction of sand sculptures. Most industrial raw materials are solids and hence come in granular form. The processes into which they feed involve their being conveyed, mixed with liquids, kneaded, or cast in molds. Many granulates actually come as fine powders, and would create enormous amounts of dust if used as received. It is therefore a standard procedure to subject such powder to granulation, a process in which the fine grains are wetted and subsequently agglomerated with the help of wetting forces to form larger granules, which are more pleasant to handle. Aside from the optimization of the resulting products, it is of major (and steadily growing) importance to minimize the energy consumption of all of the above mentioned procedures. In order to appropriately engineer these processes, a deep understanding of the mechanical properties of granular matter, at variable liquid content, is indispensable.

An example from industrial practice which is amusingly close to the construction of sand castles is the formation of black-cole cakes feeding into coking ovens, as shown in Fig. 1.3. A large fraction of the black cole, as it is mined underground, does not yield a coke which is sufficiently stable mechanically to be used in blast furnaces. Empirically, one has found that coking instead a consolidated block of fine black cole powder delivers quite satisfactory results. The black cole thus needs to be milled into a fine granulate, moistened, and formed into a 'cake' before coking. Owing to certain restrictions from the coking process, however, these cakes must be more than about 50 cm wide and no less than 6 m high. Furthermore, they must withstand their being inserted upright into the oven. It is clear that the mechanical stability of these block-shaped 'cole powder sculptures' is of central importance for the unobstructed operating sequence of the whole coking process.

Fig. 1.3   A slab of black-cole powder for feeding a coking oven.[3]

Granular matter has of course been important to mankind well before industrial processes were developed. Almost all parts of the earth's crust we can reach are covered with the granular regolith[4] on which we dwell and build our houses and cities, relying and depending on its mechanical stability. If we want to mitigate, or even reliably predict, such devastating events as land slides or mud flows, we need to study the dynamical behavior of wet granular matter in detail.

In doing so, we should be prepared to deal with an immense complexity.

---

[3]Private photograph by F. Steiner (Dillingen/Saar), on occasion of launching the first coking plant in China in 2009.

[4]From ancient greek $\varrho\acute{\eta}\gamma\mu\alpha$ = fracture and $\lambda\acute{\iota}\vartheta o\varsigma$ = stone. It denotes the fractured form of rocks and minerals produced by erosion of the bedrock.

First of all, in most granular systems not any two grains are exactly equal in size and shape. This is in strong contrast to other many-body systems in physics as, e.g., common molecular liquids, where the mutual identity of the constituents may even play a decisive role.[5] Second, a dense wet granulate is interspersed with a twisty liquid–gas interface, which connects the individual grains and exerts attractive forces onto them. The energy landscape of this interwoven liquid surface, with respect to the relative particle positions, determines most of the dynamic aspects of the material, and it is hard to imagine how this enormous degree of complexity may ever be dealt with successfully. Quite surprisingly, we will find that already quite simple concepts and model assumptions will enable us to describe a large number of important features of such materials. This concerns not only wet granular gases, where the wet grains are treated in a statistical-physics inspired fashion (much like molecules in a regular liquid) but also to wet granular 'piles' of grains, the mechanical properties of which are found to exhibit a few surprising commonalities. The development and presentation of these findings will be the main task of our journey through the dunes and canyons of granular physics.

1.2   Energy scales

In order to delineate the range of parameters we will be dealing with, it is useful to compare the various energy scales involved in wet granular systems. As the grains we are dealing with are macroscopic objects, the only attractive forces which must be taken into account are gravitation and electromagnetic forces. Throughout this book, we will assume that the grains carry no electrostatic charge. This can be easily secured for humid or wet granular systems by using water or an aqueous fluid as the wetting liquid, which provides enough electrical conductance to prevent electrostatic charging. The only electromagnetic interaction of interest is then the van der Waals force. For spherical grains of equal size, which we use as a paradigmatic model for the sake of simplicity, the van der Waals binding energy is given by [3]

$$W_{vdW} = \frac{AR}{12s},\qquad(1.1)$$

---

[5] In the theory of superfluidity, for instance, the identity of particles is essential in that it ensures the existence of a many-body wave function, which is symmetric under exchange of any two particles [1, 2].

where A is the Hamaker constant, R is the radius of the grains, and s is their surface separation.[6] The Hamaker constant is of order 10^{-20} J, for silicon oxide it is approximately 6×10^{-20} J [3].

In practice, the grains will always bear some roughness on their surfaces, which provides a natural scale for the separation of the surfaces of the idealized, spherical grain models. s will thus correspond to the depth of the surface roughness, δ. The latter is at least on the order of a micron, usually larger than that. For the purposes of the rough estimates we are aiming at here, it is sufficient to assume that δ scales with R. We set $\delta \approx 0.1R$. The resulting van der Waals energy is then independent of the grain size.

Gravitation gives rise to an energy of interaction given by

$$W_{grav} = \frac{\mathsf{G}m^2}{2R} = \frac{\mathsf{G}}{2R} \left(\frac{4\pi \varrho_s R^3}{3} \right)^2 , \tag{1.2}$$

for two spherical grains in contact, where m is the mass of a single grain, ϱ_s the mass density of the solid they are composed of, and G is Newton's constant of gravitation. Obviously, this depends very strongly on grain size.

The attractive force mediated by the liquid adsorbed close to the contact point of the grains (capillary force) is given by the circumference of the grains times the surface tension of the liquid, as we will see later. We assume here that the liquid is water, with a surface tension $\gamma = 72 \cdot 10^{-3}$ N/m. Another force of relevance is the acceleration due to gravity (on earth), which we need to multiply with a length in order to compare it to energies. As R is the only natural length scale here, this is what we choose. For providing a comparison with kinetic energy scales, we must refer to a typical velocity. We choose 1 m/s as a convenient laboratory scale velocity which appears in many typical granular physics experiments, such as granular avalanches, mixers, etc.

The results of all of our estimates are compiled in Table 1.1 and plotted in Fig. 1.4. As we can see, a particularly interesting region of grain sizes is where the capillary, inertial, and gravity forces come within the same order of magnitude. This is the case at about one millimeter grain size, which is a typical size for regular sand grains. The grey shaded region corresponds to what is officially called sand, namely clastic[7] mineral granulate with diameters from 63 μm to 2 mm [4]. The fact that van der Waals forces, gravitation, and thermal forces yield for sand a set of energies of similar

---

[6]Equation (1.1) represents the non-retarded result. In case of retardation, van der Waals forces are weaker. For more details, see Chapter 3.

[7]From ancient Greek $\kappa\lambda\acute{\alpha}\omega$, to shatter.

Table 1.1   Characteristic energy scales involved in the physics of wet granular matter.

Type of energy	Formula	Evaluation in J
van der Waals ($\delta = 0.1R$)	$AR/12\delta$	$5 \cdot 10^{-20}$
Gravitation (grains in contact)	$Gm^2/2R$	$5.3 \cdot 10^{-3} \cdot (R[\mathrm{m}])^5$
Capillary force	$2\pi R\gamma$	$0.44 \cdot (R[\mathrm{m}])^2$
Gravity on earth	mgR	$1.24 \cdot 10^5 \cdot (R[\mathrm{m}])^4$
Kinetic (1 m/sec)	$\frac{2}{3}\pi\varrho_s R^3 v^2$	$630 \cdot (R[\mathrm{m}])^3$
Thermal energy	$k_B T$	$4 \cdot 10^{-21}$

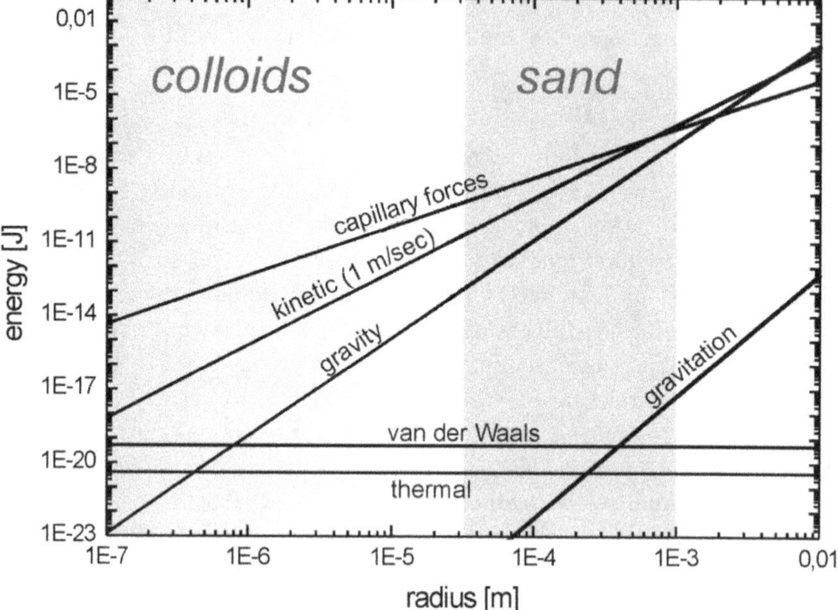

Fig. 1.4   The characteristic energy scales involved in wet granular systems, plotted as a function of grains size (radius of idealized spherical grains). Sand (dark grey shade) as a widely encountered granular system is clearly singled out due to the similarity of three easily controllable energy scales. At small grain size, capillary forces become dominant.

order (around 10^{-20} J) as well is interesting in its own right. But for our purposes, it is sufficient to mention that these are completely negligible with respect to the three other forces. Hence we directly see the reason why sand is the paradigm example for discussing wet granular matter: its grain size is just of the right order of magnitude for an interesting interplay of three relatively well controllable types of forces.

1.3 Typical questions to be asked

As an example for the character of questions that shall be asked from the physics perspective, let us consider a land slide in wet soil, such as the one depicted as a bird's eye view in Fig. 1.5. First of all, we see that we are dealing with a threshold phenomenon: some part of the slope has massively moved downwards, whereas in the adjacent parts, where conditions were certainly similar, nothing visible has happened at all. Furthermore, it is instructive to study the remains of the trail, which is clearly visible as a straight line to the left and right of the land slide region. On the slide region itself, it has been strongly deformed into a shape similar to a parabola (cf. right panel). This shows that the slope has not come down as a more or less solid piece, sliding on a fault zone deeper inside the hill, but that it rather has liquefied, and crept down like a viscous fluid. Although some individual blocks of soil are still discernible, the overall liquid-like behavior is obvious.[8]

Fig. 1.5 Left: Massive land slide at La Conchita, California, which occurred in spring 1995 after heavy rain (image courtesy of U.S. Geological Survey [5]). Right: Same image as left, but with the solid line indicating the position of the trail after the land slide. The dashed line indicates the (very probable) position of the trail before the event.

[8]Geoscientists distinguish among a large number of different types of land slides. For details, see [6].

It is tempting to suspect some kind of a discontinuous phase transition (from a solid to a liquid state) as the underlying phenomenon. This calls for an investigation with the techniques and concepts of statistical physics. A major challenge is then to find the appropriate simplifications of the system which enable a successful treatment but still retain the mechanism relevant for the behavior observed in the 'real' system.

Aside from the more practical aspects represented by land slides, phase transitions in wet granular matter are by themselves of particular interest from the point of view of statistical physics. Let us consider what happens between adjacent grains of a wet granular system when they come into solid contact. This is sketched in Figs. 1.6(a) and 1.6(b). As soon as contact is established (b), surface tension will act such as to minimize the total area of the free liquid surface. The result is a capillary bridge, as sketched in Fig. 1.6(c). When the grains are removed from each other, it is clear that this capillary bridge will rupture when the grains surfaces have reached a certain distance (Fig. 1.6(d)). Again by virtue of surface tension, and as well by wetting forces, the liquid will tend to homogeneously cover the grains again, as in Fig. 1.6(a). This process, which will take place in an agitated wet granular medium many times in many places, represents a cycle. As indicated by the arrows, most of the steps of which the cycle is comprised are irreversible. Solely the transition between c and d, which just consists in elongation of the capillary bridge, can be considered reversible (as long as the viscous dissipation in the liquid is neglected). Hence if we

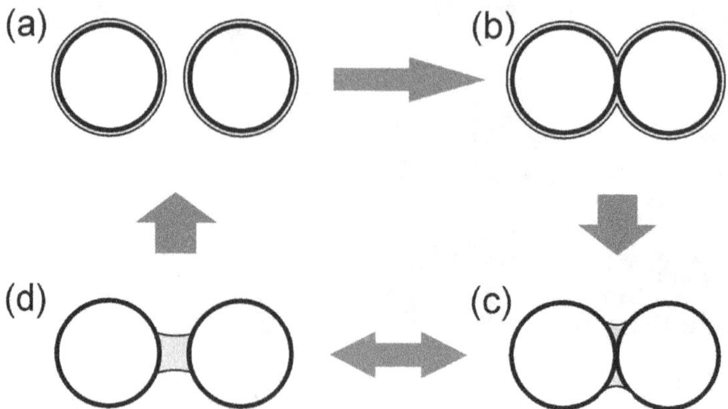

Fig. 1.6   The formation (a → b → c) and rupture (d → a) of a capillary bridge is an intrinsically irreversible process. Hence an agitated wet granular system can be viewed as a large number of coupled sub-systems each of which exhibits irreversible dynamics.

consider a wet granulate, conceptually, as a system of many interacting capillary bridges, we may say that it consists of many coupled sub-systems each of which in itself has an irreversible dynamics, or, in other words, breaks detailed balance on the microscopic scale.

This bears some analogy to systems of excitable actors, like the cells of a heart muscle, colonies of the slime mould *Dictyostelium discoideum*, or even the visitors in a soccer stadium. In all of these cases, we are dealing with many coupled actors which can be in mainly three distinct states: an excitable state, an active state, or a refractory state. Going from one to the other represents an irreversible cycle as well. In the heart muscle, the coupling of the 'actors' (the cells) may lead to fibrillation, in the *Dictyostelium* colony to the formation of a stalk, and in the football stadium to the 'Mexican wave' [7]. It is by no means clear whether in agitated wet granular media one may find indications of similar phenomena, but it is obvious that they belong to the same class of non-equilibrium many-body systems, when viewed from the appropriate angle of perspective.

That granular media are capable of generating complex patterns and structures is well known. The largest-scale examples are the collapses of primordial accretion disks into planetesimals and planets, as well as some of the fine-structure one finds in Saturn's rings (cf. Fig. 1.7). These consist of pure water ice particles with sizes from centimeters to a few meters, distributed radially over a few hundred thousand kilometers, but confined in the axial direction to less than a kilometer. Although some of the complex structure of the density distribution within the rings is due to gravitational interactions with the moons, some are direct consequences of the nonlinear interactions of the grains with each other.[9] It is interesting in its own right that there is at all a propensity in the ingredients of the universe to form structures, from the formation of galaxies to the emergence of life. We see that granular matter touches on some rather fundamental questions, and may provide suitable model systems for gaining some further insight into the fascinating field of collective phenomena far from equilibrium.

The most obvious focus of research, however, is connected to the many possible applications of granular matter. When fine powers are wetted and agglomerated to form larger grains, which can then be handled without producing lots of dust, one is dealing with a wet granular gas, and with

---

[9]It has been shown that a granulate consisting of ice particles may behave effectively as a wet granulate due to temporary melting of the ice at the points of contact when grains collide [8]. However, this is probably not important for Saturn's rings because its ice particles are much colder than the freezing temperature.

Fig. 1.7 Saturn's rings, which consist of ice particles, are another example of a granular system. While some of the inner structure of the rings is directly due to the impact of the moons (visible as the three bright dots), others are a consequence of the strongly nonlinear dynamics of the granular cloud (image kindly provided by NASA, USA).

the mechanisms discussed in connection to Fig. 1.6. It is a particularly appealing feature of wet granular materials that they are easily accessible to experiments, intimately related to a vast range of applications, and at the same time provide model systems for cutting edge physical problems. This does not only concern settings in which the relative displacement of grains is relevant, such as, e.g., in mixers or in agglomeration. The morphology of a liquid interface accommodating the geometry of a granular pile is at the heart of the physics of oil deposits in sand stone reservoir rocks, and of major relevance for oil recovery as well as for any future use of former oil reservoirs.

Aside from such straightforward ideas, the hierarchy of length scales involved in wet granular matter is inspiring the conception of novel technologies. For example, a new avenue towards self-assembling, template-assisted lithography techniques has been proposed lately. It has been demonstrated that assemblies of spherical grains can be used as a template supporting well-defined morphologies of aqueous nano-particle suspensions, which upon drying leave the nano-particles in defined shapes on a substrate [9] (cf. Fig. 1.8). These may then be further processed to form nano-wires or other building blocks for cutting-edge parallel lithography. The substrate may be either a planar support, or just the grains themselves, representing an

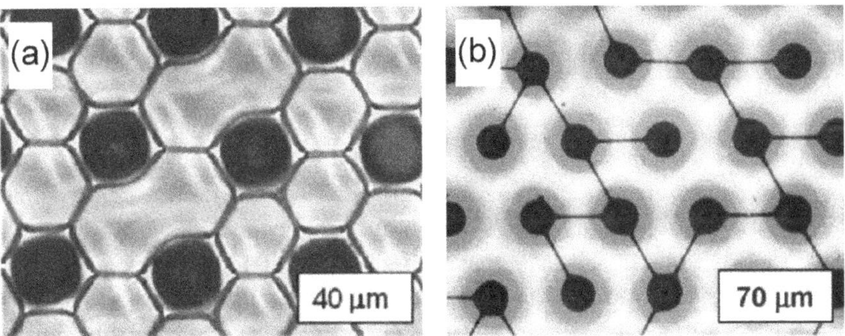

Fig. 1.8   Evaporative lithography by means of foam films within wet granular assemblies [9]. (a) Bottom view onto a glass slide on which a monolayer of spheres resides which is wetted by a liquid with a suitable surfactant and solute. As the liquid dries, the liquid forms foam films spanning between adjacent spherical particles and the glass support. (b) After the liquid has dried, the solute has assembled in thin lines, extending from one particle-glass contact to the next (reproduced with permission from [9]. Copyright (2009) American Chemical Society).

immediate extension of lithography into the third dimension. As the fundamental understanding of wet granular matter is further advanced, it is to be expected that many more such developments will surface in the future.

1.4   How we shall proceed

In order to organize our approach, it is useful to appreciate that a wet granulate represents a ternary composite, consisting of

(1) the grains, which we assume to be solid, impenetrable, and of uniform size and shape,
(2) the added liquid, which we assume to be incompressible and to wet the grain surfaces well,
(3) the gas phase, which we assume to homogeneously fill the interstitial space which is left by the grains and the liquid.

Note that this description contains already a number of considerable simplifications. First of all, as already mentioned above, there are not any two sand grains in the sculpture depicted in Fig. 1.1 which are exactly equal in shape and size, and their elastic constants are of course finite. The added liquid (water, in this case) is not perfectly incompressible, and the contact angle between the liquid and the grain surfaces will in general be non-zero.[10] In search for a model of wet granular matter which is as simple as possible, but still exhibits all important effects to be described, we will try to find ways to neglect most of these ramifications. As always in physics, it remains a central task to investigate which degree of complexity must be taken into account under what circumstances. The goal is to come up with feasible descriptions of the system, which are useful for generating quantitative and useful predictions in scientific and technological settings. We will see that a large part of the properties and phenomena connected to wet granular matter may in fact be treated successfully by means of surprisingly simple concepts.

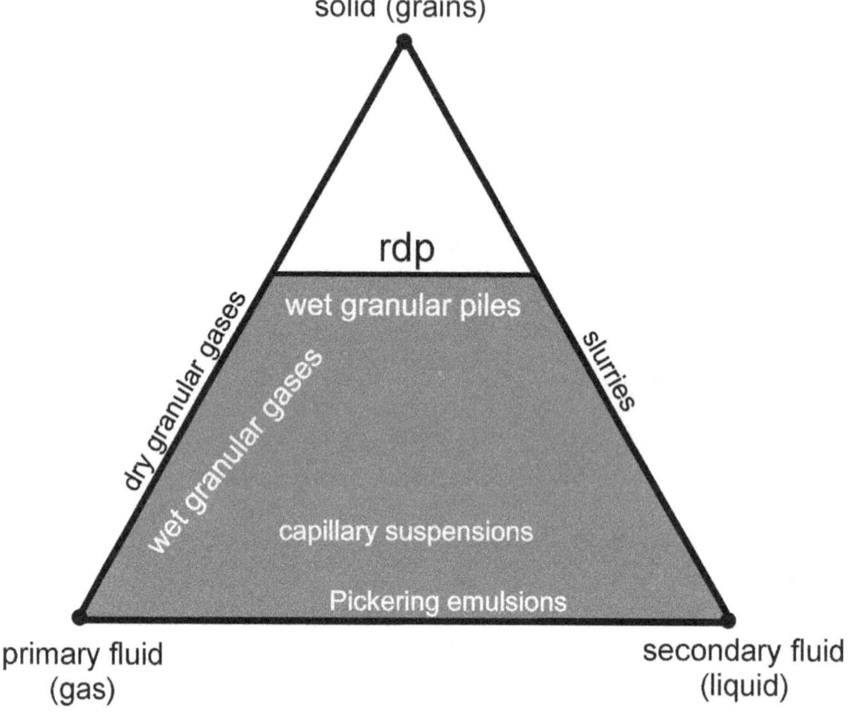

Fig. 1.9   Wet granular matter in a mixing triangle representation (rdp: random dense packing).

The topics we will discuss are laid down in Fig. 1.9, which represents a 'mixing triangle' for the three phase system consisting of the solid grains,

---

[10]As we will see, it will in most cases not even be well defined.

the interstitial gas (primary fluid), and the wetting liquid (secondary fluid). The grey shaded area indicates the accessible region, which is delimited towards the solid volume fraction by the maximum ('random dense') packing density of the grains. Immediately below this line, we find the wet granular piles, such as the sand sculpture in Fig. 1.1. All of the systems named in the diagram will be discussed, some at length, like the wet granular gases, some only shortly, because they just repeat the application of principles which are discussed in other context already.

As the ternary system bears a significant degree of complexity, its treatment requires preparation. We will therefore start by considering binary systems, which consist of the grains and just one fluid. They are found along the slanted sides of the mixing triangle. Since it makes a substantial difference whether the fluid phase is formed by a (compressible) gas or an (incompressible) liquid, either one deserves attention in its own right. The impact of this difference, as well as other important aspects of granular dynamics, will be discussed in Chapter 2, after a short introduction to the properties of the grains themselves. In Chapter 3, we will introduce basic concepts for describing the mutual interactions of all three phases, by means of a tutorial introduction to interfacial forces and wetting phenomena. Chapter 4 will be dedicated to the discussion of capillary bridges as the basic elements of wetting-induced inter-granular forces. In Chapter 5, we will apply these concepts to the dynamics of wet granular gases. Phase transitions far from thermal equilibrium will be discussed, with some emphasis on their possible role as model systems for non-equilibrium collective dynamics in general. Chapter 6 will then conceptually turn back to Fig. 1.1, as representing a dense wet granular system. The complex liquid morphology of the liquid–gas interface which develops in the interstitial space between the grains will be discussed in detail, and its relation to the mechanical properties of the pile will be derived. It will be shown how a close look at these morphologies allows to predict the mechanical properties of the pasty granulate, with a few surprisingly simple results. Aspects of liquid transport within the granular pile complete the picture. In Chapter 7, we will finally discuss not only the physics of sand sculptures, but also a few phenomena which lead us a bit outside the frame of description adopted here, but are frequently associated with wet granular systems.

Generally, we will not go very deeply into matters which have already been presented elsewhere in the textbook literature. Derivations in kinetic theory, for instance, can be looked up in the many well-written books on that topic, and we should not try a duplication here. The section *Further reading*, which can be found at the end of each chapter, is meant to guide the reader to helpful literature wherever possible and appropriate. Derivations of rather novel results, however, are presented at full length, in order to give the reader a chance to follow them in all detail if desired.

Chapter 2

Grains and Granular Fluids

Fig. 2.1 An hour glass, symbol of the flow of time and illustration of the fluid character of dry granular matter (image courtesy of Mike and Heather Fitz, UK).

Judging from the ease at which it quietly runs through the orifice of an hourglass, or forms wandering dunes in the wind like waves on the ocean, dry sand may be legitimately called a fluid. It does not maintain any shape imposed on it, and exhibits a strong tendency to form a horizontal boundary to the space above, similar to any ordinary liquid. When one tries to develop a physical description of dry sand, or any other similar granulate, it therefore suggests itself to take a similar path as with molecular fluids: as for the latter, we are dealing with a system of very many similar particles, which interact by repulsive forces in collisions, and move ballistically between these collisions. Hence the kinetic theory of dry granular gases, which we will outline now, closely resembles the theory of molecular fluids with vanishing cohesive forces. The limitations of this approach, however, are significant, and will be subsequently discussed.

2.1 Grains

2.1.1 *Kinetic theory*

Exploiting the fact that our system consists of many particles, we can make use of some standard techniques of statistical mechanics. Just as in the van der Waals theory of fluids, we start by neglecting rotational degrees of freedom, and discuss their impact later separately. Let therefore N be the number of grains, and q_i $(i = 1, \ldots, 6N)$ their center of mass coordinates and conjugated momenta. If the particles are interacting via frequent collisions, we can assume a rapid distribution of energy and momentum throughout the system, irrespective of initial conditions. The distribution of energy among the degrees of freedom of the system is then given by the equipartition theorem,

$$\left\langle q_i \frac{\partial \mathcal{H}}{\partial q_i} \right\rangle = \frac{1}{\beta_0} \quad \forall i, \tag{2.1}$$

where $\mathcal{H} = \mathcal{H}(\{q_i\})$ is the Hamiltonian of the System and the pointed brackets indicate the time average.[1] β_0 is the Lagrange multiplier of energy conservation, which appears when the distribution with highest probability is determined. In classical statistical mechanics, $\beta_0 = 1/k_B T$, where k_B is Boltzmann's constant. In the field of granular matter, one defines the granular temperature as $T_g = 1/\beta_0$, and the unit is Joule instead of Kelvin. From Eq. (2.1) we immediately obtain the well-known result that each degree of freedom which enters quadratically in the Hamiltonian, such as the position and momentum coordinates, receives an energy of $\frac{1}{2}T_g$ on average.

The analogy to molecular fluids can be pushed much farther [11, 12]. If N is sufficiently large, it appears reasonable to come up with locally averaged quantities, such as local number density, $n(\mathbf{r}, t)$, local velocity, $\mathbf{v}(\mathbf{r}, t)$, or $T_g(\mathbf{r}, t)$, which may be used in a continuum description of the system. Note that T_g refers only to the random motion [13], while $\mathbf{v}$ refers to its average over many grains in some neighborhood. In terms of these quantities, we can write down mass conservation as

$$\frac{\partial n}{\partial t} = -\nabla \cdot (n\mathbf{v}), \tag{2.2}$$

and the force balance as

$$nm \left(\frac{\partial}{\partial t} + \mathbf{v} \cdot \nabla \right) \mathbf{v} = -\nabla \mathbf{P}, \tag{2.3}$$

---

[1]A particularly clear derivation of this general result has been given by Tolman [10].

where m is the mass of a grain and $\mathbf{P}$ is the pressure tensor. The components of the latter are

$$P_{ij} = p\delta_{ij} - \eta\left(\nabla_i\mathsf{u}_j + \nabla_j\mathsf{u}_i - \frac{2}{3}\delta_{ij}\nabla\cdot\mathbf{v}\right), \qquad (2.4)$$

where δ_{ij} is the Kronecker symbol, η is the viscosity, and p is the hydrostatic pressure. For isotropic pressure, we have $P_{ij} = p\delta_{ij}$.

A viscosity shows up just as in molecular fluids, since momentum is diffusing here as well (through inter-particle collisions). The corresponding diffusivity is given by the typical velocity, $\sqrt{2T_g/m}$, times the mean free path between collisions,

$$l_{\mathrm{m}} \approx \frac{1}{8\pi R^2 n}, \qquad (2.5)$$

where R is the radius of the (assumed spherical) grains. Using the solid volume fraction, ϕ, this can be expressed as $l_m \approx R/6\phi$. The viscosity can be estimated by multiplying this diffusivity with the mass density of the granular gas, nm. This yields

$$\eta \approx \langle v\rangle l_{\mathrm{m}} nm = \frac{\langle v\rangle nm}{8\pi R^2 n} = \frac{\sqrt{2mT_g}}{8\pi R^2}, \qquad (2.6)$$

where v is the magnitude of the individual grain velocity. A more accurate result can be obtained by means of Chapman–Enskog theory [12], with the result

$$\eta = \frac{5}{64R^2 g_c}\sqrt{\frac{mT_g}{\pi}}, \qquad (2.7)$$

where R is the radius of the particles and g_c is the value of the two-point correlation function taken at particle-particle contact. A widely used expression for the latter,

$$g_c \approx \frac{2-\phi}{2(1-\phi)^3} \qquad (2.8)$$

has been given by Carnahan and Starling [14]. It provides a reasonable approximation as long as ϕ remains well below close packing. In the dilute limit ($\phi \to 0$), we find that the result of the very crude estimate of Eq. (2.6) is accurate to within about 20%.

Let us next have a closer look at the particle–particle interactions which come into play during collisions. When two grains are brought into contact, their surfaces are deformed in the contact region according to their finite elastic modulus. Assuming a spherical shape of the grains at least around the contact point, the relation between the amount of deformation and the compressive force has been first derived by H. Hertz. We will here just quote the results, detailed derivations can be found in the literature [15]. The contact area where the material is deformed is a circular region with radius $(\kappa RF/2)^{1/3}$, in which $\kappa = 3(1-\nu^2)/2Y$, and ν and Y are the Poisson ratio and Young's modulus of the grain material, respectively. If we define the separation of grain surfaces, s, as the mutual distance of the centers of mass of the grains, minus the value of that distance at contact, $-s$ can be taken as a deformation coordinate. It is then found [16] that

$$-s = \left(\frac{2}{R}\right)^{1/3} (\kappa F)^{2/3}, \tag{2.9}$$

or, if we resolve for F, the elastic force law is

$$F_{\text{el}} = \frac{1}{\kappa} \sqrt{\frac{R}{2}} (-s)^{3/2}. \tag{2.10}$$

We can finally integrate this expression to obtain an interaction potential,

$$U_{\text{el}} = \frac{2}{5\kappa} \sqrt{\frac{R}{2}} (-s)^{5/2}. \tag{2.11}$$

When the energy stored in the deformation is averaged only over the times during which an impact takes place, we find according to Eq. (2.1) an average deformation energy of $\frac{2}{5}T_g$.

So far, it may appear straightforward to derive all of the physics of granular materials from those of other many particle systems, such as regular liquids, by just re-scaling the size and mass of the particles, and replacing molecular interactions, i.e., the Lennard-Jones potential, with the potential given in Eq. (2.11). There are, however, three major differences which distinguish the grains from the molecules constituting a regular gas or liquid, which in fact turn out to be of fundamental importance: They concern the dissipative character of particle-particle interactions, the shapes of the grains, and their mass. In what follows, we will discuss these three aspects one after the other.

2.1.2 *Dissipative collisions*

We have seen above that a collision between two grains is accompanied by a temporary deformation of their surfaces around the point of first contact, which gives rise to the repulsive interaction potential in Eq. (2.11). The description by means of a potential suggests that after the collision all deformation energy is fed back into the motion of the impacting grains. A closer look, however, reveals that this is not the case. The deformation displaces the atoms around the point of contact with respect to each other, and such deformation propagates as sound in a solid body. It is thus inevitable that part of the impact energy is radiated away from the point of contact into the grains as sound pulses, which are scattered from inhomogeneities and reflected from grain surfaces. Within a matter of a microsecond or so, the energy of such pulse will have been distributed among all atomic degrees of freedom within a grain, and can never be retrieved again into the center of mass motion: it has been converted into heat. It is intuitive to assume that the intensity of the pulse will be roughly proportional to the impact energy, which immediately leads to the assumption that at each collision a certain percentage of the impact energy is lost into the atomic degrees of freedom inside the impact partners. Hence one speaks of *incomplete restitution* of the energy back into the center-of-mass kinetic energy of the impact partners.

Conventionally, one considers the momenta instead of the energies here. The relative momentum of the impact partners after impact, p_f, will be less than that before impact, p_i, and the ratio of these momenta (taken in the center of mass frame of reference) is called the *restitution coefficient, $\varepsilon :=$ p_f/p_i.* This is an important quantity for the characterization of granular materials, as it quantifies on the microscopic scale the most important fundamental difference between a granular gas and a molecular system. Table 2.1 lists several observed values of restitution coefficients for some common materials for moderate impact velocities[2] of a few cm/s.

Table 2.1   Typical values of the restitution coefficient, ε, for a few common materials at moderate impact velocities.

Materials	ε	References
Soda lime glass (both impact partners)	0.97	[17]
Nylon (both)	0.97	[18]
Steel (both)	0.95	[19]
Steel/glasss	0.95	[20]
Cellulose acetate (both)	0.87	[17]
Rubber/aluminum	0.86	[21]
Various rock samples	0.80 - 0.92	[22]
Ice (both)	0.6	[21]

---

[2]A closer look reveals that the restitution coefficient depends upon impact energy. This will be discussed in some detail in Chapter 5.

For a system with elastic collisions ($\varepsilon = 1$), such as a molecular gas, it is straightforward to compute equilibrium distribution functions, like the velocity distribution, by the standard techniques of statistical physics. As we introduce some inelasticity, which we do when setting $\varepsilon < 1$, it turns out that these techniques can still be applied in some range [12]. In order to gain a feeling for the impact of dissipative collisions on the velocity distribution, we show the analytical result [12] in Fig. 2.2 for three values of ε. The solid curve corresponding to $\varepsilon = 1$ represents the Maxwell–Boltzmann distribution. As we can see, even for quite dramatic dissipation ($\varepsilon = 0.1$) there is no qualitative change in the distribution, and we might expect that the impact of inelasticity on the granular gas can be restricted to small corrections. That this is not generally the case is revealed when we consider fluctuations in the density field, as we will do next.

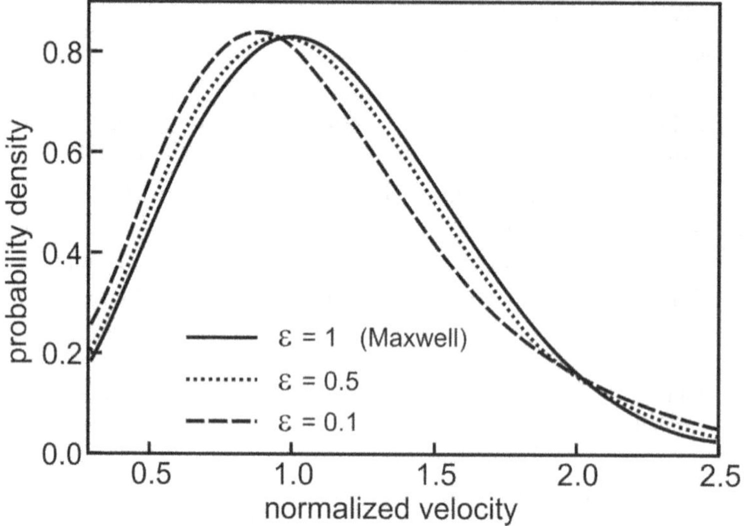

Fig. 2.2 The velocity distribution for three different values of the restitution coefficient, ε. Qualitatively, the change is not dramatic, even for values as low as $\varepsilon = 0.1$.

With the granular temperature as defined above, the energy lost in a collision scales as $\Delta E \propto (1 - \varepsilon^2)T_g$, and the collision frequency scales as $T_g^{1/2}/l_\mathrm{m}$. For the rate of cooling due to the inelastic collisions, we immediately obtain

$$\frac{dT_g}{dt} \propto -\frac{(1 - \varepsilon^2)T_g^{3/2}}{l_\mathrm{m}}. \tag{2.12}$$

This directly leads to Haff's law [23], which states that the granular temperature decays algebraically as

$$T_g \propto (1 + Ct)^{-2},\qquad(2.13)$$

in which C is some constant. Although this simple prediction has been made three decades ago, it has only recently been corroborated experimentally. The difficulty lies in the collisions with the container walls, which take place inevitably after short time, thus obscuring the cooling effect inherent to the granular gas. In order to observe the scaling of the granular temperature predicted by Eq. (2.13), observation times of several seconds are indispensable.

This has been achieved in an experiment conducted by Matthias Sperl, exploiting the merits of the Bremen drop tower. This is a slender tower, about 100 m high, the interior of which can be evacuated. Experimental setups are mounted in a sealed capsule, which is then catapulted in vacuum from the bottom of the tower vertically upwards towards its top. After a total flight time of up to 9.3 s, the capsule drops into a mash of styrofoam particles located below the launching device. This leads to a deceleration experiments can be designed to tolerate.

A data set obtained with a freely cooling granular gas is shown in Fig. 2.3. The capsule is launched at $t = -4$ s. Immediately after launching, the gas is being agitated in order to impart an average random velocity of a few centimeters per second on the grains, corresponding to a considerable granular temperature. After the agitation is stopped at $t = 0$, the gas cools down continuously during free fall of the capsule due to the inelastic collisions between the grains. The dark grey curve represents what one would expect from Eq. (2.13), showing reasonable agreement with the data. At $t > 5$ s the capsule is rapidly decelerated, resulting in erratic signals. Refined experiments have later been carried out in sounding rockets [24]. They showed even better agreement with the theoretical predictions, such that Haff's is now considered as well established.

In contrast, if there is a constant shear rate, $\nabla \mathbf{v}$, imposed on the granulate, the latter is constantly heated by the extra relative motion of the grains. Sela and Goldhirsch have shown that this gives rise to an additional term $l_\mathrm{m} T_g^{1/2} (\nabla \mathbf{v})^2$. This is readily verified qualitatively. In the prefactor, $l_\mathrm{m} T_g^{1/2}$, one recognizes the diffusivity of momentum, and hence viscosity (Eq. (2.6)). Multiplied by the square of the shear rate, $(\nabla v)^2$, it yields the granular heat source. Hence the granular temperature obeys [25, 26]

$$\frac{dT_g}{dt} \propto l_\mathrm{m} T_g^{1/2} (\nabla \mathbf{v})^2 - \frac{1 - \varepsilon^2}{l_\mathrm{m}} T_g^{3/2}.\qquad(2.14)$$

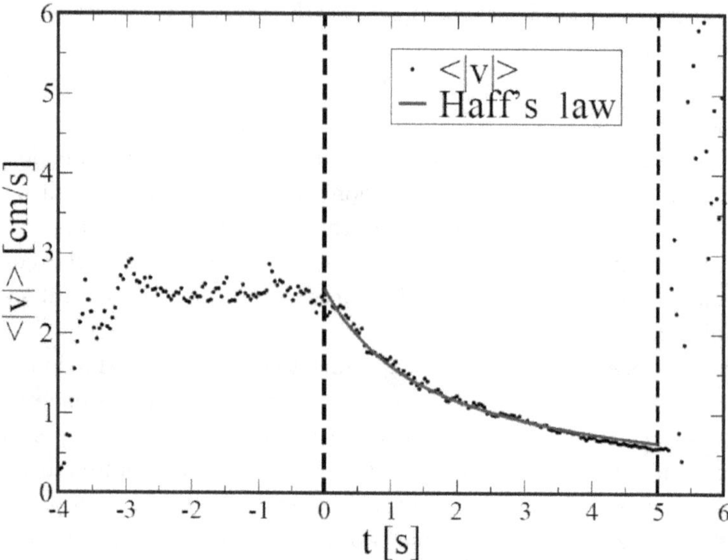

Fig. 2.3 The temporal decay of the random velocity of the grains in a granular gas, representing the granular temperature. At 'negative' times, the granulate is being agitated. Positive times exhibit the characteristic decay, after agitation has stopped at $t = 0$. The grey solid curve represents Haff's law. At $t > 5$ s, the capsule is decelerated. Image kindly provided by Matthias Sperl, DLR Köln.

This seems to lead to a stationary granular temperature, $T_g^{\mathrm{stat}} = (l_\mathrm{m} \nabla \mathbf{v})^2 / (1 - \varepsilon^2)$.

However, this is not a *stable* state [27, 28]. Consider a fluctuation corresponding to a locally increased density. Grains which enter this region will have more frequent impacts, and thus lose energy more rapidly than in regions of smaller density. The grains thus leave this region significantly slower than they entered it. Hence there will be a net flow of particles towards the high density region, and a clustering instability takes place [27–29]. As a result, the system breaks its translational symmetry of the homogeneously dense gas, and forms distinct clusters, as shown in Fig. 2.4 from a simulation [27].

From a fundamental point of view, one immediately appreciates the intimate connection between the dissipative impacts between the grains on the one hand, and the cluster formation on the other. Since the overall entropy of the system must increase with time, the large scale symmetry can only be broken if heat is dissipated into some 'invisible' degrees of

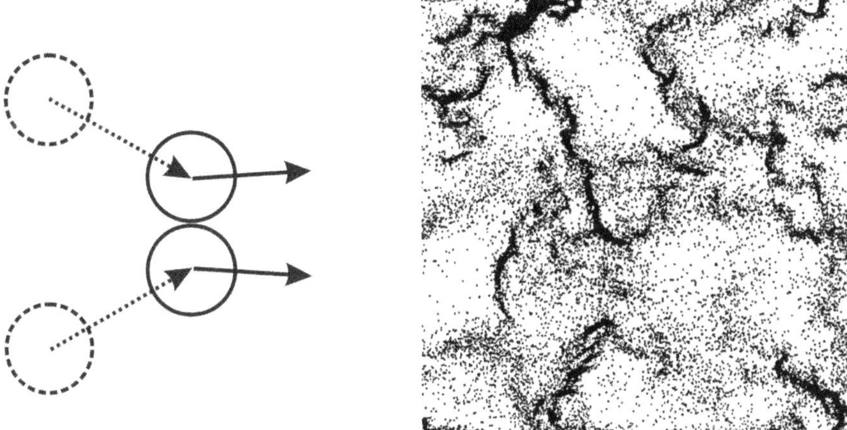

Fig. 2.4 The impact of inelastic collisions (sketch to the left) on spatial density fluctuations. The right panel shows the result of simulations of 40.000 particles (reprinted Fig. 3 of [27]. Copyright (1993) American Physical Society).

freedom, in this case the atomic degrees of freedom of the grains. Alluding to the Saturn rings in Fig. 1.7, we can say that since all structure in the universe (including ourselves) can develop only with the help of dissipative processes, granular matter may serve as a well accessible model system for the study of structure formation far from thermal equilibrium. We will see later that this approach can be carried quite far, in particular in the case of wet granular systems.

There is a particularly illustrative way to demonstrate the connection between dissipative inter-particle collisions and an apparent (large scale) violation of the second law of thermodynamics, which would predict a homogeneous distribution of particles. Into a container which is mounted on an electromagnetic shaker, one places more than enough grains to completely cover the container bottom. When the container is vertically vibrated, the grains will at first glance behave similar to the molecules of a gas, filling the container with approximately homogeneous density. If, however, one places a partitioning wall on the container bottom, with a height comparable to the average jump height of the grains, one will find that after some time almost all grains assemble in one of the two partitions, leaving the other empty [30–33]. Figure 2.5 shows an experiment with several partitions at four different times after vibration had been started. One of the compartments received all of the grains in the end, leaving all other compartments empty. The physical reason of this striking effect is obvious: a grain falling

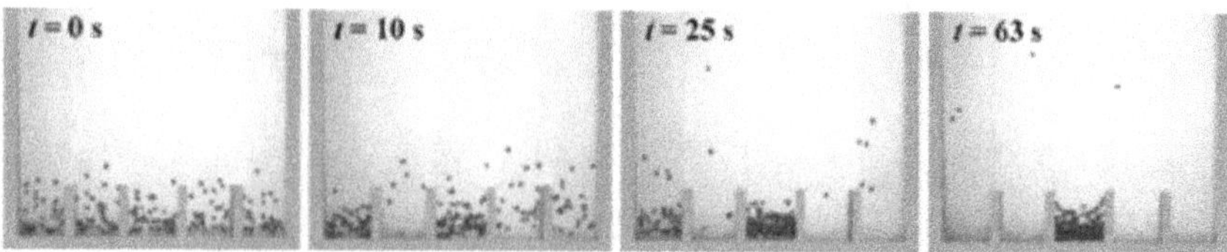

Fig. 2.5   Experimental demonstration of the tendency of granular gases to form clusters. A container which is partitioned into five compartments by walls of finite height is vertically shaken, such that the erratically moving grains can overcome the walls occasionally. Four snapshots are shown at $t = 0$ s, 10 s, 25 s, and 63 s after vibration has started. After about a minute, all grains have collected in one compartment (image courtesy of Ko van der Weele).

on a section of the container bottom which is covered with grains will not rebound, while on the bare container bottom it clearly will, with much better chances to overcome a partition. The fact that the grain does not rebound from the surface of a granular pile is of course directly related to the dissipative character of individual inter-particle collisions, but is greatly amplified by the presence of many grains and their mutual contacts.

Let us now return to the continuum description we have introduced above, and see how the clustering effects just discussed show up in that framework. First we note that mass conservation and force balance alone (Eqs. (2.2) and (2.3)) do not yet close the equations, since the viscosity, which enters in the pressure tensor, depends on (granular) temperature (Eq. (2.7)). We thus need another equation which correctly accounts for the flow of 'thermal' energy. Just as the equations for density and velocity, it can be borrowed from standard theory of molecular fluids, but this time we will see that there is an important modification. The equation for the change in temperature reads [12]

$$\left(\frac{\partial}{\partial t} + \mathbf{v} \cdot \nabla\right) T_g = -\frac{2}{3n}\left(P_{ij}\nabla_j v_i + \nabla \cdot \mathbf{q}\right) - \zeta n T_g, \qquad (2.15)$$

where $\mathbf{q}$ is the vector of heat flux. Although we will make no more use of the latter, we give the pertinent equations for the sake of completeness. We have [27]

$$\mathbf{q} = \frac{25\pi^{1/2} R\varrho_s}{64} T_g^{1/2} \nabla T_g. \qquad (2.16)$$

The first expression on the r.h.s. of Eq. (2.15) is the standard expression for the divergence of energy flow. The last term, however, is absent in the theory of elastically colliding particles. It describes the loss of energy in particle collisions into the internal atomic degrees of freedom of the particles. ζ is a 'cooling coefficient' which is zero in an elastic gas, but for inelastic gases is proportional to $R^2\sqrt{T_g/m}$ [12]. This last term is the only structural difference between the equations describing molecular fluids and those describing granular gases, but its impact on the behavior of the solutions may be dramatic, as one can anticipate from Figs. 2.4 and 2.5. More examples will be presented towards the end of this section.

So far we have neglected rotational degrees of freedom altogether, although according to Eq. (2.1) they receive the same average kinetic energy as the translational ones. When introducing the restitution coefficient, we have tacitly assumed that it is only the translational impact energy which is partly dissipated into the internal degrees of the grains. This is of course not strictly true: if two rotating grains come into contact, such that there is a relative tangential velocity of their surfaces, some of the rotational energy will be transferred to atomic degrees of freedom as well. This is just another way of saying that there is a finite amount of friction between the grain surfaces, which is in fact generally the case for all known materials.

Although the microscopic physics of friction is enormously complex and manifold (and by far not fully understood) there are a few rather simple, classical laws which are fulfilled in a wide range of cases. They have been developed long ago by Leonardo da Vinci, Guillaume Amonton, Charles-Augustin de Coulomb, and others. We will restrict the discussion here to these general observations, as the microscopic physics of friction is largely irrelevant to granular matter. If one considers the tangential force acting on a solid contact, $F^{\parallel}$, as a function of the load force, $F^{\perp}$, one finds that

$$F^{\parallel} \approx \mu F^{\perp}, \qquad (2.17)$$

independent of the area of contact. μ is a numerical constant and is commonly called the friction coefficient. This reflects a simple experience. If the flat support on which a solid body rests is gradually tilted, there is a certain critical angle at which the body starts to slide down the slope. The magnitude of that critical tilt angle depends on the materials used for the support and the solid body, but not on the size and shape of the latter. The tangent of the critical angle is equal to the friction coefficient. Hence μ lies typically between 0,2 and unity, and depends only on the materials used.

In particular, the ratio $F^{\parallel}/F^{\perp}$ is found to be largely independent of the relative velocity of the contacting surfaces. Although this appears simple enough at first glance to lead to a tractable theory of granular systems with friction, a closer look immediately reveals its malice. If we write the tangential force as a function of sliding velocity, $v^{\|}$, we have

$$F^{\|} = \mu F^{\perp} \, \mathrm{sgn}(v^{\|}). \tag{2.18}$$

Despite its simplicity, this is a strongly nonlinear function, and it is anything but trivial to derive its impact on the collective behavior of a pile of grains.

We will discuss the various properties of friction in some more detail in Chapter 4, when we consider the various interaction forces which are practically relevant between dry and wet grains. Here it is sufficient to say that as far as the kinetics of granular gases is concerned, friction is just another channel through which macroscopic kinetic energy is transferred *irreversibly* to atomic degrees of freedom within the grains.

2.1.3 *Aspects of grain size*

In strong contrast to a molecular gas, the mass of each individual particle is many orders of magnitude larger than that of an atom or molecule (a sand grain contains on the order of 10^{18} atoms). As a consequence, a grain is too heavy for being lifted noticeably from its support by thermal energies. If we take silicon oxide as an example (of which most sand is composed), we are to deal with a density of $\varrho_{\mathsf{SiO}} \approx 2.6 \ \mathrm{g/cm}^3$. It is readily verified that for grains with a radius larger than a micron, thermal energies will not be sufficient to lift a grain by its own radius (cf. Table 1.1). Granular materials, however, are usually envisaged as being composed of grains with many microns in size. Figure 2.6 shows cumulative size distributions (diameters) for a few types of sand which are frequently referred to. By convention, sand is defined as a clastic[3] sediment with grains sizes between 63 μm and 2 mm.

As long as it is not agitated mechanically, any granulate will thus form a condensed phase at the bottom of its container, kept together by gravity alone, without any cohesion being necessary. This is in contrast to regular liquids, which form condensed phases only by attractive interaction between their particles.

Another consequence of the large mass of the particles is the enormously large number of internal (atomic) degrees of freedom contained in each grain. We have already mentioned the irreversibility of energy transfer into these degrees of freedom. Let us now illustrate this by means of a simple gedanken experiment. If we take an hourglass like the one in Fig. 2.1 and shake it in our hands, just enough to 'fluidize' the sand which elsewhere rests

---

[3]From ancient greek $\kappa\lambda\acute{\alpha}\omega$, to break, to shatter.

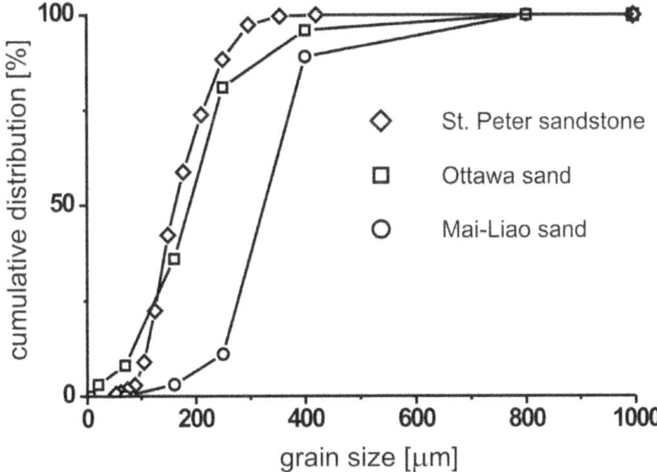

Fig. 2.6 Cumulative size distributions of three types of sand widely used for reference. On that scale, the cumulative size distribution of the ballottini shown in Fig. 2.7 would appear as a sharp step.

on the bottom of the lower glass bulb, we introduce typical translational velocities of several cm per second. The granular temperature corresponds only to the random motion, and will be of the order of $T_g \approx m \, (1 \text{ cm/s})^2$. If we insist to compare the granular temperature to the 'thermal' temperature of the atomic motion within the grains, we have to use Boltzmann's constant. We arrive at a temperature of the irregular center-of-mass motions of the grains in the Tera-Kelvin regime. The dissipative nature of the inter-granular collisions intimately couples this Tera-Kelvin heat bath to the atomic degrees of freedom within the grains, which represent a heat bath at room-temperature. This extreme non-equilibrium aspect can be seen to be at the heart of a large fraction of the striking properties of dry granular matter [12, 34, 35]. The obvious fact that even vigorously shaken granulates remain largely at ambient temperature illustrates the vast number of atomic degrees of freedom within each grain, compared to which the center of mass degrees of freedom are *extremely* few.

2.1.4 *Heterogeneity in size and shape*

Another glance at Fig. 2.6 shows that there is typically a whole range of grain sizes in a granular sample, as represented by the finite widths of the steps of the cumulative size distributions. This width, divided by the average grain size of the sample, is a dimensionless number called the poly-dispersity of the granulate. It has important consequences on the packing geometry of granular piles, which will be discussed in Section 6.1.2. Here we just refer to what we stated in the introduction: in a pile of sand, like the one shown in Fig. 1.1, there are not any two grains with the same size and shape. Let us take a moment to illustrate this statement. If we are satisfied with 'optical' accuracy, i.e., a resolution of about one micron, and consider grains with a 'radius' of 200 microns, we need $4\pi R^2 = 4\pi \times 4 \times 10^4 \approx 5 \times 10^5$ pixels to fully characterize the shape of a grain's surface. If we restrict the slope of the surface roughness to the interval $[-1; +1]$, we have at least $3^{5 \times 10^5}$ possible shapes which are 'different' on the chosen level of accuracy. This is to be compared to thoroughly 10^{10} grains in the pile, which make about 5×10^{19} possible pairings. Hence in about 10^{238541} sculptures like the one shown in Fig. 1.1, there is (statistically) just a single one which contains two grains of equal size and shape. This in enormously improbable and can be safely disregarded.

We can also ask the opposite question: how inaccurate do we have to characterize the shape, in order to have a chance for finding two 'identically' shaped grains in the pile? If we use spherical harmonics, and allow one byte (255 possible values) for each harmonic, we find that we need just eight harmonics to make it improbable to find to identical shapes. Hence for any accuracy which allows to account reasonably well for the rough grain surface, the model of having only indistinguishable, 'individual' grains is a very good approximation to reality. Obviously, we are not dealing with an ensemble of identical particles, as we do in most regular fluids. It will therefore be important to choose the questions to be posed to the system such as to make the answers as independent as possible of this inherent 'geometric polydispersity', i.e., of the variability of grain shapes.

Figure 2.7 shows a few micrographs of sand grains, as well as glass beads (also called ballottini). Clearly, the degree of roundedness is strongly variable. While optical micrographs give a qualitative idea of the shape and roughness (Figs. 2.7(a) and 2.7(b)), scanning electron microscopy is better suited for assessments of the surface structure due to its higher resolution. Examples of quantitative measurements of surface roughness profiles by means of scanning force microscopy will be shown in the next chapter.

As a direct physical consequence of geometric polydispersity, granular systems never crystallize. Crystallization is a prerequisite to systems consisting of particles of identical size and shape, and even in such systems it may be sometimes hindered, as in glass-forming molecular liquids. While

Fig. 2.7 (a) Optical micrograph of sand from Marina Beach, Chennai, India (courtesy of Nirmal Thyagu, Göttingen). (b) Optical micrograph of sand from Lüderitz bay, Namibia (kindly provided by Sabine Hamm, Göttingen). (c) Scanning electron micrograph of sand grains, revealing more of the surface structure. (d) Optical micrograph of commercially available glass beads (ballottini) as frequently used as model systems for granular matter physics (a, b, and d: images by Guido Schriever and Kris Hantke, Göttingen. c: image kindly provided by Kim Dalby, University of Copenhagen).

crystallization can be helpful in many situations, such as scattering experiments, it can also lead to complex side effects, like creeping dislocations, which give rise to transport phenomena which are difficult to interpret. We therefore should take it as a blessing that we do not have to worry about crystallization in granular matter, and always consider our systems as geometrically well randomized. It is then straightforward to separate this randomness conceptually into three levels:

(1) the random set of center of mass coordinates of the grains,
(2) their orientations in space (two Euler angles per grain),
(3) the roughness of the grain surfaces.

These three levels of consideration directly lead to three different stages of treating the different aspects of granular systems. When we are interested in the positions of the grains, we will refer to their centers of mass as their location coordinates, and consider them as ideally smooth, spherical bodies to first order approximation. On the next stage, we may take into account that the shape of the grain deviates from a sphere. In many cases, it turns out to be sufficient to assume that this shape is convex. Finally, we appreciate that the surface is not smooth, but always bears a certain roughness. For 'natural' materials, like sand, this roughness has its origin in the production process of the grains, which is dominated by fracture and weathering. But even in the case of artificial, 'idealized' granulates (e.g., 'ballottini', Fig. 2.7(d)), the grains become rough by their mutual collisions and wear. There are many models at hand for describing such roughness, and we will have to see which are appropriate for the length scales to be typically encountered in the case of granular materials.

Since it is not clear *a priori* how we might conceptually separate 'shape' from 'roughness', we shall provide a method how these two notions can be clearly distinguished. We make now explicit use of our restriction to convex shapes. Let the function $g(\vartheta, \varphi)$ denote the distance of the grain surface from its center of mass.[4] Then we approximate the grain surface by the surface of a convex body, $\mathcal{B} = \{\mathbf{r} = (r, \vartheta, \varphi) | r \leq c(\vartheta, \varphi)\}$ (cf. Fig. 2.8). Convexity implies that for any two points $\mathbf{r}_1$ and $\mathbf{r}_2$ in $\mathcal{B}$, all points lying between $\mathbf{r}_1$ and $\mathbf{r}_2$ belong to $\mathcal{B}$ as well. Hence $c(\vartheta, \varphi)$ must enclose the straight line *between any two points* in $\mathcal{B}$. Furthermore, in order for c to represent the shape of the grain, we require the integral

$$\int_{\vartheta,\varphi} \frac{(g-c)^2}{1+(\nabla c)^2} \longrightarrow \min. \tag{2.19}$$

to be minimal. The denominator in the integrand makes sure that we minimize the residual difference between $g(\vartheta, \varphi)$ and $c(\vartheta, \varphi)$ measured *normal* to the surface of the convex model grain. We call $c(\vartheta, \varphi)$ the shape of the grain. The vertical distance between g and c, expressed by the function $f(\vartheta, \varphi) = (g-c)/\sqrt{1+(\nabla c)^2}$, is what we call its roughness.

Given this hierarchy of structure on different scales, it is clear that in experiments we need to control the shape and roughness of the grains occasionally, in order to exclude spurious side effects due to the irregular shape of the grains or determine their effect. A straightforward option is

---

[4]This implies the assumption that there be no 'overhang' in $g(\vartheta, \varphi)$, i.e., that the grain is not 'too' rough.

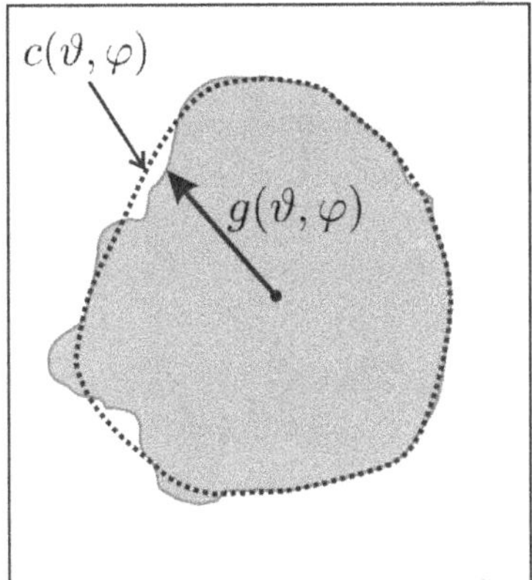

Fig. 2.8   Separating shape from roughness. The dashed line indicates the surface of a convex body, which is chosen such as to minimize the vertical distance to the true grains surface (see text).

to use glass beads, as they are commercially available for technical purposes like sand blasting, or as fillers for reflective coatings of street signs. Figure 2.7(d) shows a micrograph of a typical sample. A frequently chosen technique for their generation is to blow glass powder through a hot flame. The glass particles melt and acquire a spherical shape by virtue of their surface tension. Owing to this process, commercial samples are not mono-disperse, their size distribution rather exhibits a typical width of 5–15% of the average grain size (polydispersity 0.05–0.15). This is not a drawback, however, since this is enough to prevent crystallization of the granular packing [36, 37], which we should try to avoid anyway, for the reasons mentioned above. At the same time, the shapes are quite close to a sphere, and therefore rather well defined.

Another advantage of using glass beads in experiments is their chemical similarity to quartz, since this is the material of which almost all sand on earth is composed. This is a direct consequence of the chemical stability of SiO_2, which is exploited in the wide spread use of glassware in chemical laboratories. Other abundant minerals, like feldspars or clay minerals, rapidly dissolve by chemical weathering. Shales or lime stone are readily milled down to finer grain size (silt). Hence the quartz alone remains in the form of grains of appreciable size. This is an additional (geochemical) justification of the use of glass beads as a model system in experiments on granular dynamics.

Sand grains which have experienced transport over large distance by wind, such as those from the deserts, are usually quite well rounded. Sand from rivers, which is still much closer to the place where it emerged from rock by erosion, has more pointed and rough grains. Consequently, if one chooses to use silicon oxide grains for experiments, one has a good chance to control effects from roughness and grain shape irregularity by using both glass beads and quartz sand samples from different sources, and compare the results. In this way, effects due to grain shape can be efficiently separated from other effects. We will see later that the effects due to surface irregularities of the grains are anyway surprisingly small in most aspects of relevance.

2.1.5   *Some phenomenological consequences*

After having laid out the main commonalities and differences between granular gases and regular fluids, it is illustrative to list a few peculiar phenomenological properties of (dry) granular systems which serve to underpin the differences. These properties can be seen as more or less direct consequences of two main ingredients of granular physics, as discussed above. These are

- the cooling coefficient introduced in Eq. (2.15),
- the friction between adjacent grains, Eq. (2.18).

As far as this is possible, we will connect each phenomenon to the mechanism which is responsible in the first place.

Probably the most prominent effect which is commonly associated with granular media is the Brazil nut effect. This term refers to the observation that if one opens a package of mixed breakfast cereals containing various kinds of 'particles', like oat flakes, raisins, and nuts, one usually does not encounter an intimate mixture of species, but rather finds the largest nuts (e.g., Brazil nuts) lying on top of the pile. Quite generally, when a formerly well-mixed granulate consisting of differently sized grains is being shaken for some time (such as the cereal package during transport), the larger grains tend to rise to the surface and stay there.

Several mechanisms seem to contribute to this surprising, and surprisingly robust, phenomenon. One is the excitation of convective currents in the granular pile from the shaking due to ratcheting effects [38, 39]. Another has to do with the buoyancy of a shell of small particles around a large one [40]. The peculiar mechanisms of force propagation within the granular pile play their part as well. We want to stress here that all effects contributing to the Brazil nut effect are due to the dissipative character of the interactions of the grains. Friction leads to complex collective ratcheting effects, and thus convection in the pile. Inelasticity gives rise to the mentioned buoyancy effects around large objects. Owing to the enormous complexity of the interplay of all these effects, there is as yet no closed theory of the Brazil nut effect which allows precise predictions, although it is of great importance in handling granular materials.

Consequently, the response of granular mixtures to mechanical agitation is still a lively field of research in the engineering sciences. An experimental setup which is used frequently in this field setups in this field is the rotating drum, as sketched in Fig. 2.9. It consists of a cylindrical container which rotates about its axis of symmetry and is closed on both ends with transparent plates for optical access. A granulate is filled into the drum, and the angle Θ the pile surface makes with the horizontal is measured as a function of time while the drum rotates.

For a usual viscous fluid, one would expect Θ to approach a constant value if the angular frequency of the rotation of the drum is constant. In contrast, the granular pile produces, when the rotation is sufficiently slow, a series of avalanches in which Θ is rapidly reduced, with intermittent phases of quiescence where the pile stays at rest with respect to the drum, such that Θ increases gradually as the drum turns. These quiescent phases exist by virtue of the inter-granular friction, while the avalanches are governed mainly by dissipative collisions. The resulting temporal variation of $\Theta(t)$ is sketched on the right-hand side of Fig. 2.9. Its maximum value, Θ_M, at which an avalanche sets in, and the so-called angle of repose, Θ_R, at which

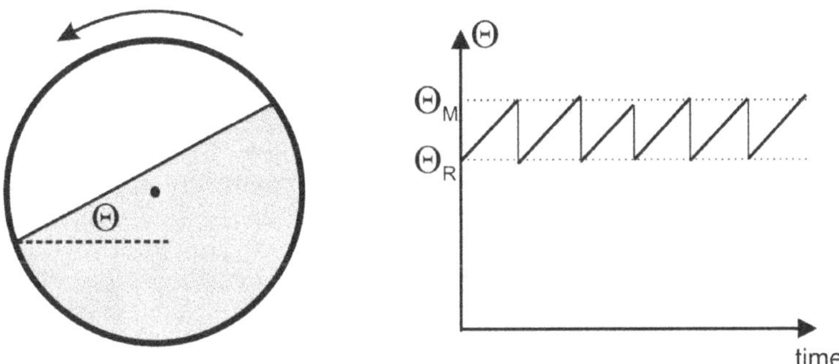

Fig. 2.9 The rotating drum is a standard setup for characterizing the avalanching properties of a granulate, i.e., its maximum angle of stability, Θ_M, and its angle of repose, Θ_R.

the avalanche comes to rest, are important quantities for characterizing a granulate. They depend strongly, e.g., upon the roughness of the grains and on humidity [41].

As the rotation speed is increased to velocities larger than those typically encountered in a granular avalanche, there will be a rather stationary flow field of granulate within the drum. The transition from the hysteretic behavior which oscillates between Θ_M and Θ_R to the stationary flow has been studied in detail. It shows characteristic features of a first-order phase transition when the rotation speed is varied [42, 43].

It is not surprising that as a collection of different species of grains are filled into the drum, even more complex scenarios arise. To stress the potential complexity one may have to deal with, Fig. 2.10 shows the result of similar experiments with grains of different color and size in differently shaped drums. Clearly, there is a plethora of complex patterns possible [44]. They yield a glance at the many different aspects of granular flow which may play a role when grains are moved with respect to each other [45].

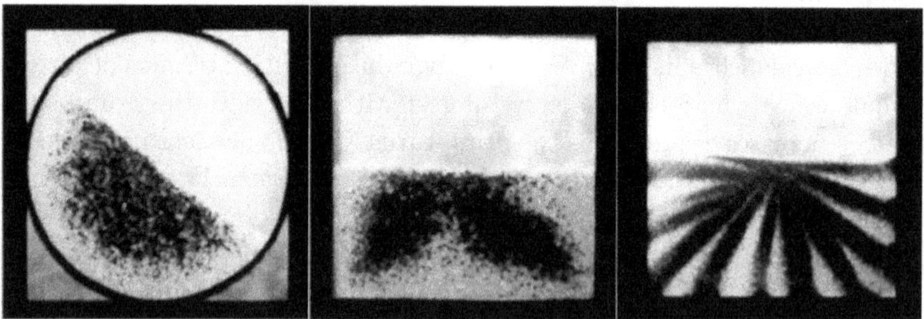

Fig. 2.10 Segregation patterns in various experimental setups involving grains of different size or weight (copyright (1999) National Academy of Sciences, USA [44]).

It will thus be a central challenge to ask questions to our systems such that the answers are as independent as possible of those microscopic mechanisms which lead to such immense degrees of complexity. As it will turn out, it is indeed possible in many cases to keep the underlying physics a lot simpler, in particular in the case of wet granular systems.

2.2 Granular fluids

So far, we considered the properties of individual grains and their mutual interactions. This is appropriate for settings in vacuum, where there is no matter involved other than the grains themselves, such as Saturn's rings, the asteroids, or any proto-planetary accretion disc. We are now ready for the next step of complexity, which is to explicitly include a fluid which fills the interstitial space between the grains. This is equivalent to considering a binary system, which in the language of Fig. 1.9 is located along the slanted edges of the triangle. It consists just of the grains and a single fluid, the physical properties of which have to be included in the treatment. This will be developed in the present section.

To get into the matter, let us first acknowledge that in almost every situation of practical relevance, the grains are imbedded in a fluid 'carrier phase'. In the case of sand dunes in the desert, it is the surrounding air which fills the space between the grains, thus enabling small animals to breathe even when they dug themselves deep into the sand. When the air is in motion, such as in a gentle breeze, it may lift individual grains and carry them for several meters. This process is called saltation, and is an essential ingredient in the formation process of dunes. In case of a sand storm, grains can be carried form the sahara desert to northern Europe. If one has a sufficiently deep understanding of the interaction of the wind with the sand grains, one can from the shapes of sand dunes infer the temporal pattern of wind strength and directions. This is of particular value for the exploration of regions as remote as other solar planets, or their moons, where access is limited to optical inspection from spacecraft [46, 47]. As an example, Fig. 2.11 shows so-called barchan[5] dunes, or just barchans, on Earth and on Mars. This type of dune is indicative of a limited supply of sand, and a fixed wind direction, which is indicated in the images by the white arrows.

If the sand is submerged in water, as in the coastal areas of the oceans, the physics is clearly different. The difference in the mass densities of the grains and the fluid is smaller, the viscosity of the fluid is larger, and the sedimentation times are thus much larger than in the case of air. However, many phenomena are strikingly similar, like the formation of ripples on the surface of the sediment bed, which are ubiquitously known from sand desert settings as well as from shallow-water littoral zones. It suggests itself to assume that by proper rescaling, the physics of under-liquid and under-gas granular systems can be mapped upon each other. In many cases this turns out to be true, in particular when the compressibility of the gas phase can be neglected. Several research groups have used this fact to

---

[5]From turcmenian *barchan* [bar′xa:n] = sickle dune.

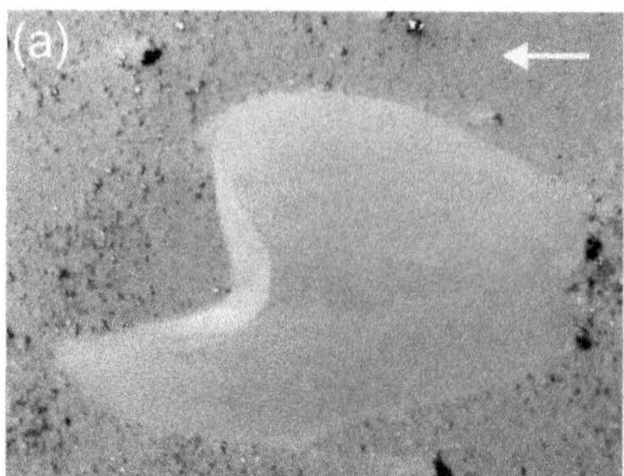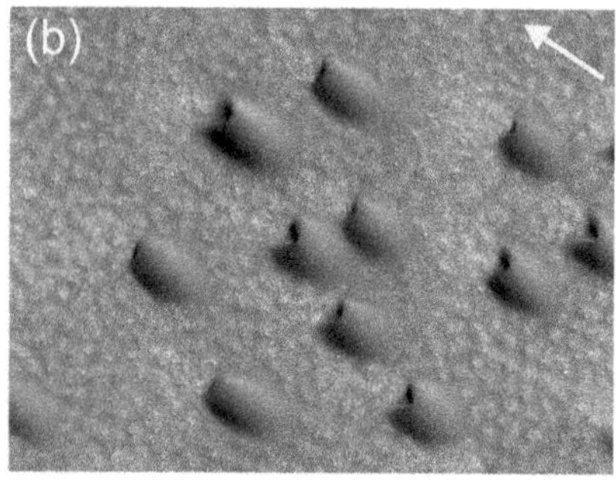

Fig. 2.11   (a) Top view of a barchan taken with a camera from a kite (image kindly provided by Philippe Claudin, ESPCI, Paris). (b) Barchans on Mars (courtesy of NASA, University of Arizona, USA). As barchans are indicative of constant wind direction (arrows), the latter can as well be inferred for the marsian setting.

set up laboratory-scale model experiments for the formation of sand dunes [48–50]. There are, however, circumstances under which this rescaling is not appropriate. In what follows, we will discuss a few scenarios which are selected to exemplify either case. For the sake of coherence with the above discussion of dune structures, we stay with examples from geoscience.

2.2.1   *Buoyant clouds*

Powder snow avalanches can be considered as particularly dramatic granular dynamics phenomena. The knocking-off of a tiny amount of snow by, say, a bird, may be sufficient to trigger the downhill rush of many hundreds of tons of material, which crush and bury everything on their way. Such avalanches differ from the ones observed at the slopes of granular piles, such as in the rotating drum (cf. Fig. 2.9), in that the main transport takes place within the air above the slope, which is heavily laden with snow particles raised from the slope. This is why powder snow avalanches are also referred to as *buoyant clouds*.

Another example of buoyant clouds which is of great practical as well as scientific relevance is the occurrence of so-called turbidity currents in the oceans. These are massive granular avalanches moving down the slopes of the continental shelves, which mainly consist of granular sediments washed into the sea from the rivers. Turbidity currents can carry enormous amounts of material over many kilometers, such that they can be abundantly traced in geological outcrops of sedimental rocks, so-called turbdites. They have thus become an important means of assessing conditions earlier on earth,

and are known from basically all stages of earth history after land masses
had formed. The basic mechanism of their formation can be easily understood and will be discussed in what follows. It applies, *mutatīs mutandīs*,
as well to the powder snow avalanches mentioned above, but for the sake
of specificity we stay with the turbidity currents here.

Consider the leading edge of the under water sedimental deposits (i.e.,
sand) of a river delta, a few kilometers towards the open sea, where the
sand makes a slope corresponding to its angle of repose, Θ_R (cf. Fig. 2.12).
Assume that this critical angle has been reached everywhere on the slope.
If somewhere a little bit of sand is stirred up, perhaps by a fish rummaging
the ground for food, it forms a suspension locally around the disturbance.
Since the specific weight of the sand grains is larger than that of water,
the density of the suspension will as well exceed that of the surrounding
liquid, such that there is a gravitational downhill-slope force which induces
a flow along the slope (cf. Fig 2.12(a)). If this is strong enough to stir up
more sand out of the slope, a runaway instability may occur (Fig 2.12(b)),
which eventually affects the whole slope below the place of the original
disturbance.

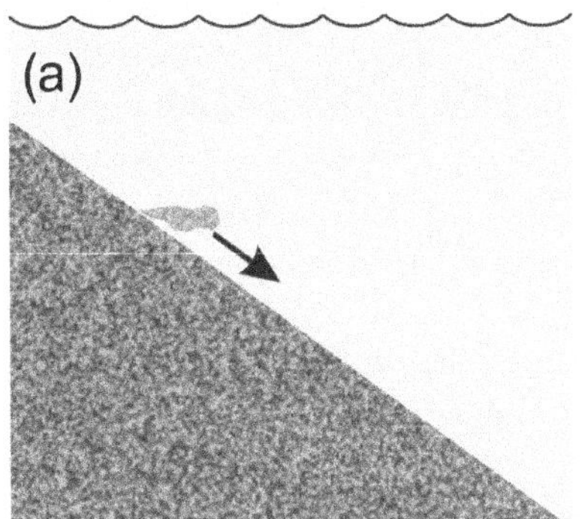

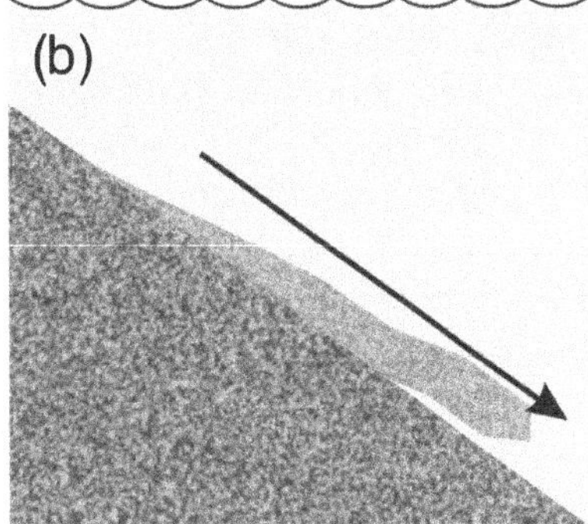

Fig. 2.12 Schematic representation of the development of a turbidity current, on the
slope of the continental shelf towards the deep sea. (a) A small local perturbation forms
a suspension of higher density. (b) This suspension flows down the slope by virtue of
gravity and stirs up more and more material on its way.

In order to appreciate the robustness of this mechanism, let us set up
a very simplified, linear dynamical model. If sediment is stirred up to
form a suspension layer of height h with solid volume fraction ϕ, we can
immediately write down the Navier-Stokes equation of that layer,

$$\varrho_l \partial_t v = g \sin \Theta_R \, (\varrho_s - \varrho_l)\phi + \eta \Delta v, \tag{2.20}$$

where ϱ_l and ϱ_s are the mass densities of the liquid and the solid grains, respectively, v is the flow velocity field, and Δ is the Laplace operator. For the sake of simplicity, we do not only neglect variations along the flow direction (which avoids the nonlinear $v\nabla v$ term), but also reduce the problem to its temporal behavior by considering only an averaged flow velocity of the layer. Thus we can write

$$\partial_t v \approx G\phi - \frac{\eta}{\varrho_l}\frac{v}{h^2}, \tag{2.21}$$

with the abbreviation $G = g\sin\Theta_R(\varrho_s - \varrho_l)/\varrho_l$. The stirred-up granulate will settle in some characteristic time τ_s, and we express the sedimentation process accordingly in the linearized form $\partial\phi \approx -\phi/\tau_s$. For the uptake of granulate into the liquid, we set $\partial_t\phi = f(v)$ with $f(0) = 0$. We expect $f(v)$ to increase steeply when v exceeds a certain threshold, since there can be no sediment uptake for a whole range of (too) small velocities. Introducing $\tau_0 = h^2\varrho_l/\eta$, we then arrive at the dynamical system

$$\partial_t\begin{pmatrix}\phi\\v\end{pmatrix} = \begin{pmatrix}f(v) - \phi/\tau_s\\G\phi - v/\tau_0\end{pmatrix}, \tag{2.22}$$

which describes the dynamics in the 'phase space' spanned by v and ϕ. The corresponding phase portrait is shown in Fig. 2.13. The bold solid curve is given by $\phi = \tau_s f(v)$ and indicates the locus of all points where $\partial_t\phi = 0$ (nullcline of $\partial_t\phi$). Any curve with this shape will have an intersection with a straight line through the origin, such as the dashed. This represents $\phi = v/G\tau_0$ and thus the locus of all points where $\partial_t v = 0$ (nullcline of $\partial_t v$). The point of intersection is the fixed point of the system, where both $\partial_t v$ and $\partial_t\phi$ vanish. The phase flow in the (v, ϕ) plane is sketched as arrows, qualitatively on the basis of Eq. (2.22). The dotted curves indicate the invariant manifolds belonging to the fixed point, i.e., the trajectories which would finally end up at the fixed point for either $t \to \infty$ or $t \to -\infty$.

Without explicitly carrying out the stability analysis of the fixed point, or even solving for the trajectories, we clearly see what happens. When some granulate is stirred up, with the liquid remaining largely at rest, an initial condition is set which corresponds to some point on the ϕ axis. Depending on whether this lies below or above ϕ_c, the results will be dramatically different. If the initial perturbation is weak and leads to some point below ϕ_c, the trajectory will pass below the fixed point, and approach the origin for $t \to \infty$. The system thus returns spontaneously to the quiescent resting state. If the initial perturbation is strong enough to surpass ϕ_c, the trajectory will pass above the fixed point, and thus take off into the

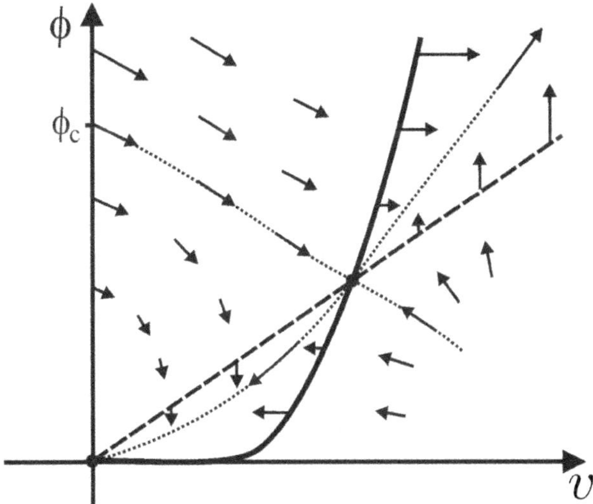

Fig. 2.13   Phase portrait of a buoyant cloud, following the simplified description adopted in Eqs. (2.20) through (2.22). The initial disturbance leads to some starting point on the positive ϕ-axis. Depending on its position, the trajectory will either return to the origin, or take off towards infinity. Solid curve, nullcline of $\partial_t \phi$. Dashed line, nullcline of $\partial_t v$. Dotted, invariant manifolds.

direction of unbound flow velocity. This quite robust mechanism has given rise to many deadly accidents in quarry lakes, where people are likely to encounter situations like the one sketched in Fig. 2.12 when bathing over unconsolidated sedimentary beds.

In buoyant clouds, the grains move essentially with the carrier fluid. As long as the flow velocities are small as compared to the velocity of sound in the fluid (i.e., for small Mach number[6]), the compressibility of the carrier can be neglected, and even a gas can be treated as an incompressible liquid. Hence in these cases there is no qualitative difference between a gaseous and a liquid as the carrier.

2.2.2   *Filling an earthquake fissure*

This changes as soon as the relative motion of the fluid with respect to the grains plays an essential role. In order to investigate this in some detail, we will discuss another example of geological relevance, in which

---

[6]The Mach number, Ma, is defined as the typical velocity in the system divided by the speed of sound in the involved fluid.

the interstitial fluid may be either gaseous or liquid: we shall investigate in which manner a granular material, like soil, or the sand of a desert or beach, will fill a fissure which opens up instantaneously in the bedrock below. This sometimes happens during earthquakes, when the bedrock yields to stress which has built up over time due to tectonic activity. Somewhere below the ground, at the so-called hypocenter of the earthquake, the rock fails first, and the lateral boundaries of the fissure rapidly propagate away from the hypocenter. This proceeds roughly at the speed of sound in the rock, i.e., at several kilometers per second. Since the typical length scale of such fissures is kilometers, we expect that the process of fissure opening takes place within fractions of a second, or at most a few seconds.

When the ground opens, there is not any air yet in the fissure, and whatever stands close to its rim will be violently sucked inside. If an overlayer of a clastic sediment, such as sand, is present above the cracking bedrock, it will completely fill the fissure, forming a so-called clastic dyke. It is far from trivial to determine how, and how fast, the fissure will be filled with the overlying material. In particular, we may anticipate that this process will strongly depend on whether the sediment is dry (e.g., desert) or wet (the floor of a lake or ocean). In present day tectonic events, this is of minor relevance, since it concerns only regions well below ground level, which are hardly accessible. However, as time passes by and the bedrock becomes eroded, clastic dykes generated in tectonic events long ago may finally crop out at the newly generated surface. Based on a thorough understanding of granular flow, one may thus gain valuable information about tectonic events in geological history from investigations of outcropping clastic dykes.

There are, in fact, a number of beautiful outcrops of such fossil tectonic fissures. Some of the best preserved can be found in Namibia, at the slopes of the Great Gamsberg (GPS coordinates: S23°20′30″ E16°13′30″) [51]. The Gamsberg[7] is a mesa plateau at an elevation of 2347 m above sea level, which can be seen in Fig. 2.14. It is one of the last remains of a vast peneplain which had formed from the Permian through early Jurassic ages. In the Gamsberg area, the bedrock, which now forms the mountain, was granite stemming from the times of the formation and shaping of Gondwanaland (Damara orogeny). After the granite had been flattened by meandering rivers, climate changed to arid conditions in the late Jurassic age. Most of Namibia was covered by a desert, and a thick layer of sand

---

[7]The name has nothing to do with the German word 'Gams' for chamois. The correct transcription would be 'Xams -', where the 'X' indicates a snapping sound characteristic of the Nama language.

Fig. 2.14 (a) The Great Gamsberg, sacred mountain of the Nama, indigenous people in Namibia and dedicee of this book (viewed from the Hakos farm house, photo taken by the author). The Little Gamsberg to the left appears higher due to perspective, but is approximately the same height. Both mountains are remains of a vast peneplain. The plane was covered with a thick layer of sand (Etjo formation), which now forms the flat sand stone top. (b) A fossil earthquake fissure, cropping out almost vertically at the Gamsberg slope, striking approximately north-south. Its width (distance between white arrowheads) is about 5 cm. It is composed of quartzite (sand consolidated by diagenesis), while the surrounding rocks are granite (photos by the author). (c) The flatness of the Gamsberg top is best appreciated by satellite imagery (courtesy of NASA). The slanted white arrow indicates the position of the outcrop shown in (b).

was deposited above the plane granite surface. This shows up today as a thick (roughly 30 m in the Gamsberg area) layer of sand stone which forms the flat top of the Gamsberg.

At first glance, it may appear surprising that the sand stone seems so much more resistant to weathering than the proverbially hard granite basis. The reason is that in spite of its mechanical stability, granite is not very resistive against chemical weathering. It consists of a dense fabric of millimetric crystals of quartz, feldspars, mica, and amphibols. The feldspars are readily decomposed to more or less soluble clay minerals, similarly the amphibols and mica. As already mentioned above (Section 2.1.4), it is only the grains of quartz, as is the chemically most resistive of the rock-forming minerals, which finally survive the weathering process, and roll down the slopes as a characteristic loose gravel. This is why the steepness of the slopes of the Gamsberg, as obvious from Fig. 2.14(a), is close to the angle of repose of a dry granular pile (cf. Figs. 1.2 and 2.9).

When one climbs the slopes of the Gamsberg, one finds vertical plates of sand stone embedded in the granite matrix. They extend from the top of the mountain to up to hundreds of meters below the top, and have thicknesses ranging from a few millimeters to about one half meter [52]. An example is shown in Fig. 2.14(b). The sandstone filling is indicated by the horizontal white arrows. The distance between the arrowheads represent the thickness of the clastic dyke, which consists of low porosity quartzite, i.e., sandstone lithified to a solid rock material through quartz overgrowth on the grains into the pore space. The surrounding rock is granite. These dykes are believed to be the fillings of fissures which formed in the granite bedrock when Gondwanaland broke apart about 135 My ago. The rift zone forming between what today are South America and Africa exerted large extensional forces on the granite, which in response broke apart forming a series of fissures striking roughly north-south [52]. The sand, which at those days was lying on the peneplain as a loose granulate, was sucked into the fissures and later consolidated to the sand stone cropping out today at the Gamsberg slopes. Figure 2.14(c) shows a satellite top view of the great Gamsberg, with the slanted arrow indicating the approximate position of the outcrop depicted in Fig. 2.14(b). More easily accessible outcrops, as described already by Wittig [52], can be found along the dirt road (visible as the faint meandering line at the top right of Fig. 2.14(c)) leading to the Gamsberg plateau. The extraordinary flatness of the latter can be well appreciated from the satellite top view presented in Fig. 2.14(c). We shall return to this remarkable feature further below.

Several hundred kilometers north of the Gamsberg, one finds another well-accessible outcrop of clastic dykes which can be interpreted as fossil earthquake fissures. Somewhat north of the Brandberg, there is a vertical cliff exposing the uppermost parts of the Gai-As formation bedrock (GPS:

S20°46′50″ E14°4′30″), an old marine shale stemming from Permian times [53]. On top of these Gai-As rocks, one encounters a bright sand stone layer which represents the analogue of the Gamsberg top sand stone, belonging to the same (so-called Etjo[8]) formation. It exhibits extensive cross bedding, in a way which clearly identifies it as fossilized sand dunes from an old desert, and there is so far no evidence for water having been present when the sediment was still a loose granulate. Below this sand stone layer, several vertical fissures can be found in the dark-brown shale bedrock, which are filled with the bright yellow sand stone corresponding to the Etjo overlayer. They present evidence that the Gai-As shales were similarly exposed to the extensional forces during late Jurassic rifting, and formed fissures similar to those found at the Gamsberg slopes. As an illustrative example of the dry climate in this northern part of today's Namibia, Fig. 2.15 shows one of many fossil sand dunes found on the top plateau of the Waterberg (S20°40′30″ E17°4′0″), which belongs to the same formation.

Fig. 2.15 A fossil sand dune on the top plateau of the Little Waterberg in central Namibia. The slanted bedding, which is characteristic of a wind blown dune (cf. Fig. 2.11), is clearly visible (photograph by the author).

---

[8]Named after Mt. Etjo (S21°7′ E16°27′).

In contrast to these dry conditions, there is some evidence for water having been present at the top of the Gamsberg before the sand stone had consolidated. This can be seen from fossils like those depicted in Fig. 2.16. Figure 2.16(a) shows a fossilized mollusc trace on the surface of a piece of sand stone found at the Great Gamsberg. It is well known that molluscs are likely to appear where sufficient amounts of ground water intrude into dessert scenarios, even aeolian sand desert like the Sahara, far away from the oceans [54, 55]. Remains of gypsum roses, as evidenced by the imprints shown in Fig. 2.16(b), are considered indicative of variable levels of saline ground water within the sediment [56, 57]. Gypsum roses are known to grow, in part displacively, within the sand deposit from the mineral content of ground waters even far away from genuinely saline mileus such as the oceans [58–60]. Curved crystal facets, which are also visible in Fig. 2.16(b), are furthermore indicative of the presence of organic material [61, 62].

Can we now, from a detailed analysis of the filling material of the fissures, determine whether the fluid carrying the sand was just air (as in

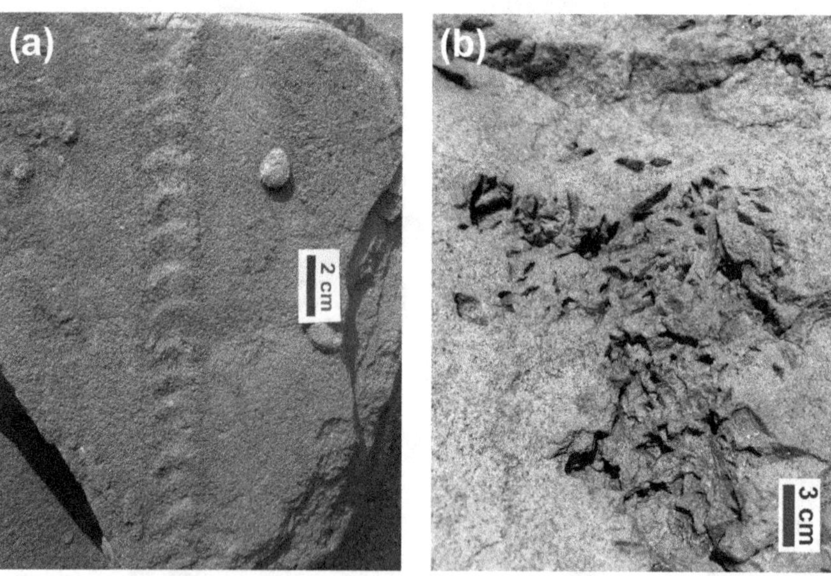

Fig. 2.16   (a) Fossilized trace of a mollusc on the surface of a piece of sandstone found at the Great Gamsberg (courtesy of the Wehner farm and Reinhold Wittig. Photo by the author). (b) Imprints of gypsum roses in quartzite samples from the top of the Gamsberg plateau. The predominance of the characteristic spar shape is clearly discernible, as well as the sometimes curved facets, characteristic of gypsum roses in the presence of organic matter (photo by the author).

the case of a desert), or water (as in the case of a lake)? This would be very valuable for helping to unravel, e.g., the geologically rather complex situation at the Great Gamsberg[9] [52]. In what follows, we will thus try to predict what exactly will happen in these two cases, and whether insight into granular flow can lead us to determine, from the appearance of the fissure filling material, whether there was a desert or a lake (or at least considerable amounts of ground water) on top of the Gamsberg granite when Gondwana broke apart.

2.2.3 *Granular flow with gaseous carrier*

If the sand overlayer is dry, the carrier fluid is just the surrounding air. Since the atmospheric pressure was quite constant through earth's history, we can assume that the physical properties of the air when the fissures were created was about the same as today. We now assume that at time $t = 0$, a cleft of width b opens instantaneously in the bedrock. Both the sand grains and the gas molecules will now tend to fill the vacuum in the fissure. Initially, the grains are at rest, while the gas particles move according to their thermal velocity distribution, whose average corresponds roughly to the velocity of sound. We can thus safely assume that initially there will be only a current of gas through a matrix of densely packed sand grains which are more or less static, corresponding to the situation sketched in Fig. 2.17, with $l = 0$. This will lead to a characteristic distribution of gas pressure within the sand layer after a short transient. The pressure gradient will then, together with the acceleration due to gravity, set the grains into motion, such that the fissure will finally be filled with sand.

We are interested in the characteristic time scales at which this happens, and in the velocities at which grains are driven into the fissure. It is well conceivable that as the atmospheric gas is sucked into the evacuated fissure, a scenario quite similar to sand blasting may arise, where grains are impinging on the solid walls at high velocities. This would lead to multiple grain fracture as soon as velocities around 100 m/s are reached. These would then show up in optical microscopy of thin sections of the fissure filling material as broken grains, and could be used for identifying the filling process. On the other hand, the overlying sand may constitute enough of a seal to prevent rapid inflow of gas, resulting in a much less

---

[9]For the sake of brevity, we have in fact to greatly simplify the geological setting at the Great Gamsberg, which is in several aspects still far from understood. We restrict the discussion here to what is necessary to appreciate the importance of granular flow.

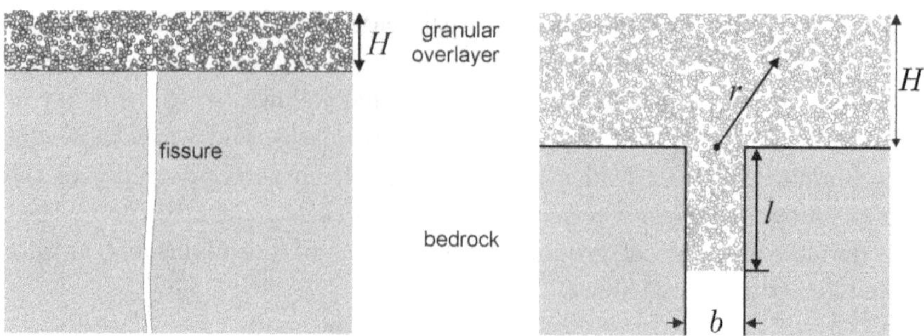

Fig. 2.17 Left: Sketch of the situation considered. Right: Closeup of the left sketch, showing some of the quantities used in the calculations.

dramatic process. Sand would then rather trickle into the fissure as it does in Fig. 2.1, with no breakage of grains to be expected.

We start by considering the gas flow within the matrix of densely packed grains. As it is obvious from the sketch in Fig. 2.17, the flow is effectively two-dimensional, and we can treat the problem in cylindrical coordinates, (r, φ). Furthermore, if $b \ll H$, what we henceforth assume, we can safely disregard all φ-dependence. For the description of the flow, we thus use the average flow velocity, $v(r)$, of the gas within the granulate, which is defined such that if $n(r)$ is number of gas molecules per unit volume of sand, nv is the number of molecules flowing through a unit area of sand pile per unit time. Darcy's law then tells us that

$$\nabla p = \alpha \eta v(r), \tag{2.23}$$

where p and η are the pressure and viscosity of the gas ($\eta \approx 2 \cdot 10^{-5}$ Pas for air), and α is a geometrical constant characteristic of the packing geometry, with the dimension of an inverse area. A well established expression for α is [63]

$$\alpha = \frac{A\phi^2}{R^2(1 - \phi)^{3.6}}, \tag{2.24}$$

where R is the typical grain radius, $\phi \leq 0.6$ is the packing density of the sand pile (volume fraction of the grains), and $A \approx 33$ for sand with $R \leq 1$ mm. Inserting numbers yields $\alpha \approx 530/R^2$. Combining the equation of state of the interstitial air,

$$p = \frac{nk_BT}{1 - \phi}, \tag{2.25}$$

with Eq. (2.23), we obtain

$$\frac{k_BT}{(1 - \phi)} \partial_r n(r) = \alpha \eta v(r). \tag{2.26}$$

For the pressure distribution which builds up within the granular pile, we first seek a stationary solution of Eq. (2.26). In this case, the total current of gas through any cylinder surface around the fissure entrance, J, must be the same, due to mass conservation. We thus have $J = nv(r)\pi r = const.$ This yields

$$\frac{k_B T}{2(1-\phi)}\partial_r n^2 = \alpha\eta\frac{J}{\pi r}, \tag{2.27}$$

which can be directly integrated over r, with the result

$$n(r) = \sqrt{\frac{2\alpha\eta J(1-\phi)}{\pi k_B T}\ln\frac{r}{r_0}}, \tag{2.28}$$

where r_0 represents an integration constant. If n_0 is the gas density in the atmosphere above the sand layer $(r = H)$ and n_f is the density at the entrance of the fissure $(r = b/\pi)$, we can eliminate the unknown r_0 and obtain

$$\ln\frac{\pi H}{b} = \frac{\pi k_B T}{2\alpha\eta J(1-\phi)}\left(n_0^2 - n_f^2\right). \tag{2.29}$$

Since there is vacuum in the fissure, it is clear that $n_f \ll n_0$ at all times of relevance, such that we can safely neglect n_f^2 with respect to n_0^2. This finally yields

$$J \approx \frac{\pi k_B T n_0^2}{2\alpha\eta(1-\phi)\ln\frac{\pi H}{b}}, \tag{2.30}$$

which is the number of molecules entering the fissure per unit length and unit time. For the time it would take to fill an entire fissure of depth D with gas at a pressure of one atmosphere, we readily obtain

$$t_{\text{fill}} = \frac{\alpha\eta DH}{p_0}Q\left(\frac{b}{H}\right), \tag{2.31}$$

with the abbreviation

$$Q(x) = \frac{x}{\pi}\ln\left(\frac{\pi}{x}\right). \tag{2.32}$$

In order to be more specific, we shall henceforth consider a fissure of width $b = 10$ cm, depth $D = 10^3$ m, and grains with a radius of $R = 200$ μm covering the cracking rock to a thickness of $H = 10$ m. We then find that t_{fill} is a few minutes.

This will speed up considerably as soon as the granulate is set in motion, since it adds its flow velocity to the gas velocity. We can at least come up with an upper limit of the gas inflow velocity by assuming the grains to be frictionless. Each streamline reaching down to the fissure entrance sees a pressure drop determined by the hydrostatic pressure of the granular layer, plus the gas pressure drop. The velocity at which the granulate will stream into the fissure, u_f, is then given by Bernoulli's law,

$$\frac{1}{2}\varrho u_f^2 = \varrho g H + p_0 - p_f, \tag{2.33}$$

where $\varrho = \phi \varrho_s$ is the mass density of the granulate. This leads to

$$u_f = \sqrt{2gH + 2\frac{p_0 - p_f}{\varrho}} > \sqrt{2gH}. \tag{2.34}$$

The expression on the right-hand side is equal to the final velocity reached when dropping an object from a height H. As this corresponds to about 14 m/s for $H = 10$ m, the fissure would be filled with granulate within about a minute.

However, since the gas is streaming *relative* to the grains towards the fissure, it will always be ahead, such that the pressure in the fissure will approach atmospheric pressure much more rapidly than suggested by Eq. (2.31). A steady state is reached when the gas pressure in the fissure has become large enough to sustain a zero gas velocity while the granulate is streaming in. The corresponding velocities are readily derived. Due to mass conservation, we have

$$u(r) = \frac{bu_f}{\pi r}, \tag{2.35}$$

for the granular flow velocity field. The gas flow relative to the granulate is just $v = -u$, such that following Eq. (2.23) we can write

$$\nabla p = -\frac{\alpha \eta b u_f}{\pi r}. \tag{2.36}$$

From this we immediately obtain, by integration over r, the pressure drop from the entrance of the fissure back to the outer atmosphere,

$$p_f - p_0 = \frac{\alpha \eta b u_f}{\pi} \ln \frac{\pi H}{b}. \tag{2.37}$$

Introducing the characteristic velocity

$$c = \frac{\alpha \eta H}{\varrho} Q\left(\frac{b}{H}\right), \tag{2.38}$$

we find

$$u_f^2 + 2cu_f - 2gH = 0, \tag{2.39}$$

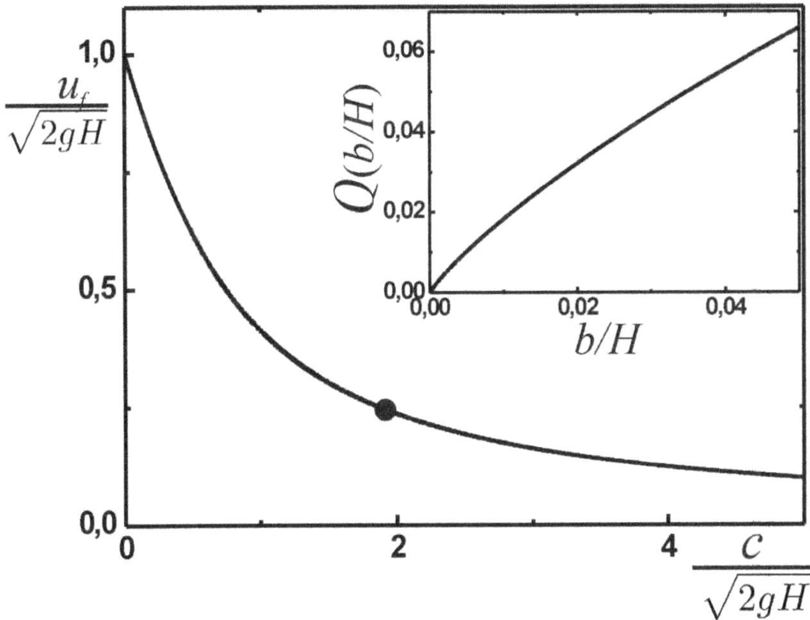

Fig. 2.18   The steady state velocity of grains at the fissure entrance, u_f, as a function of the characteristic velocity scale c. The inset shows the variation of $Q\,(b/H)$, which is used repeated times in this theory. The black dot indicates the example we consider here with $b = 10$ cm, $H = 10$ m, and $R = 200$ μm. The latter enters via the characteristic velocity c (Eq. (2.38)).

and hence

$$u_f(R, b, H) = \sqrt{2gH} \left[\sqrt{1 + \frac{c^2}{2gH}} - \frac{c}{\sqrt{2gH}} \right], \tag{2.40}$$

which is plotted in Fig. 2.18. Note that c depends on R through α. The black dot indicates the conditions corresponding to our example. From Eqs. (2.24) and (2.38), it is obvious that the finer the grains, the smaller will be u_f in the steady state.

We see that in the final steady state influx regime, the grains are entering the fissure at a velocity considerably smaller than a free fall from a height H. At no stage do we find a scenario reminiscent of sand blasting, and in particular towards the end of the filling process, the grains will rather trickle gently into the fissure than being violently sucked in. Even if a grain would fall freely over a long distance, its velocity would saturate at a value determined by the viscosity of the gas in the fissure. As the viscosity of a gas is generally independent of pressure once the mean free path is smaller than the dominant length scale (here: the grain diameter), the fall would be as mild as in standard atmosphere once the pressure in the fissure exceeds about 10 Pa. As a consequence, we expect *no* grain fractures to show up in thin sections of fissure filling material if the fissure was filled under dry conditions.

2.2.4 *Granular flow with liquid carrier*

Let us now turn to a situation where the interstitial space of the granulate is filled not with a gas, but with a liquid. A dense suspension of grains in a liquid is called a *slurry*. We will try to predict how a tectonic fissure is filled with a slurry immediately after it has formed. In order to formulate the problem, it is helpful to estimate first the typical Reynolds number, Re, which indicates the relative importance of inertial and viscous forces. It is defined as

$$\mathsf{Re} = \frac{\varrho u b}{\eta},\tag{2.41}$$

where ϱ is the mass density of the fluid streaming at velocity u, η its viscosity, and b a characteristic length scale, which again is the width of the fissure.

It should be well noted that we are treating the slurry here in terms of Newtonian fluids, which appears far from obvious. In fact, it has only recently been shown with some rigor that the rheology of suspensions at appreciable shear rate can be well modelled as Newtonian fluids [64–66].

It is clear that the viscosity of the slurry depends strongly on the volume fraction, ϕ, of the suspended grains. A well established expression which is sufficiently exact for our needs is the well-known Krieger–Dougherty law [67], which reads

$$\eta = \eta_0 \left(1 - \frac{\phi}{\phi_j}\right)^{-2.5\,\phi_j},\tag{2.42}$$

where η_0 is the viscosity of the carrier liquid (about 10^{-3} Pas for water) and $\phi_j \approx 0.64$ is the maximum attainable value of ϕ, at which the granular pile jams. If the sediment has this density, it cannot flow at all. Hence for a flow to take place, the density needs to be significantly below this value. We assume for the moment that the sand overlayer has attained a sedimentation density, ϕ_s, which is a few percent below ϕ_j. If we assume $\phi_j - \phi_s \geq 3\%$ [68], we have $\eta < 200\eta_0$ by insertion in Eq. (2.42). For our

10 cm wide fissure and velocities of a few meters per second, we then have
Re $\approx$ 1000, such that we can safely neglect viscous effects and focus on
inertial terms alone. Larger initial packing fractions will be treated further
below.

Again we first consider the velocity field in the granular overlayer (e.g.,
the bottom of a lake or ocean). As both the water and the surrounding
atmosphere are adding to the pressure which drives the slurry into the
fissure, we introduce a new quantity, $\tilde{H}$, which is defined as the depth of a
layer of slurry which at its bottom would have the same hydrostatic pressure
as actually prevails at the entrance of the fissure. If H is the thickness of the
bottom sand layer of a lake or shallow ocean, it is clear that $\tilde{H} > H$. It will
be helpful to assume that there be no noticeable friction of the fluid with
the bedrock floor (full slip condition). This may seem a harsh assumption
at first glance, but at high Reynolds numbers, the shear zone is confined to
a boundary layer which is localized close to the walls.[10] We will therefore
in effect have to deal with full slip conditions, justifying a purely cylindrical
symmetry in the description. Furthermore, finite slip can be treated in an
approximate way and would lead to not more that a slight quantitative
correction.

If we accept this assumption, we can write the Navier-Stokes equation
in the completely 'cylindrical' form

$$\varrho(\partial_t + u\nabla)u = -\nabla p. \tag{2.43}$$

For the velocity field, mass conservation implies again

$$u = -\frac{bu_f}{\pi r}, \tag{2.44}$$

with $u_f = dl/dt$, as it can be read off Fig. 2.17. Note that we assume plug
flow of the slurry, which again rests upon the observation that Re is large.
Inserting Eq. (2.44) into Eq. (2.43) yields

$$\frac{\varrho b}{\pi}\left[\frac{1}{r}\frac{du_f}{dt} + \frac{bu^2}{\pi r^3}\right] = \partial_r p. \tag{2.45}$$

This can be directly integrated over r, which leads to

$$p(r) = \frac{\varrho b}{\pi}\left[\frac{du_f}{dt}\ln\frac{r}{\tilde{H}} - \frac{bu^2}{2\pi r^2}\right] + \varrho g\tilde{H}. \tag{2.46}$$

In the first term on the right-hand side, we have chosen the integration
constant as $-\ln\tilde{H}$ to account for the fact that the acceleration becomes

---

[10]The thickness of the boundary layer scales as $1/\sqrt{\text{Re}}$, according to L. Prandtl's boundary layer theory [69].

negligible as $r \to \tilde{H}$. Furthermore, we neglected b against $\tilde{H}$ in the hydrostatic pressure drop (rightmost term in Eq. (2.46)).

Within the fissure, mass conservation and incompressibility together assure that u is spatially constant as long as b is. Thus the gradient term in Eq. (2.43) vanishes here. However, it is important to explicitly include the gravitational body force, ϱg, which we tacitly treated as a hydrostatic pressure background above. We then obtain

$$\partial_z p = \varrho \left[\frac{du_f}{dt} - g \right] = \frac{p_f}{l}, \tag{2.47}$$

where z is the vertical coordinate along the fissure. Since u must be constant with respect to z, p can only vary linearly. If l is the depth to which the fissure has been filled, we have

$$\frac{dp}{dz} = \frac{p(z=0) - p(z=-l)}{l}. \tag{2.48}$$

Next we consider the pressure at the entrance of the fissure, p_f. If we demand that the pressure of the fluid in the overlayer, Eq. (2.46), and the pressure of the fluid within the fissure are identical where they meet at $r = b/\pi$, we obtain a differential equation for the temporal development of the flow velocity in the fissure,

$$l\varrho \left[\frac{du_f}{dt} - g \right] = \varrho g \tilde{H} - \frac{\varrho b}{\pi} \frac{du_f}{dt} \ln \left(\frac{\tilde{H}\pi}{b} \right) - \frac{1}{2} \varrho u_f^2. \tag{2.49}$$

Introducing the dimensionless quantities $\lambda = l/\tilde{H}$, and $\tau = t\sqrt{g/\tilde{H}}$, we obtain

$$\lambda(\lambda'' - 1) = 1 - \frac{1}{2}\lambda'^2 - \lambda'' \frac{b}{\pi \tilde{H}} \ln \frac{\pi \tilde{H}}{b}, \tag{2.50}$$

where the primes denote differentiation with respect to the dimensionless time variable, τ. This can be further simplified setting $\Lambda := \lambda + Q\left(\frac{b}{\tilde{H}}\right)$ and $B = 1 - Q\left(\frac{b}{\tilde{H}}\right)$, which finally leads to

$$(\Lambda'' - 1)\Lambda = B - \frac{1}{2}\Lambda'^2. \tag{2.51}$$

For considering the dynamics in phase space, we define a momentum-type variable, $M := \Lambda'$. We can now write Eq. (2.51) as a first-order differential equation for the flow in (Λ, M) space,

$$\partial_\tau \begin{pmatrix} \Lambda \\ M \end{pmatrix} = \begin{pmatrix} M \\ 1 + \frac{1}{\Lambda}(B - \frac{1}{2}M^2) \end{pmatrix}. \tag{2.52}$$

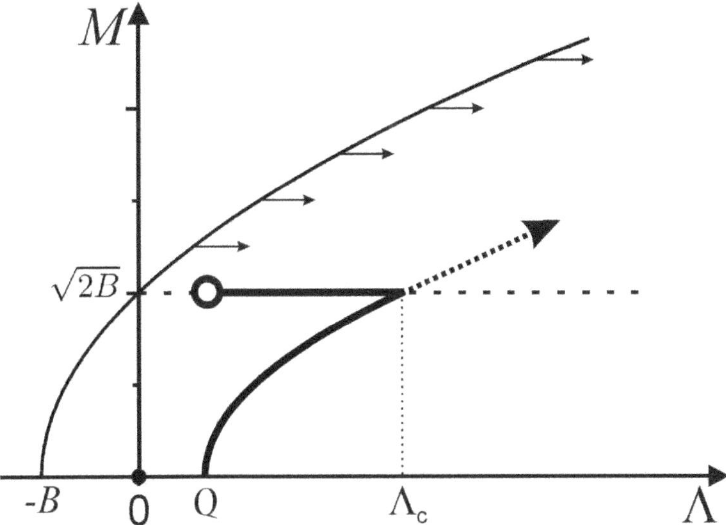

Fig. 2.19 Phase plot of the solutions of Eq. (2.51) in the plane spanned by Λ and $M = \Lambda'$. The thick solid curve indicates the trajectory starting at $l(t = 0) = 0$.

We first discuss the phase flow field at a few important points. On the Λ-axis ($M = 0$), the trajectories are generally pointing vertically upward ($\Lambda' = 0$). A horizontal phase flow pattern, $M' = 0$, is found along $M = \sqrt{2(\Lambda + B)}$, which is plotted as the thin solid curve in Fig. 2.19. As it will turn out below, it is of particular interest to investigate at which point the acceleration of the column, du_f/dt, drops below the acceleration due to gravity. Setting $du_f/dt = g$, we have $\frac{g}{H} l'' = g$, and therefore $\lambda'' = \Lambda'' = M' = 1$. This yields $M = \sqrt{2B}$, corresponding to the dashed horizontal line in the phase plot. Since there is zero pressure (vacuum) in the fissure below the column and there is no pressure drop along the column when $du_f/dt = g$, it follows that the pressure at the fissure entrance is zero as well. Since it is bound to drop further as the acceleration keeps decreasing, it will become negative. The inevitable result is *cavitation*: the column will break apart at the fissure entrance, forming voids which are only filled with water vapor at the saturated pressure of a few millibar.

As a consequence, the trajectory will never really enter the region above the dashed line. Instead, it will jump back along that line (since the streaming velocity will remain constant due to inertia), to the point marked by the bold open circle, above the point where it first started on the Λ axis ($\Lambda = Q$). From this time on, cavitation will occur again and again in rapid sequence, with the trajectory fluctuating about the open circle in the phase plot. Isolated patches of slurry are thus entering the fissure, and are accelerated further according to g as they proceed downward. It is of interest at which length of the column, l_c, cavitation will occur first. This corresponds to the point of intersection off the main branch of the trajectory with the dashed line.

In order to derive the shape of the trajectory in the (Λ, M) plane, we divide the differentials of Λ and M by one another and obtain

$$\frac{d\Lambda}{dM} = \frac{\Lambda M}{\Lambda + B - \frac{1}{2}M^2}. \tag{2.53}$$

Dividing by Λ and multiplying by dM, it is readily seen that the substitutions $y = \ln \Lambda$ and $q = \frac{1}{2}M^2$ directly lead to

$$\frac{dq}{dy} + q = e^y + B, \tag{2.54}$$

which is linear in q and can thus be solved by standard methods. The general solution of the homogeneous system (setting $e^y + B$ to zero) is obviously $q_{\text{hom}} = a\exp(-y)$, where the constant a is still to be determined. A special solution for Eq. (2.54) is easily guessed and reads

$$q_s = \frac{1}{2}e^y + B. \tag{2.55}$$

Re-substituting Λ and M into $q(y) = q_{\text{hom}} + q_s$, we finally obtain for the trajectory

$$M = \sqrt{\Lambda + 2B + 2a/\Lambda}. \tag{2.56}$$

The constant a is determined from the condition that $M(\lambda = 0) = 0$. For $\lambda = 0$ we have $\Lambda = Q = 1 - B$. We directly obtain from Eq. (2.56)

$$a = \frac{Q}{2}\left[Q - 2\right]. \tag{2.57}$$

This determines the trajectory, which is shown in Fig. 2.19 as the bold solid curve.

Cavitation takes place if $M = \sqrt{2B}$, i.e., when $\Lambda + \frac{2a}{\Lambda} = 0$. This yields

$$\Lambda_c^2 = Q\left[2 - Q\right] \approx 2Q, \tag{2.58}$$

and thereby, owing to the properties of Q (cf. inset of Fig: 2.18)

$$l_c \approx \sqrt{b\tilde{H}}. \tag{2.59}$$

Hence cavitation sets in when the length of the column corresponds to about the geometric mean between the fissure width and the height of the sand overlayer. This is generally much less than the depth of the sand layer, and we can conclude that the cavitation scenario dominates the whole process, except the very first moment when the column has just entered the fissure.

Filling will thus proceed by isolated loads of slurry falling more or less freely into the fissure. At the fissure entrance, the fluid rapidly reaches a stationary inflow velocity, u_{stat}, which can be obtained form the condition $M = \sqrt{2B}$ (cf. Fig. 2.19). This yields

$$u_{\text{stat}} = \sqrt{2g\tilde{H}\left(1 + Q\left(\frac{b}{\tilde{H}}\right)\right)} \approx \sqrt{2g\tilde{H}}, \qquad (2.60)$$

which again is the velocity an object falling freely from a height $\tilde{H}$ would reach at the bottom. The flow velocity at the entrance of the fissure is thus similar as in the dry case, but somewhat larger, since $\tilde{H} > H$. Furthermore, in contrast to the dry granulate, the slurry will experience the full acceleration due to gravity down into the fissure. Since the presence of water strongly supports the formation of cracks in silicon oxide, as is commonly exploited in glass cutting devices, we may conclude that *if* we find a significant amount of broken grains in thin sections from an outcropping earthquake fissure, this would indicate the filling to have taken place under wet conditions.

2.2.5 *Dilatancy*

We can narrow down our conclusions even further. What do we expect if sedimentation has led to a solid volume fraction which is large enough to prevent the slurry from flowing as easily as we so far assumed? This is in fact what we should expect if the sediment had ample time to form a compact pile. Looser packings are possible, but usually require exceptionally rough grains (which would show up in thin sections of fossil fissure samples) or the presence of metabolism products from, e.g., bacteria or algae, providing sufficient adhesion between the grains to preclude effective compaction. In most cases, we should thus expect that the granulate is *not* able to flow at the density of its sedimentation. If such granulates are *forced* to flow, it directly follows that they need to reduce their packing density. This is in fact a widely observed phenomenon in granular systems, and is called *dilatancy*. The volume fraction above which dilatancy occurs when the granulate is being deformed is called *dilatancy onset*, and will be denoted by ϕ_{do}.

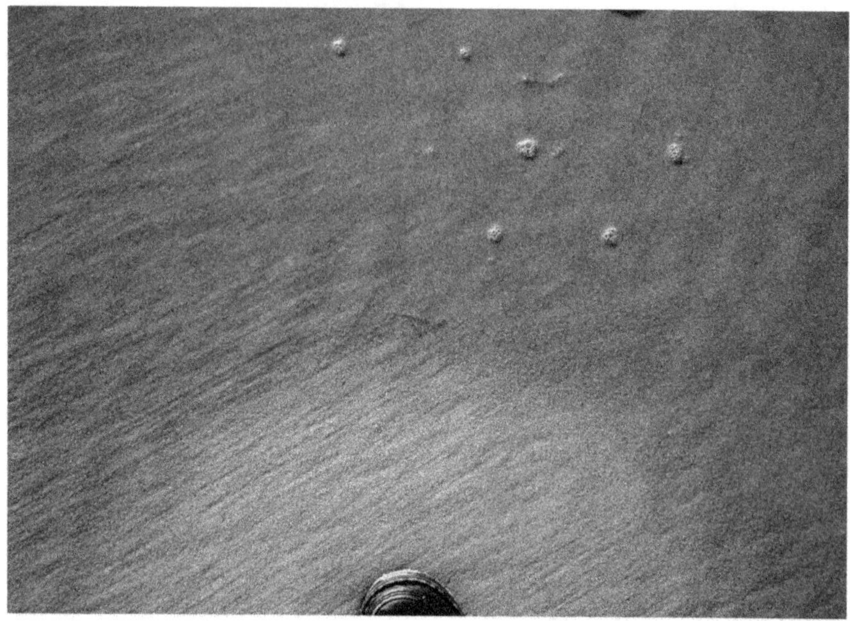

Fig. 2.20   A simple field experiment on dilatancy. If one steps on liquid-saturated sand at a beach, the surface of the sand becomes dry around the foot (center of lower edge of picture), due to the retraction of the fluid deeper into the ground (photo by the author).

A well-known consequence of dilatancy can be observed at the beach (cf. Fig. 2.20). The sand had a long time to consolidate, and can be considered to be rather densely packed. In the surf zone, the sand is fully saturated with sea water, i.e., the interstitial space is free of air, but the liquid interface coincides with the surface of the sand packing. When one walks over the sand, one observes that it seems to become dry around the foot whenever it presses on the ground. In Fig. 2.20, this shows up as the brighter area near the bottom of the picture. At first glance, this is counterintuitive, because one might expect the water to be pressed out of the ground instead. However, since stepping on the ground shears the granular packing out of its optimal configuration, the interstitial volume must increase, since the overall volume does, while the total volume of the grains of course stays constant. Since the water is incompressible, its volume remains constant as well, and so it must recede into the ground.

Modern imaging techniques allow to investigate this effect on the microscopic scale. By means of X-ray microtomography, a three-dimensional

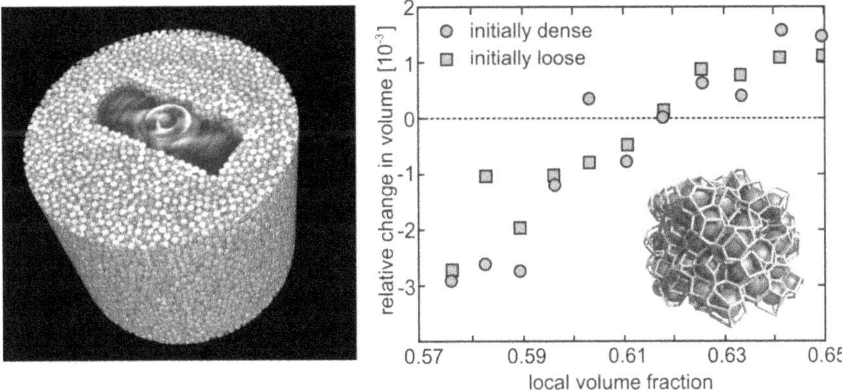

Fig. 2.21   Left: Rendering of three-dimensional X-ray tomography data of a pile of glass beads in a cylindrical sample vial. In the center, a paddle is visible to which a disordered layer of glass beads has been glued. Moving this paddle deforms the packing geometry of the pile. Right: The relative change in volume of Voronoi cells. A Voronoi cell of a glass bead is the set of points closer to the center of that bead than to the center of any other bead. The inset illustrates the Voronoi cells by virtue of their edges. Obviously, the change in Voronoi volume (or, equivalently, the local volume fraction), only depends upon the local volume itself, but not on preparation history (dense or loose packing). Data kindly provided by J.-F. Metayer and M. Schröter, Göttingen.

image of a granular pile can be generated.[11] The left panel of Fig. 2.21 shows a rendering of such three-dimensional data, obtained for a pile of glass beads in a cylindrical glass tube with an inner diameter of about one centimeter. In the center of the sample one can see a paddle sticking into the pile of beads, which can be used to deform the packing in a controlled manner. With suitable algorithms, one can identify each individual bead and determine the position of its center, as a function of displacement of the paddle. It is possible to determine the local packing densities around each grain while the pile is deformed by slowly pulling the paddle out of the sample.

The local packing fraction is determined by means of the so-called Voronoi tesselation, which is illustrated in Fig. 2.22. The Voronoi cell belonging to a grain is defined as the set of points in space which are closer to the surface of that grain than to the surface of any other grain. For ideally spherical grains of equal size, this is identical to the well-known Wigner-Seitz cell, determined with respect to the centers of the spheres. The local volume fraction can be defined as the inverse of the volume of the Voronoi cell.

---

[11] A detailed description of X-ray micro-tomography will by given in Chapter 6.

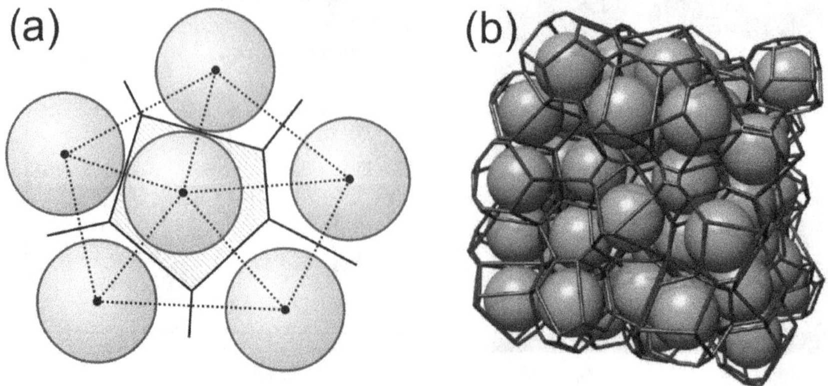

Fig. 2.22 (a) Construction of the Voronoi cell, which is defined as the set of points in space which are closer to the surface of that grain than to the surface of any other grain. (b) Rendering of the Voronoi cells of a three-dimensional random packing of spherical grains.

The right panel of Fig. 2.21 shows the results for small deformations, with the paddle having been moved only by a fraction of a particle diameter. At many times during the deformation process, the set of all center of mass positions of the beads has been determined, as well as the corresponding Voronoi volumes assigned to each bead. The plot shows that each Voronoi volume dilates or shrinks depending only on its actual size, but independent of the preparation of the pile. This shows that dilatancy is an intrinsic property of the granular material, and is not contingent upon preparation protocols or sample size. The packing density at which the data cross through the ϕ-axis is called dilatancy onset, $\phi_{\mathrm{do}} \approx 0.62$. As long as $\phi_s < \phi_{\mathrm{do}}$, we do not have to take dilatancy into account. This corresponds to the calculation above (Section 2.2.4), where we assumed the granulate to flow as a slurry with Newtonian rheology. Here we want to find out what happens if $\phi_s > \phi_{\mathrm{do}}$, such that the granulate needs to expand when shear is applied.

Let us derive a mathematical formulation of dilatancy which is suitable for being inserted into our equations for earthquake fissure filling dynamics. We introduce the parameter

$$\delta_{\mathrm{dil}} = \frac{\phi_s - \phi}{\phi}, \tag{2.61}$$

which describes the amount of dilatancy which incurs when the slurry starts flowing. Mass conservation of the granulate entails that at the fissure entrance, where the granulate is fully dilated, the flow velocity of the grains must be elevated by a factor $\phi_s/\phi = (1 - \delta_{\text{dil}})^{-1}$ with respect to its value without dilatancy. At the same time, the velocity of the interstitial liquid will be decreased, because the interstitial volume is increased by a factor $(1 - \phi)/(1 - \phi_s)$. As a result, there must be a relative motion of the interstitial liquid with respect to the grains with a certain relative velocity, v_{rel}. We easily find, to first order in δ_{dil},

$$v_{\text{rel}} = -u\frac{\delta_{\text{dil}}}{1 - \phi_s}. \tag{2.62}$$

Far away from the fissure entrance, where there is no dilatancy, the velocity of the granulate and the interstitial liquid must be equal, since a relative displacement would require a pressure drop scaling with the total system size. Focussing thus on the vicinity of the fissure entrance, we can combine Eq. (2.62) with Darcy's law, Eq. (2.23), and obtain

$$\partial_r p = -u_f \alpha \eta_0 \frac{\delta_{\text{dil}}}{1 - \phi_s}\frac{b}{\pi r}, \tag{2.63}$$

which can be integrated over r to yield the total extra pressure drop due to dilatancy,

$$p_{dil} = -u_f \alpha \eta_0 \frac{\delta_{\text{dil}}}{1 - \phi_s}\tilde{H}\, Q\left(\frac{b}{\tilde{H}}\right). \tag{2.64}$$

Including this term in the equation of motion, Eq. (2.49), leads directly to

$$\lambda(\lambda'' - 1) = 1 - A_{\text{dil}}\lambda' - \frac{1}{2}\lambda'^2 - \lambda''Q, \tag{2.65}$$

with the abbreviation

$$A_{\text{dil}} = \frac{\alpha \eta_0}{\rho g}\frac{\delta_{\text{dil}}}{1 - \phi_s}Q \tag{2.66}$$

in place of Eq. (2.50). It is now of interest to see which of the terms is dominant. From the last paragraph we can see that since u_f is at most of order $\sqrt{g\tilde{H}}$, λ' will never become much larger than unity. On the other hand, it is readily appreciated that A_{dil} is of order one hundred, or even more; in any case where there is noticeable dilatancy, it will be large as compared to unity. We therefore can safely neglect the λ'^2 term in Eq. (2.65) against the A_{dil} term. In place of Eq. (2.51), we then obtain

$$(\Lambda'' - 1)\Lambda = B - A_{\text{dil}}\Lambda'. \tag{2.67}$$

60   *Wet Granular Matter: A Truly Complex Fluid*

As before, we discuss its solutions in phase space,

$$
\partial_\tau \begin{pmatrix} \Lambda \\ M \end{pmatrix} = \begin{pmatrix} M \\ 1 + \frac{1}{\Lambda}(B - A_{\mathrm{dil}}M) \end{pmatrix}, \tag{2.68}
$$

which is plotted in Fig. 2.23. The trajectories take off vertically from the Λ-axis as before, and flow horizontally to the right wherever $M' = 0$, i.e., along the line $M = (\Lambda + B)/A_{\mathrm{dil}}$ (thin solid line in Fig. 2.23). As before, the acceleration of the column falls below the acceleration due to gravity when $M' = 1$. This yields $M = B/A_{\mathrm{dil}}$ for the locus of points in phase space where we expect cavitation. It is indicated by the dashed line in Fig. 2.23. Note that since B is of order unity and A_{dil} is of order one hundred, we are now exploring a range of velocities which are much smaller than before (Fig. 2.19). More specifically, the velocity at which cavitation (detachment of the column) sets in is smaller by a factor $B/A_{\mathrm{dil}}\sqrt{2B} = \sqrt{\frac{1}{2}B}/A_{\mathrm{dil}}$.

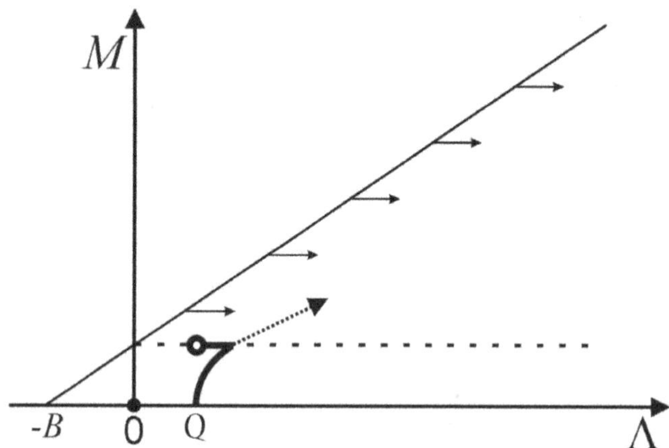

Fig. 2.23   Phase plot of the solutions of Eq. (2.67). The ordinate spans a much smaller range as compared to Fig. 2.51.

The stationary velocity is of order $\sqrt{g\tilde{H}/A_{\mathrm{dil}}}$, which is roughly two orders of magnitude slower than before. The granulate enters the fissure at a velocity of only several centimeters per second, and does so as a paste which is no longer saturated with liquid, due to dilatancy. The interstitial space between the grains is partly filled with water, and partly with its saturated vapor. This is just the ternary material from which the sand castle in Fig. 1.1 has been constructed. As we know, this is a stiff matter which does not easily flow at high rates. Filling the earthquake fissure will take many hours in this case, and we should not expect cataclastic flow, i.e., no breakage of grains. Consequently, *if* we find a significant amount of broken grains in thin slices of the fossil fissure quartzite, we can conclude that there was water present in the sediment as the earthquake occurred, *and* the sedimentation density was low, pointing to the presence of sticky organic material, and hence microbial activity in the interstitial space between the grains.

2.3   The Great Gamsberg: A fossilized oasis from the Karoo

In fact, a glance of thin sections of Gamsberg sediment samples yields enlightening insight here. Figure 2.24 shows two micrographs of quarztite thin slices taken from two different places. In Fig. 2.24(a), a sample from the top quartzite plate is presented [70]. In this technique, rock samples are cut into thin slices which are subsequently milled down to a thickness of about 30 microns. At this thickness, the optical anisotropy of crystalline quartz is just sufficient to let quartz grains appear in various shades of grey between crossed polarizers, depending on their angular orientation relative to that of the polarization filters. We see that all grains in the sample are well rounded, as expected from a well-sorted aeolian quartz sediment such as the Karoo desert sand. Furthermore, the pore space is seen to be entirely filled, with its hue corresponding to one of the surrounding grains. This is indicative of epitactic overgrowth of crystalline quartz on the respective

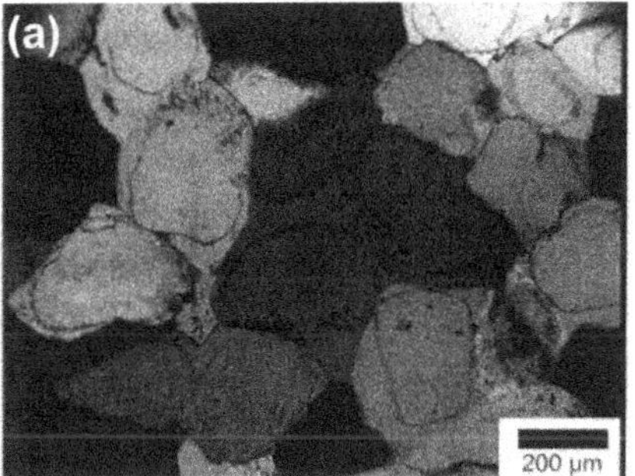

Fig. 2.24   Thin slices of quartzite samples from the Great Gamsberg. (a) Top quartzite plate, taken from [70]. The high degree of lithification is evident, which hints at the presence of interstitial liquid bearing organic material [71]. (b) Earthquake fissure (courtesy of Reinhold Wittig). The abundance of fractured grains is obvious, providing evidence for cataclastic flow in the filling process.

grains. Note that excessive precipitation of crystalline silicate is known to correlate with the presence of organic material [71].

2.3.1　*The earthquake fissure quartzite: Evidence of water*

Figure 2.24(b) shows a thin slice from a quartzite sample taken from one of the earthquake clefts (cf. Fig. 2.14(b)). In marked contrast to Fig. 2.24(a), it exhibits a predominance of fractured grains, with otherwise similar diagenesis, i.e., interstitial pore space virtually completely filled with epitactic overgrowth of crystalline quartz on the fractured grains. According to what we have said at the end of Section 2.2, the abundance of fractured grains in the fissures suggests that the process of filling of the earthquake fissure has been cataclastic, pointing to a liquid as the carrier, and to only limited compaction before the process. The latter may be interpreted as being due to the presence of biological matter, the stickiness of which prevents compactification towards densities which might give rise to dilatancy when flow of the slurry sets in. This suggests that during the time of the formation of the clastic dykes, the sand sediment on top of the Gamsberg was not a dry sand layer, but was saturated by water, which also contained biological material. This could have been ground water, or the sand sediment even was the bottom sediment of a lake.

In order to distinguish between these two possibilities, we need more evidence, which we can gain from the large scale morphology of the Gamsberg plateau. As Fig. 2.14 already suggests, the plateau is amazingly flat. Furthermore, the panoramic view in Fig. 2.25 suggests that there is almost no deviation from a perfect plane, aside from the usual brush, debris and pieces of rock lying on its surface. The slight tilt of this plane with respect to the perfect horizontal, which has been determined to be of the order of a milliradian, is well within what may be expected from large scale subsidence or tectonic corrections of the position of the craton the mountain is sitting on.

On the other hand, the thickness of the quartzite top plate varies considerably across its extension. At its boundaries, the outcropping quartzite exposes a thickness (i.e., vertical distance between the top plane and the contact between the quartzite and the granite basis) ranging from just one meter at some spots on the SE rim to about 30 m at the NW rim, and even more towards the west. Hence we find that the interface between the top quartzite plate and the supporting granite deviates much more from a perfect plane than its top surface. The quartzite top plate appears to have

Fig. 2.25   Panoramic view of the Great Gamsberg top plate, evidencing its unusual flatness. In fact, the deviation from a perfectly horizontal plane merely consists (aside from the brush an loose rocks) in a slight tilt, amounting to less than ten meters of elevation difference over its entire length of about 3 km (photo by the author).

a lens shape, with an almost perfect plane on the top, but with a thickness considerably varying along its lateral extension.

This immediately suggests that the morphology of the Gamsberg top plate represents the shape of a lake, with its perfectly flat upper (water) surface, and the lower surface being determined by the topography of the underlying rock. Hence we arrive at another question concerning the physics of wet granular matter: if a lake is being filled with sand, e.g., because it is surrounded by a desert which sends large sand dunes wandering over the lake, will the surface of the resulting sand deposit represent the surface of the former lake?

2.3.2   *Experimental support for the model*

Let us contemplate for a moment why this might indeed be the case. Dunes, such as those being shown in Figs. 2.11 and 2.15, are formed and being transported by wind blowing sufficiently strongly to carry sand grains over some distance. As a result, sand dunes are known to wander considerable distances as time proceeds, following the direction of the wind. If there is a lake on the trajectory of a sufficiently large dune, the sand freight of the latter will fill the lake. We might anticipate (and will indeed corroborate in Chapter 4) that the water will glue the grains together by virtue of its surface tension, such as to prevent the wind from carrying further away any grains which have become wet. Although we can not (yet, but cf. Section 7.1) anticipate how far capillary ascension may suck the water up into the granular pile being blown over the lake, it may appear possible that the wind will carry along all grains which remain above the water level (i.e., the water table within the dune), but leave those in place which have made contact with the water. After the dune will have passed the lake, a body of wet sand will then result whose interstitial space is completely filled with water and whose upper surface represents the former surface of the lake.

As this would be an immediate explanation for the shape of the Gamsberg top quartzite plate, it appears valuable to test this idea by means of an experiment. A photograph of the setup is shown in Fig. 2.26. The central (stainless steel) chamber visible in the center of the image has a height of 60 cm and a lateral extension of 80 cm. Its width, as well as the height

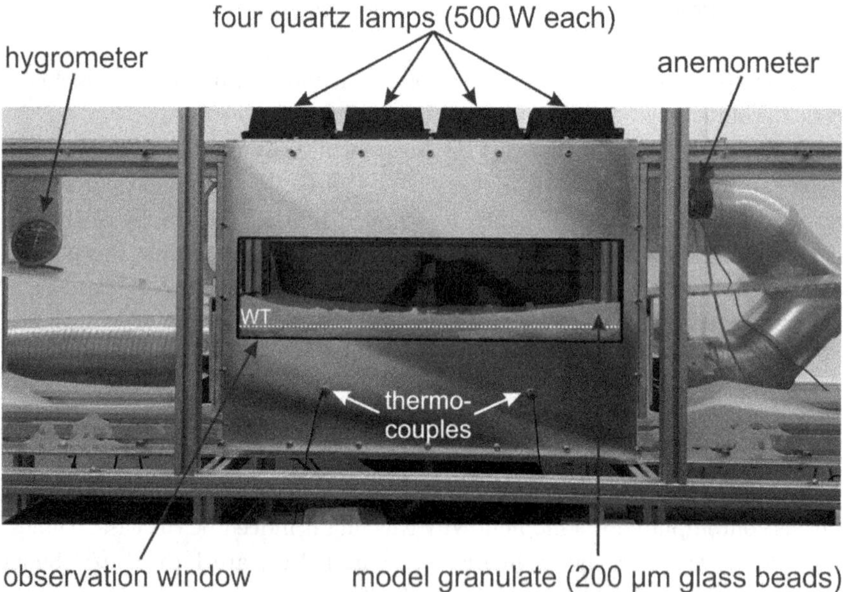

Fig. 2.26 Experimental setup for investigating the interaction of a water table (WT, white dotted line), a granulate (200 microns glass beads, white deposits), and wind (actively blown by a fan). The water table was adjusted by connecting the central chamber with a flexible hose to a bucket of water whose vertical position could be varied. Typical wind velocities were about 7 m/s. With the quartz lamps (top) switched on, humidity settled to about 20%.

of the window, is 20 cm.[12] As the model sand, we use standard ballottini with an average diameter of 200 microns. The bottom of the chamber is connected by a flexible hose to a bucket (not shown) which is half way filled with water, and whose vertical position can be adjusted. In this way, the water table within the granulate can be adjusted at will. In the picture, it is indicated by the white dotted line (WT).

To the left and to the right of the central chamber, there is an entrance (left) and an exit (right) air chamber which have plexiglass covers in the front and in the back. In the entrance chamber one can discern a hygrometer, an anemometer in the exit chamber. Two thermocouples are used to monitor the temperature of the material below the water table (WT). Behind the setup, one can see the air tubes through which air is blown in a closed cycle by means of a powerful fan. On the top, four quartz lamps can be seen which are used to mimick solar irradiation, with a total power of 2 kW. Typical wind velocities were about 7 m/s, humidity was measured to be about 20% when all lamps were switched on. This is well in line with typical conditions encountered in desert scenarios.

After blowing for 22 h with a wind velocity of 7 m/s, the morphology of the surface of the granular bed was as depicted in Fig. 2.27. In Fig. 2.27(b), the water table is indicated by the white arrows and the faint dark line drawn on the glass window. We can see that while capillary ascension effects, which may vary laterally due do the wide size distribution of the grains, lead to a quite rugged topography of the resulting wet granulate surface after wind blow. However, its topographic deviation from the water table nowhere exceeds a few centimeters (the higher parts of the granular pile towards the glass windows are due to the finite container size and can be disregarded). More evidence is provided in the lower part of the figure, which shows height profiles, both longitudinal and transversal, taken from of the resulting wet granular body in Figs. 2.27(a) and 2.27(b), relative to the water table. We see that the deviation from the water table is nowhere greater than 3 cm.

As it appears, the idea that the wind will carry along all grains which have not been touched (appreciably) by the water is corroborated by the experiment. This suggests that indeed the consequence of sand dunes being blown over a lake may be a body of wet (saturated) sand whose upper surface, up to an accuracy of a few centimeters, coincides with the water surface of the (former) lake. A condition for this to happen is that during

---

[12]The window is a glass plate of 8 mm thickness. Its presence is evidenced from the mirror image of the author taking the photo.

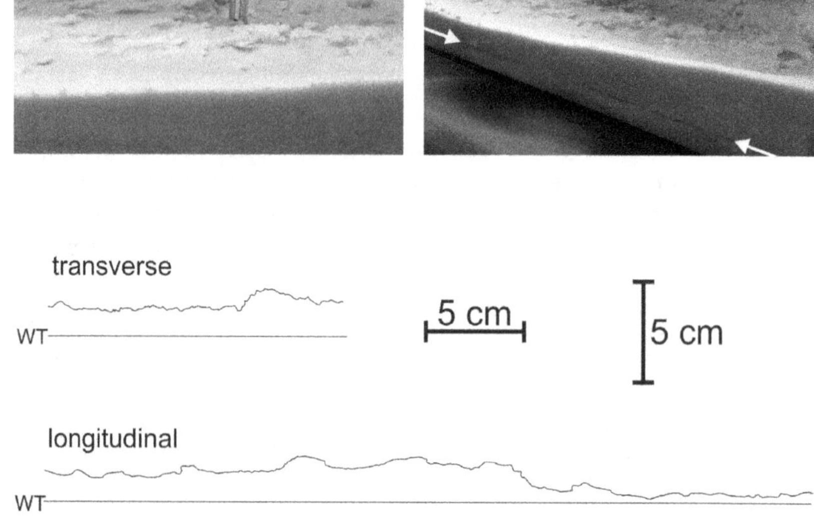

Fig. 2.27   Experimental results for a wet granular bed exposed to blowing wind. After 22 h of wind (7 m/s, 30°C, 20% humidity), morphologies shown in (a) and (b) are typically encountered. Bottom: profiles taken from the granular bed morphologies. WT: water table. The surface profile obviously follows the (horizontal) water table within few centimeters.

the whole process, the height of the water table remains constant. This is possible if there is a continuous source of water filling the lake and, at the same time, its water table is determined by the height of a saddle point in the rock basis, over which any surplus water drains away. If this is fulfilled, the granular body which finally forms will precisely resemble the shape of the water body at maximum height. This strongly supports the idea that the top quartzite plate of the Gamsberg is just that: the fossil remains of a lake which has been filled with sand, by sand dunes wandering over it.

2.3.3   *The Gamsberg: A geological narrative*

Taking all of the above together, including our studies of granular flow, we are now in a position to sketch an approximate history of the origin of the Gamsberg, as we see it today. It is sketched schematically in Fig. 2.28 in six

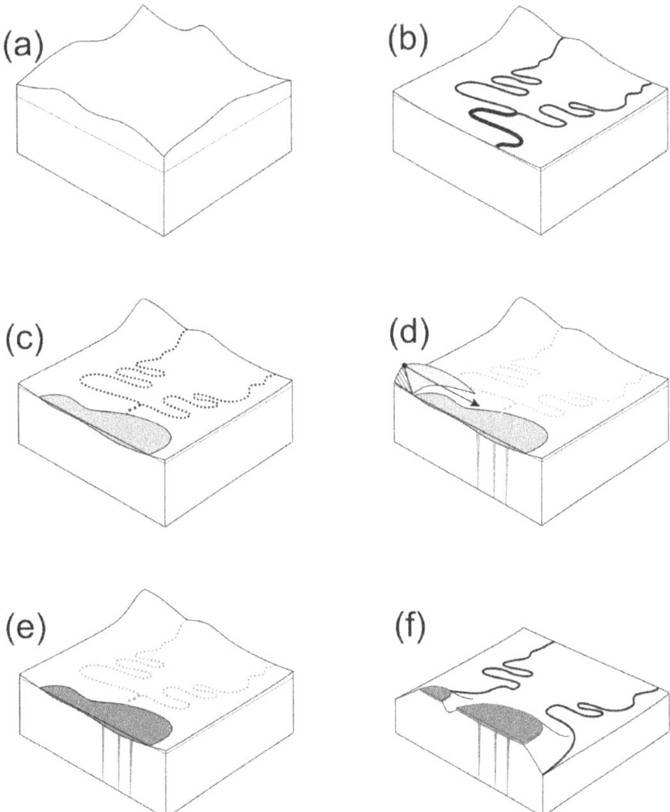

Fig. 2.28 A pictorial history of the emergence of the Gamsberg. (a) Starting point: the Damara mountain range, close to the south pole at that time (the thin dotted lines indicate a horizontal plane for reference). (b) At the end of the Dwyka glaciation, massive melt waters give rise to the formation of a vast peneplain. (c) As a peneplain is never an exact plane, water accumulates in shallow depressions forming lakes. (d) As climate grows more arid, aeolian sediment (sand) forms which accumulates in dunes. These are wandering over the lakes (arrow). Extensional tectonic forces (breakup of Gondwana) lead to clastic dykes (earthquake fissures). (e) Advection of silicates enables strong lithification of the saturated granular body, forming a solid quartzite plate. (f) As stronger erosion sets in again, the granite is removed wherever there is no protection through the (inert) quartzite. The Little and the Great Gamsberg are formed.

steps. It starts off with the Damara mountain range, which was based upon late precambrian granite (Fig. 2.28(a)) [72]. In the later Mesozoic (Permo-Carboniferous), it was close to the southern pole, such that it was largely covered by massive ice sheets, a period known as the Dwyka glaciation. As plate tectonics moved the Gamsberg area farther away from the pole, temperatures started rising, heralding the end of the Dwyka glaciation. Mighty meltwater streams then milled down parts of the mountain range, forming a vast peneplain in the granite, Fig. 2.28(b).

Such peneplains are generally rather flat, and close to horizontal, as a consequence of the meandering of the rivers which lead to their formation. However, they are of course by no means mathematically exact planes; there are shallow depressions, slight elevations, and saddle points in between. As a consequence, temporary freshwater lakes formed in depressions, as the water supply gradually decreased from massive melt water supply (at the end of the glaciation) to occasional supply from rain within the respective catchment (Fig. 2.28(c)). For each of these lakes, the highest possible stationary water level is given by the height of the lowest adjacent saddle point, over which it drains any surplus water further along, towards the ocean. A depression a few tens of meters deep, where the Gamsberg later stands, is shown here schematically.

Further aridization then lead to the formation of aeolian sand, mainly from the quartz granulate which must have been present as the chemically inert remnant of the eroded granite. Because of its limited quantity, it probably accumulated into barchan dunes (Section 2.2). When a barchan (or any dune) migrates across a freshwater lake (Fig. 2.28(d)), the latter is being gradually filled with sand, and all sand that comes to lie above the filling mark dries up and is blown away, just as observed in the experiment discussed above. This creates a sand surface that represents, with an accuracy of a few centimeters (cf. Fig. 2.27), the water surface of the lake. A necessary condition for this scenario is that there is still enough water supply to have the depression filled to the maximum height (i.e., the height of the saddle point over which it drains surplus water) most of the time. The time span needed to complete the process sketched in Fig. 2.28(d) must have been the one where the earthquake fissures (Fig. 2.14(b)) have formed. During this time span, one may say the lake which later formed the Gamsberg was an oasis in a Karoo desert. The mollusc trace shown in Fig. 2.16(a) might belong to this time.

As time proceeded, continued advection of dissolved silicates from the eroded granite peneplain lead to epitactic growth of quartz on the sand grains, into the interstitial space of the saturated granulate. Hence a completely lithified quartzite plate formed, whose shape was identical to that of the water body of the lake which had initially formed. Since the dissolved silicate was in direct (diffusive) contact with the interstitial space of the granulate in the earthquake fissures, lithification took place there as well (Fig. 2.28(e)).

Later the arid conditions alleviated again, and flooding became more frequent. The granite surrounding the inert quartzite slab started to be eroded away, such that the edges of the quartzite plate became exposed. As they were very thin, they broke off into small pieces which were distributed over the plate during floods. It is these rocks in which gypsum rose casts like the ones shown in Fig. 2.16 are abundantly found on the Gamsberg plateau. The main part of the quartzite slab, where it was many meters thick, formed a solid stencil which prevented the erosion of the granite below. Hence the latter was eroded away only outside the quartzite plate area, thus giving rise to the (double) mesa mountain we see today. In this picture, we may assume that the Gamsberg is a fossilized oasis (Fig. 2.28(d)) from the Karoo desert.

2.4 Conclusions

We have seen that many concepts of the kinetic theory of molecular fluids apply quite well to granular systems, even though the grains are not identical and the typical energy scales exceed thermal energies by far. We can use concepts like pressure, temperature, and viscosity just as in a regular fluid. However, the inherent dissipation of energy into the many atomic degrees of freedom of the grains adds new aspects to the dynamics, which may appear subtle, but at times lead to substantial consequences, like an apparent violation of the second law of thermodynamics. These ramifications are particular to granular systems, and identify them as systems far from thermal equilibrium.

For binary systems consisting of grains and a 'carrier fluid', we have used examples from geosciences to illustrate the interplay of the rheological properties of the fluid and the peculiarities of the granular system. We showed that a situation as complex as the filling of an earthquake fissure leads to equations of motion which can be solved analytically for both gaseous and liquid carrier fluids, and both with and without dilatancy. The observed variability in the outcomes is in fact striking, and can potentially be used to draw detailed conclusions on subtle differences in the fluid properties, like the initial packing density or the friction between the grains. In fact, we could draw immediate conclusions on the geological history of a natural monument in central Namibia, the Great Gamsberg, which in light of our reasonings appears to be a fossilised oasis of the Karoo desert, from the times when Gondwana broke apart into what are today Africa and South America.

Further reading

We had to be very brief on the kinetic theory of granular gases. There is an extensive review by Isaac Goldhirsch [73] and the well written book by Brilliantov and Pöschel [12]. Both are, at different levels of completeness, perfectly suited to fill this gap for everyone interested in more details. There are as well reviews available covering the most recent developments in granular gases [74, 75].

The physics of the binary granular systems, like slurries and powder clouds, is of course a lot richer than it was presented here. There is meanwhile a quite complete picture as to the rheology of slurries in terms of some important dimensionless numbers, such as the inertial number,

$$I = \frac{\text{strain rate}}{\text{inertial rate of rearrangement}}, \tag{2.69}$$

and the viscous number,

$$J = \frac{\text{viscous stress}}{\text{confining stress}}. \tag{2.70}$$

A careful selection of recent papers can bring the reader swiftly to the cutting edge of current wisdom [64–66]. Concerning the particular aspect of dilatancy, the compilation of pertinent literature found in [68] should be very useful.

There is a vast body of literature about the general phenomenology of dry granular matter. From the classical papers of Bagnold [76–78], who pioneered granular science as a modern physics topic, to the many exciting contemporary articles on the dynamics of dunes [46, 50, 48, 79], there are lots of interesting and sometimes unexpected effects to learn about. A particularly appealing line of thought is the application of knowledge about dune formation and morphology to remote scenarios on other planets, where it can lead to valuable insight into climatic conditions [46, 47]. To get into the matter, it is certainly useful to read through available reviews [29, 80–82].

For those who are interested to learn more about avalanches and turbidity currents, a few papers [83–86] with the literature cited therein should make a reasonable start.

Chapter 3

Wetting

Fig. 3.1 A drop of dew sitting if a leaf of *Cotinus coggygria*, making almost no contact with the leaf surface. This is due to the interplay of the wetting forces with the microstructure of the epi-cuticular wax of the leaf (Image by Jutta Wolf, Ulm).

Having dealt so far with just a binary system, let us now turn to the case where both a liquid and a gas are present in the interstitial space between the grains. The surface tension of the liquid thus enters the stage, and we will readily see that (and how) it requires a dominant role in the play. We will have to deal with the complex geometrical structure of the solid surfaces which are wetted by the liquid. The surface of even a perfectly spherical grain is non-planar, requiring special attention. Moreover, the inner surface of a random pile of such grains exhibits a geometry of mind-boggling complexity. Moreover, the omnipresent surface roughness poses a significant additional challenge to successfully predict the interplay of the liquid surface with the system of grains. The drop of dew shown on the picture above illustrates the substantial effect the interplay between wetting forces and substrate geometry can have. Although the contact angle of water with the epi-cuticular wax of the leaf is only around ninety degrees, the drop beads off almost perfectly, which is due to the microscopic structure of the rough crystalline wax layer on the leaf surface. Roughness can in fact promote both wetting and non-wetting, and we will see that the appropriate treatment of roughness is of substantial importance in wet granular materials.

In order to acquire some feeling for the interaction of a pile of grains with a wetting liquid, we start by imagining a pile of dry grains to which we gradually add small amounts of liquid. The best defined way to achieve this is to add increasing amounts of the vapor of the liquid to the gas phase. As a consequence, liquid will homogeneously adsorb on the grain surfaces everywhere in the pile. Initially, this will form a molecularly thin film which follows the topography of the grain surface, including the roughness on it. Later, when more liquid has adsorbed, it will have filled the troughs and crevices of that roughness, and the grain surfaces appear completely wet. As more and more liquid accumulates, the liquid–gas interface will have to accommodate the twisty maze of interstitial pores between the grains, before finally all of the gas phase is being expelled in favor of the liquid phase.

It does not come as a surprise that on the way from the adsorption of the first molecules to the expulsion of the last bits of gas, a large number of interesting and quite complex phenomena are encountered. Figure 3.2 gives a rough idea of the complexity of liquid morphologies we will have to deal with. It shows cross-sections of three-dimensional snapshots taken by X-ray micro-tomography while a wetting liquid (oil phase, white) is expelled from a pile of glass beads (grey disks) by means of injection of a non-wetting liquid (water plus contrast agent, black). In order to provide the reader with the necessary tools to tackle such complex structure, we will in this chapter outline the basic concepts of wetting, and apply them to a few paradigmatic cases pertinent to wet granular matter.

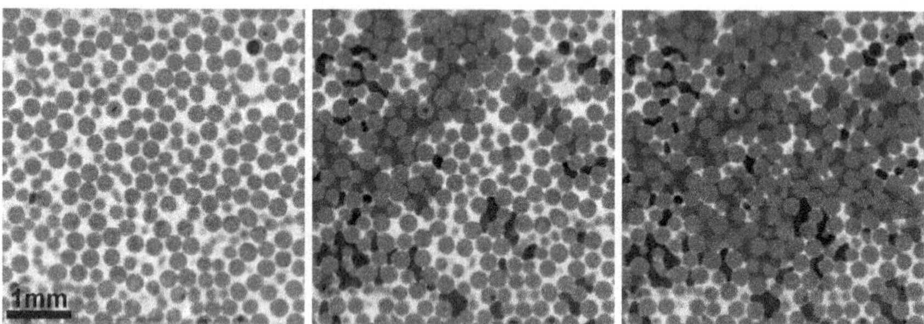

Fig. 3.2   Two-dimensional sections through piles of glass beads at variable content of wetting liquid, taken by means of X-ray-micro-tomography. The grey disks are sections through the individual glass beads. White: wetting liquid (oil phase; secondary fluid). Black: non-wetting liquid (aqueous phase, including contrast agent; primary fluid). The oil content of the sample decreases from left to right, as the oil phase is gradually expelled by injection of aqueous fluid. The average diameter of the beads is 300 μm (image courtesy of Ralf Seemann, Saarbrücken).

3.1 Planar substrates

In order to introduce the concepts of wetting, it is useful to forget for a moment the complex topography of the solid–gas interfaces encountered in wet granular media. With an ideally flat solid substrate in mind, we now discuss the basic interaction forces between an adsorbed liquid and this substrate, and the resulting liquid–gas interfacial morphology thus emerging.

3.1.1 *van der Waals forces*

Aside from gravitation, the most ubiquitous force acting between uncharged bodies is the van der Waals force. The most general way to explain the mechanism of its occurrence is to consider the quantum-mechanical zero-point energy of the electromagnetic field in the vicinity of the bodies under consideration. It turns out that the frequencies, ν_i, of the eigenmodes of the electro-magnetic field always shift to lower values as the bodies are approached to each other. Hence the energy of the zero-point fluctuations, $\frac{1}{2}h\nu_i$, are reduced as well, resulting in an attractive force between the two bodies [3].

A less formal picture of this mechanism is to consider the two fluctuating electro-magnetic dipoles of the bodies separately. The field of one instantaneous dipole of one body induces a dipole moment in the other body, the field of which will in return lower the energy of the first one. As the field strength of the dipolar field varies as the inverse third power of the distance for point-like particles, it is clear that the induced dipolar interaction, named after its discoverer *van der Waals*, varies as the inverse sixth power of the distance between the particles.

If the separation of the particles becomes comparable to the wavelength of light corresponding to their electronic resonances, retardation effects come into play, and the decay is proportional to the inverse seventh power of the distance. Since the resonances of most molecules of relevance are in the ultraviolet, a few tens of nanometers are sufficient for retardation effects to become relevant.

Owing to its ubiquity, the impact of van der Waals interactions can be traced everywhere, starting from the condensation of a gas into a liquid, due to the mutual van der Waals attraction of its molecules, to the interaction of the resulting liquid phase with a solid surface. To discuss these effects with some focus on the systems to be treated here, it is useful to start by considering the interaction between a solid surface and a single molecule.

As we have seen above, the van der Waals interaction energy between two atoms or molecules scales as the inverse sixth power of their distance. The interaction energy between a solid wall and a molecule at distance d will involve an integration over the half space filled by the solid. By virtue of dimensions, the result of the integration over three spatial coordinates is that the total interaction energy scales as the inverse third power of d. At larger distances, a crossover to a scaling with the fourth power of d takes place due to retardation, but this will henceforth not be discussed.

It should be noted in passing that the electromagnetic field is singled out here only because of the large speed of light. In principle, e.g., acoustic fluctuations contribute as well. However, as the frequency of the modes, and thereby the zero-point energy, scales directly with the corresponding propagation velocity, their contribution is negligible as compared to the electro-magnetic contributions. It should be noted at this point that the electromagnetic field is singled out here only because of the large speed of light. In principle, e.g., acoustic fluctuations contribute as well. However, as the frequency of the modes, and thereby the zero-point energy, scales directly with the corresponding propagation velocity, their contribution is negligible as compared to the electro-magnetic contributions.

3.1.2   *Adsorption isotherms*

For the comparably simple case of purely van der Waals attraction, it is instructive to discuss what we would expect for the behavior of a thin liquid film adsorbing at the free substrate-gas interface. This corresponds to pressures below the saturated vapor pressure, p_s, because any pressure above p_s corresponds to the bulk liquid, and hence to infinite 'film' thickness. Let us therefore, in a gedanken experiment, have a solid substrate mounted in a thermostated container which contains the adsorbate at a pressure below saturation. Now the pressure shall be gradually increased, such that it will finally approach saturated vapor pressure. What can we anticipate for the excess adsorbate accumulating at the substrate surface?

It is clear that a film of noticeable thickness will form only if the pressure is already very close to saturation. A corresponding situation can be prepared in the way sketched in Fig. 3.3. The substrate is thought of as dipping into the bulk liquid. At a height z above the surface of the latter, the pressure is slightly reduced due to gravity, according to Boltzmann

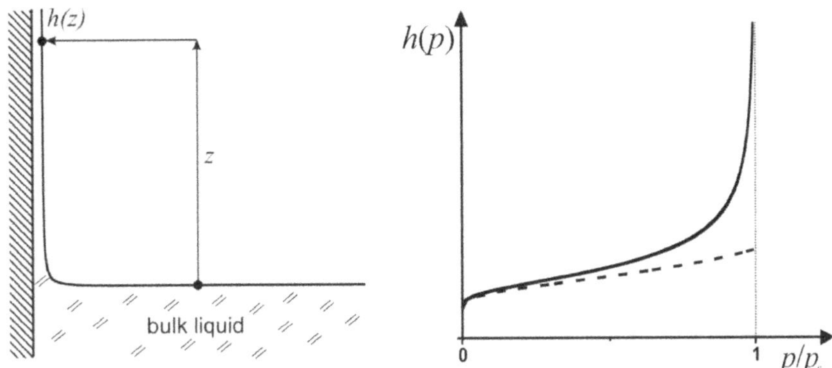

Fig. 3.3 Sketch of a gedanken experiment to derive the form of a van der Waals adsorption isotherm. The solid substrate (hatched) is thought to dip into the bulk liquid, but at some height z the pressure is below saturation due to gravity (barometric formula).

statistics in disguise of the barometric formula,

$$p(z) = p_s \exp\left(-\frac{m_{\mathrm{m}} g z}{k_B T}\right), \tag{3.1}$$

where m_{m} is the mass of a single molecule. The thin liquid film adsorbed on the solid substrate is at equilibrium with the bulk. Hence the energy we would have to afford to lift a molecule from the bulk liquid surface to the height of z (cf. Fig. 3.3) must be exactly recovered when moving the molecule sideways to the surface of the film. The energy change connected to the horizontal arrow is solely the van der Waals energy, which is to be afforded for replacing a half space of liquid, up to a distance h from the surface, with the solid. This is given by

$$E_{vdW} = \frac{\alpha_s - \alpha_l}{h^3}, \tag{3.2}$$

where the constants α_s and α_l indicate the strengths of the van der Waals interaction of the gas molecule with the solid and the liquid phase, respectively. Setting this equal to the gravitational energy in the barometric formula above, we immediately arrive at a relation between the gas pressure and the thickness of an adsorbed film which is in equilibrium with the gas phase:

$$h(p) = \frac{\alpha_s - \alpha_l}{k_B T}\left(\ln\frac{p_s}{p}\right)^{-\frac{1}{3}}. \tag{3.3}$$

This is plotted in Fig. 3.3 as the solid curve, representing the van der Waals adsorption isotherm, or Frenkel–Halsey–Hill isotherm. For small pressures, a film of only molecular thickness is adsorbed to the solid, but as saturation is approached, the film thickness diverges algebraically. This corresponds to the formation of an infinitely thick coverage with bulk liquid at saturated vapor pressure.

However, as everyone knows from the turbidity of the mirrors in the bathroom after one has taken a shower, this is not what will usually happen. Instead, the condensing liquid will tend to form individual droplets on the solid substrate, the free surface of which makes a finite angle with the solid. The large number of such droplets give rise to diffuse scattering, and thus to turbidity, which so effectively impedes the practical use of the mirror. It is clear that the environment of the mirror surface is very close to liquid–gas equilibrium due to the presence of the droplets. On the other hand, there is no macroscopically thick liquid coverage between the droplets. The corresponding adsorption isotherm looks like the one plotted as the dashed line in Fig. 3.3. Although the amount of adsorbed liquid will certainly increase monotonously with gas pressure, it must attain a finite value at saturation. It turns out (and will become clearer below) that all wetting scenarios fall into the two classes indicated in Fig. 3.3: either the film thickness diverges continuously at saturation, in which case the contact angle the liquid makes with the substrate is zero; this is called *complete wetting*. Or the film thickness remains finite as saturation is approached from below, in which case the contact angle is finite. This is called *incomplete wetting*,[1] or partial wetting.

3.1.3 *The contact angle*

The first thorough investigation on the interaction of a liquid with a solid substrate owes to Thomas Young, who in his 1805 paper [87] described, just in words and without use of a single formula, the geometry of the interface between a liquid and its vapor close to a solid surface. There have been important contributions later by other researchers, like Carl Friedrich Gauss and, most notably, Athanase Dupré, which finally led to a concise formulation of Young's results as

$$\cos\theta_Y = \frac{\gamma_{sg} - \gamma_{sl}}{\gamma},\qquad(3.4)$$

---

[1] In the literature, some more sophisticated terms, like semi-partial wetting, are sometimes encountered. These refer to unnecessary (and frequently unprecise) distinctions, and will not be used in this book.

where θ_Y, which is called Young's contact angle or the Young–Dupré angle, is the angle the liquid surface makes with the solid immediately at the surface of the latter. γ_{sg} and γ_{sl} are the excess free energies of the solid–gas interface and the solid–liquid interface, respectively. γ_{lg}, or just γ for short, is the liquid surface tension. Equation (3.4) has become known as Young's equation, or the Young–Dupré equation. The corresponding situation is sketched in Fig. 3.4(a).

We discuss interfaces in terms of excess *free* energies because we are dealing with thermal many-body systems here. The equilibrium configuration of the interfaces are achieved by means of continuous redistribution of the molecules, which is mediated by thermal motion. Hence what we see as, e.g., the spherical shape of a drop is the result of the system's search for the most likely spatial distribution of molecules under the given constraints, such as number of molecules, temperature, pressure, etc. In thermodynamic nomenclature, what we see is the configuration which corresponds to the minimum of the total free energy of the system, which includes molecular interaction energies as well as entropic contributions. As far as the interfaces are concerned, we hence have to consider the extra free energies which are to be afforded for creating these extra interfaces.

Let us now try to rationalize Eq. (3.4) from the sketch in Fig. 3.4(a). Imagine we have zoomed in on the three-phase contact line, at which the solid, the liquid and the gas phase meet, sufficiently close that the liquid surface can be considered as planar, i.e., the curvature of the drop surface

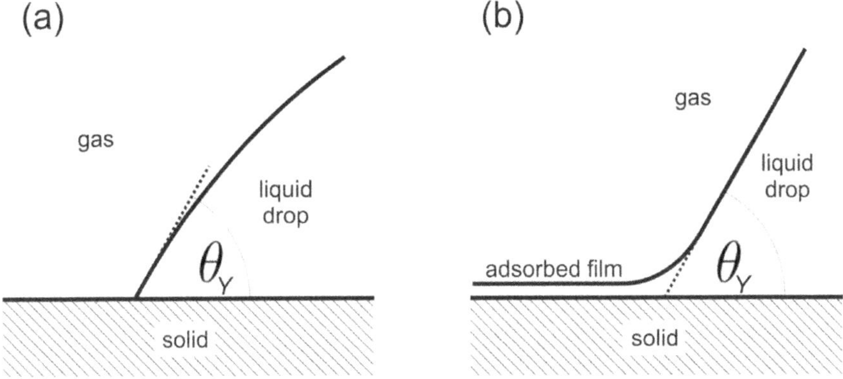

Fig. 3.4   (a) The rim of a liquid drop sitting on a solid surface. At the contact line, the solid, liquid, and gas phases meet (three-phase contact line), making a characteristic angle θ_Y with the substrate. (b) Closing up on the contact line reveals the fine structure of the liquid–gas interface and the microscopically thin liquid film next to the drop.

can be neglected. The sketch represents a cross-sectional view, such that the three-phase contact line is just a point, and the liquid surface is represented by a line, hitting the substrate at an angle θ_Y. Thermal fluctuations will lead to tiny displacements of the liquid surface due to the molecular exchange between the liquid and the gas phase. In thermodynamic equilibrium, such shift of the liquid surface, and hence of the three-phase contact line, will leave the total free energy of the system constant, since the latter must be extremal (namely minimal) in equilibrium. If we assume the three-phase contact line is shifted by a distance δx, the change in free energy of the system associated with this shift is proportional to $\delta x(\gamma_{sl}-\gamma_{sg}+\gamma\cos\theta)$, as one can immediately read off Fig. 3.4(a). If we demand this expression to vanish, we directly arrive at Eq. (3.4).

Contact angles measurement (goniometry) usually proceeds by optically imaging the vicinity of the three-phase contact line, followed by suitable image processing. Figure 3.5 presents such images for three different settings. Figure 3.5(a) shows a sessile drop on a flat solid substrate, representing the classical goniometer setup.[2] The drop is deposited on the horizontal substrate through a capillary, which is visible in the center of the image. In Fig. 3.5(b), capillary bridges are visible between spherical basalt particles. Figure 3.5(c) shows an example of goniometry at the interface between two immiscible fluids. In particular the setups in Figs. 3.5(b) and 3.5(c) are of practical relevance for experiments with wet granular matter, since the grains material often comes only in granular form, and is not available as large flat samples suitable for the standard setup shown in Fig. 3.5(a).

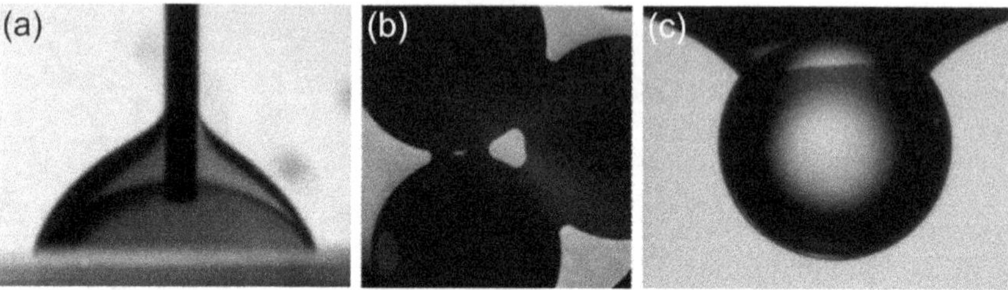

Fig. 3.5 Contact angle measurement geometries. (a) standard goniometer setup; (b) basalt beads wetted by water; (c) glass bead (1 mm diameter) at the interface between an aqueous solution of ZnI_2 and a mixture of dodecane with 8% bromdodecane. These liquids are frequently used in X-ray tomography because of their high X-ray contrast.

---

[2] *Goniometry* (from ancient greek) means 'angle measurement'.

3.1.4 *The effective interface potential*

The picture developed above suggests that in the case of partial wetting, there are some parts of the substrate which are covered by liquid and others which remain completely dry. This can, however, not be the full truth, as Fig. 3.3 suggests. We thus redraw Fig. 3.4(a) as shown in Fig. 3.4(b), which closes up further on the three phase contact line, such that the adsorbed layer, which is only a few molecular layers thick, can be seen. It suggests itself to describe the position of the free liquid surface not anymore with a contact angle, but with a function, $h(x)$, indicating the vertical position of the liquid–gas interface as a function of distance along the substrate, x.

Before we proceed, however, we should first define clearly what we mean by the position of that interface. It should be expected (and is in fact true) that the adsorbed material may, within the film, acquire a structure which is different from its bulk condensed phase under the same pressure and temperature; the adsorbed 'liquid' can in fact even be a solid film [88]. Furthermore, the thermally excited capillary waves which are present on the interface between the adsorbate and the gas bestow a certain fuzzyness to the latter. Hence there is a need for a clear cut definition of the film thickness, h, which in the case of non-trivial liquid profiles will be treated as the function $h(x)$. This can be achieved by considering the density of the adsorbate as a function of the distance from the substrate, as sketched in the left panel of Fig. 3.6.

Now let us consider the quantity

$$\xi = \int_0^\infty (n(z) - n_g)dz. \tag{3.5}$$

If there was only the vapor phase and no liquid adsorbed on the substrate, ξ would vanish. It thus represents the material which is present due to wall in excess of the gas, and is called the Gibbs excess. We can now define the adsorbed film thickness by just dividing ξ by the bulk liquid density. Another possible definition which relates to the actual density profile would be to take the inflection point of $n(z)$, where the liquid crosses over towards the gas phase. Since the density variations within the film are usually minute, there is no noticeable numerical difference between the two definitions, and we will not explicitly refer to one or the other. Most of the time, we will use the latter (more geometric) definition, and assume the density to be constant throughout the adsorbed film. In any case, we envisage the density profile of the film to be represented by a sharp step at $z = h(x)$, from the liquid density (for $z < h$) to the gas density (for $z > h$).

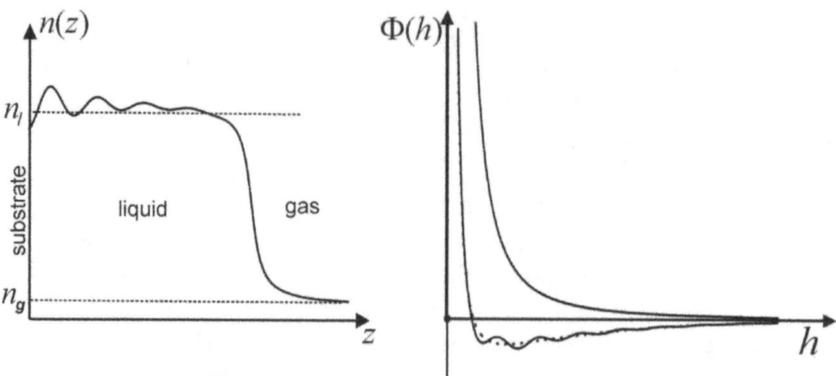

Fig. 3.6 Left: Typical density profile of a film of liquid adsorbed onto a solid substrate. Close to the 'wall', a certain degree of layering is observed, sometimes to the point that the first few monolayers are solid. At the liquid–gas interface, the density varies over a distance corresponding to the bulk correlation length of the liquid, which is of just molecular size, except very close to the critical point of the liquid. Outside the film, the density has decayed to the gas density. Right: Typical behavior of $\Phi(h)$ for various types of interactions (see text for details). The oscillations at small distances correspond to the molecular scale oscillations in the density as sketched in the left panel. The long range tail decays to below thermal energies within few tens of nanometers.

In the literature, this model is widely used and known as the sharp kink approximation. It provides an excellent account of the physics of wetting as long as the system is not considered close to the critical point of the liquid.

Let us now consider the total interfacial free energy per unit area of a substrate together with a film adsorbed on it. If the film thickness were infinite, this would just be the sum of two interfacial free energies, the surface tension of the adsorbed liquid, which is just the excess free energy per unit area of liquid–gas interface, plus the interfacial free energy of the substrate–liquid interface. If we now reduce the film thickness, such that the two parallel interfaces start to interact with each other, there will be corrections to the total free energy. We thus write

$$\mathcal{F}(h) = \gamma_{sl} + \gamma + \Phi(h), \tag{3.6}$$

such that all corrections due to the mutual interaction of the interfaces are absorbed into the new quantity $\Phi(h)$. The latter is called the *effective interface potential* and is a very powerful means for describing wetting phenomena.

Before we proceed with applications of $\Phi(h)$, let us discuss a few typical contributions to this quantity. In case of a pure van der Waals interactions, the excess free energy of a liquid film of thickness h on a solid substrate can be obtained by integration of Eq. (3.2) from zero to h. As this is one more integration over a length, the result is proportional to h^{-2}. It is usually written (by convention) in the form

$$\Phi_{vdW} = \frac{-A}{12\pi h^2},\qquad(3.7)$$

where A is called the Hamaker constant of the system. Note the minus-sign here, which is in accordance with the widely accepted definition.[3] If the liquid tends to wet the substrate, $A < 0$, resulting in the form plotted in Fig. 3.6 as the upper solid curve.

If the van der Waals interaction between the liquid with itself is stronger than between the liquid and the solid, A is positive. Although at very short distance there are almost always other interactions which lead to the adsorption of at least one molecular layer, the long range tail of $\Phi(h)$ lies then below the h-axis, such as sketched as the dashed curve in Fig. 3.6. Furthermore, short range structures such as the oscillations shown in Fig. 3.6 are likely to lead to a similar structure in $\Phi(h)$, as indicated by the lower solid curve. In such cases, where there are multiple minima in $\Phi(h)$, the adsorption isotherms as depicted in Fig. 3.3 will look more complex at intermediate pressures. Adsorption then proceeds, at least in some range of pressures, in single molecular layers ('layering'). However, the overall classification into complete and incomplete wetting, as given above, still holds. As the typical thickness of the adsorbed film is up to a few nanometers, this is as well the distance over which the van der Waals force is noticeable.

3.1.5 *The interface displacement model*

To demonstrate the convenience and generality of the description of wetting phenomena by means of $\Phi(h)$, we turn to discussing the case of adsorption from the gas phase, which we have done more qualitatively already above. When the pressure of the surrounding gas is below saturated vapor pressure, there is an extra penalty for forming the condensed (liquid) phase, which is proportional to the deviation of the chemical potential from coexistence. This is given by

$$\Delta\mu = k_B T \ln \frac{p_s}{p}.\qquad(3.8)$$

---

[3]Some authors use the opposite sign convention, such that one must be careful when interpreting the sign of published Hamaker constants.

Off coexistence, the free energy of an adsorbed film is then given by

$$\mathcal{F}(h,p) = \gamma_{sl} + \gamma + \Phi(h) + h(n_l - n_g)\Delta\mu, \tag{3.9}$$

which is sketched for purely van der Waals interaction in Fig. 3.7(a) as the solid curve, with the effective interface potential being indicated as the dashed curve. The dotted asymptote corresponds to the rightmost term in Eq. (3.9). It is obvious that for any finite and positive $\Delta\mu$, the global minimum of $\mathcal{F}$ will be at finite h. This is the film thickness which forms in equilibrium. As $\Delta\mu \to 0$, this global minimum will gradually shift towards higher h, and finally diverge towards infinity. This corresponds to the isotherm we have discussed above for the case of complete wetting (Fig. 3.3). Note that the global minimum of $\mathcal{F}$ tends to infinity for $\Delta\mu \to 0$ whenever the global minimum of Φ lies at infinity, i.e., when $\Phi(h)$ is positive for all finite h.

If this is not the case, we have a situation like the one sketched in Fig. 3.7(b). If Φ is negative for some h, there must be a global minimum of Φ at finite h, since $\Phi \to 0$ for $h \to \infty$. As we see, this leads to a global minimum of $\mathcal{F}$ at finite h for all positive $\Delta\mu$. Consequently, the adsorbed film thickness, which corresponds to this global minimum, remains finite for all pressures up to saturated vapor pressure. We thus see that there is a clear correspondence between the structure of the effective interface potential and the details of the wetting scenario. In particular, there is complete wetting if and only if the global minimum of Φ is at $h \to \infty$, and there is incomplete wetting otherwise.

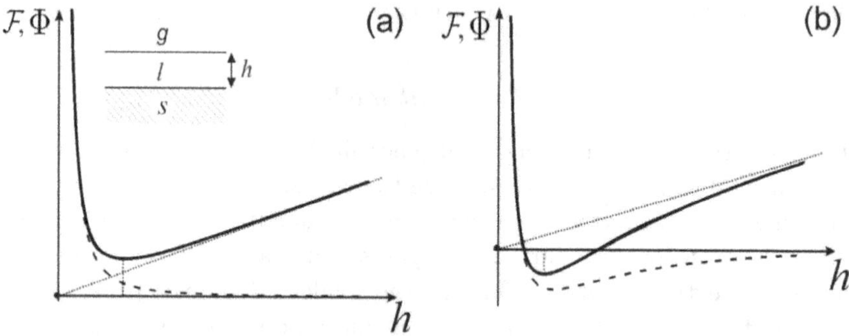

Fig. 3.7 Construction of the adsorption isotherm from the effective interface potential, Φ. The equilibrium film thickness is given by the minimum of the free energy, which contains the effective interface potential as well as the penalty for forming the condensate below the saturated vapor pressure. (a) Construction for purely van der Waals interaction and complete wetting. (b) Construction for incomplete wetting, which is characterized by a finite adsorbed film thickness at coexistence.

It suggests itself to seek a similar connection to the equilibrium contact angle. The existence of a finite contact angle requires the presence of a three-phase contact line, at which the solid substrate, the gas phase, and the liquid phase meet. The position of the liquid–gas interface relative to the substrate is then not anymore spatially constant, but must be allowed to vary laterally, $h = h(x)$ (interface displacement model). A liquid–gas interface making an angle of θ_Y with the substrate, for instance, would then be described as $h(x) = x \tan \theta_Y$. For the sake of simplicity, we start by assuming that $\theta_Y \ll 1$, which allows for some mathematical simplifications. This is well justified since for wet granular matter, the liquid must wet the solid phase quite well (small Θ_Y).

For a spatially varying position of the liquid–gas interface, the total free energy of the system is readily written down. It reads

$$\mathcal{F} = \int \left[\gamma_{sl} + \gamma \sqrt{1 + \left(\frac{dh}{dx} \right)^2} + \Phi(h(x)) \right] dx, \qquad (3.10)$$

and is to be minimized for the equilibrium profile the system will attain. With the above mentioned approximation (i.e., shallow profiles $h(x)$), we can expand the square root in a Taylor series and keep just the first term. Dropping furthermore all additive constants, which are irrelevant for finding the minimum of $\mathcal{F}$, we readily arrive at

$$\int \left[\frac{\gamma}{2} \left(\frac{dh}{dx} \right)^2 + \Phi(h) \right] dx \longrightarrow min. \qquad (3.11)$$

It is instructive to compare this to Hamilton's principle in classical mechanics. It is immediately clear from the form of Eq. (3.11) that the shape of the liquid meniscus described by the solution of Eq. (3.11) corresponds to the motion of a particle of mass γ in a potential $U = -\Phi$, when x is replaced with time, and h with the position of the particle.

For the effective interface potential which is sketched as the solid curve in Fig. 3.8(a), an imagination of the corresponding liquid interface profile can be obtained as follows. On the potential $U = -\Phi$ which is sketched as the dashed curve, a particle with mass γ is put such that it lies on the maximum, as indicated by the black dot. This corresponds to the equilibrium film thickness (vertical dotted line), which is encountered far away from any bulk liquid structure, such as, e.g., a droplet sitting on the substrate. Now imagine the mass receives an infinitesimally small kick to the right, such that it creeps away from the maximum, leaving it in an accelerated fashion. It will then pass through the minimum of U and continue

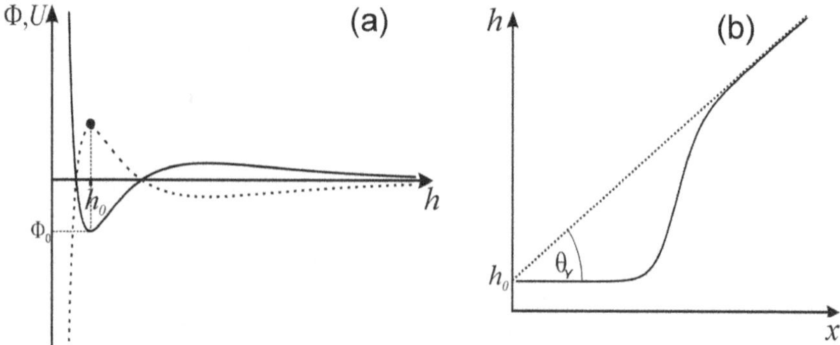

Fig. 3.8 (a) The mechanical analogue for the construction of the liquid film profile at the three-phase contact line. The solid curve represents the effective interface potential, Φ. The trajectory of a mass point (black dot) starting from the maximum of $U = -\Phi$ (dashed curve) towards larger h can be mapped upon the liquid profile, $h(x)$. (b) The profile of the liquid–gas interface near the three-phase contact line, according to the effective interface potential sketched in (a). This is to be seen as a detailed view of Fig. 3.4(b). Depending on the shape of $\Phi(h)$, there need not be an inflection point in $h(x)$.

to the right, where it will finally attain a constant velocity which is given by 'energy conservation', $\frac{\gamma}{2}h'^2 = -\Phi_0$ (the prime indicates differentiation with respect to x). The corresponding profile of the liquid–gas interface is sketched in Fig. 3.8(b), where the contact angle is indicated as well as the equilibrium film thickness, h_0. The complex structure of the liquid profile, which extends to film thicknesses where the van der Waals interaction is still noticeable, is often referred to as the van der Waals tail of the contact line.

The typical range of extension, h_{vdW}, of van der Waals tails away from the substrate can be estimated from a comparison of the involved energy scales. From Eq. (3.7) we then obtain

$$h_{vdW} = \sqrt{\frac{|A|}{2\pi\gamma}}. \tag{3.12}$$

A is typically of order 10^{-20} J, such that h_{vdW} is on the order of a few nm.

The profile displayed in Fig. 3.8(b) represents a closeup of the liquid–gas interface near the three-phase contact line, say, of a droplet sitting on the substrate. To the right, the interface leaves the substrate with an angle θ_Y, which is Young's contact angle. To the left, we have a homogeneous film with thickness h_0. The presence of a droplet, which determines the system to be at liquid–gas coexistence, entails $\Delta\mu = 0$. Hence the rightmost term in Eq. (3.9) vanishes, such that the minimum of the free energy coincides with the global minimum of the effective interface potential. This must be at finite h (and thus below the h axis) for the equilibrium film thickness, h_0, to be finite. It is immediately clear that otherwise the construction sketched in Fig. 3.8(a) will not work. We see that a minimum of $\Phi(h)$ at finite h is necessary and sufficient for the contact angle to be finite.

From Eq. (3.11), we obtain an equation for the effective interface profile in a straightforward manner by inserting the Lagrangian

$$\mathcal{L} = \frac{\gamma}{2}\left(\frac{dh}{dx}\right)^2 + \Phi(h), \tag{3.13}$$

into the Euler–Lagrange equation,

$$\frac{d}{dx}\frac{\partial\mathcal{L}}{\partial h'} - \frac{\partial\mathcal{L}}{\partial h} = 0. \tag{3.14}$$

This leads directly to

$$\gamma\frac{d^2h}{dx^2} - \frac{d\Phi}{dh} = 0, \tag{3.15}$$

which has the form of Newton's equation of motion, and can be used to calculate the shape of the liquid profile once the effective interface potential is known.

Multiplying on both sides by dh/dx and integrating over x, we obtain the analogous expression to energy conservation,

$$-\Phi(h_0) = \frac{\gamma}{2}\frac{d^2h}{dx^2}\Big|_{x\to\infty}. \tag{3.16}$$

This finally yields

$$\tan\theta_Y = \frac{dh}{dx}\Big|_{x\to\infty} \approx \sqrt{\frac{2|\Phi(h_0)|}{\gamma}}. \tag{3.17}$$

The '$\approx$' shall remind us that this still rests upon the assumption that θ_Y is small. We see that the contact angle is directly connected to the depth of the global minimum of the effective interface potential, $|\Phi(h_0)|$.

3.1.6 *Curved interfaces and the Laplace pressure*

In the vicinity of a curved liquid surface, there is an extra pressure, the Laplace pressure, which is due to that curvature. It can be derived by considering a droplet with radius R and calculating the energy required to increase its volume by a small amount. The Laplace pressure is the

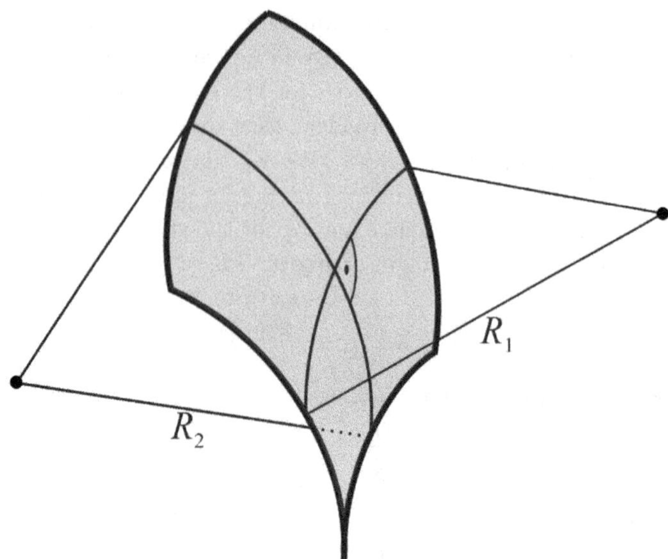

Fig. 3.9 Definition of the two principal radii of curvature of a general curved surface. Depending on where the center of a circle (solid dots) lies with respect to the surface (inside or outside the body to which the surface describes the boundary), the corresponding radius of curvature is counted positive or negative.

derivative of the surface free energy, $4\pi R\gamma$, with respect to the volume. We readily obtain

$$p_L(R) = \frac{2\gamma}{R}. \tag{3.18}$$

For more general curved surface shapes, one defines the two principal curvatures at the point of interest [3]. This is done as sketched in Fig. 3.9. At the point at which its curvature is to be evaluated, the surface can approximated by two tangent circles intersecting at right angles in the tangent plane of the surface. It can be shown that the expression

$$\mathsf{H} = \frac{1}{2}\left(\frac{1}{R_1} + \frac{1}{R_2}\right) \tag{3.19}$$

is independent of the orientation of the two circles in the tangent plane. H is called the mean curvature of the surface, at the point of intersection of the circles. Comparison with Eq. (3.18) shows that the latter is appropriately generalized by

$$p_L = 2\gamma\mathsf{H}. \tag{3.20}$$

If the surface is given as a function, $f(x, y)$, the mean curvature can be expressed (Monge parametrization) as

$$\mathsf{H} = \frac{1}{2} \nabla \left(\frac{\nabla f}{[1 + (\nabla f)^2]^{3/2}} \right). \tag{3.21}$$

For shallow profiles, this reduces to $\mathsf{H} = \frac{1}{2}\Delta f$, where Δ is the Laplace operator. Hence for the important case of shallow profiles, we have

$$p_L = \gamma \Delta f. \tag{3.22}$$

This allows for an interesting revisit to Eq. (3.15). In the first term, we now recognize the Laplace contribution to the pressure in the film. The second term has of course the dimension of a pressure as well. In fact, the expression

$$\Pi(h) = -\frac{d\Phi}{dh}, \tag{3.23}$$

is called the *disjoining pressure* and can be used alternatively to the effective interface potential. In the physical chemistry literature, it is frequently even preferred. This said, Eq. (3.15) reveals itself as a balance between the local Laplace pressure, stemming from the interface curvature, and the disjoining pressure, which is due to the interfacial interactions. It should finally be noted that we can establish a direct relation between the van der Waals constants α_i in Eq. (3.2) and the Hamaker constant of the system by identifying $\Pi_{vdW} = n_l E_{vdW}$.

The character of the equation of motion for the interface profile becomes even more apparent if we lift the assumption of shallow profiles for a moment. We then have, instead of Eq. (3.13),

$$\mathcal{L} = \gamma \sqrt{1 + \left(\frac{dh}{dx} \right)^2} + \Phi(h(x)), \tag{3.24}$$

just omitting the additive constant, γ_{sl}. Inserting this into Eq. (3.14) leads directly to

$$\frac{\gamma h''}{(\sqrt{1 + h'^2})^{\frac{3}{2}}} - \frac{d\Phi}{dh} = 0, \tag{3.25}$$

where the prime indicates differentiation with respect to x. Comparing the first term with the exact expression for the mean curvature given in Eq. (3.21) immediately reveals the role of the first term as the Laplace pressure, as in Eq. (3.15).

Even in this case of larger contact angles, we can use the analogue to energy conservation in order to obtain a (this time exact) formula for the contact angle. We construct a Hamiltonian as usual from

$$\mathcal{H} = \frac{\partial \mathcal{L}}{\partial h'} h' - \mathcal{L}. \tag{3.26}$$

If we demand this to be a 'constant of motion' (i.e., to be constant along the interface profile), it must be equal in particular for $h = h_0$ and for $h \to \infty$. This leads to

$$\cos \theta_Y = 1 - \frac{\Phi(h_0)}{\gamma}, \tag{3.27}$$

as the exact expression, which for small θ_Y indeed approaches Eq. (3.17).

3.1.7 *The contact angle away from coexistence*

In a wet granular medium, we have to deal with sometimes strongly curved liquid surfaces, and with situations away from liquid–gas coexistence. The Kelvin equation,

$$\mathsf{H} = \frac{n_l k_B T}{2\gamma} \ln \frac{p_s}{p}, \tag{3.28}$$

provides a direct connection between the curvature of the liquid surface, H, and the pressure of the surrounding vapor in units of the saturated vapor pressure. Although the contact angle has been defined in the Young–Dupré equation through quantities alluding to coexistence, it is intuitively clear that the contact angle will not differ greatly away from coexistence. A sessile drop, which is commonly used to measure contact angles, has a convex shape and is thus, strictly speaking, 'above' saturation ($p < p_s$, according to Eq. (3.28)). In addition, the large body of literature on capillary condensation suggests, both theoretically and experimentally, that the concept of a contact angle applies very well also for pressures well below saturation.

We can illustrate the pertaining physics with the help of Fig. 3.7(b). The global minimum of Φ is at finite h (dashed line) and therefore below the h-axis, thus representing a finite contact angle. The slope of the dotted line indicates the deviation from coexistence. A negative slope corresponds to a pressure above coexistence, with the global minimum of $\mathcal{F}$ at $h \to \infty$ (bulk liquid). If we insert this into Eq. (3.14), we immediately appreciate from the construction we used in Fig. 3.8(a) that the asymptote will not be straight as in Fig. 3.8(b), but slightly curved downwards, and come back toward the x-axis after some distance. This is the one-dimensional analog of a sessile droplet, coexisting with a thin film of thickness h_0. If, however, the slope of the dotted line in Fig. 3.7(b) is positive, we are at pressures below saturation. In this case, the asymptote in Fig. 3.8(b) will be curved upwards, as expected from a liquid meniscus in a slit pore at capillary condensation.

It is obvious from the construction that the depth of the minimum of $\mathcal{F}$, and thus the contact angle according to Eq. (3.27), changes only by a very small amount, because the minimum at h_0 lies so close to the origin. In the language of the Kelvin equation, we have to go to meniscus curvatures of order $1/h_0$ to change Φ_0 appreciably. For all curvatures relevant to granular matter, including roughness on the grains surfaces, the change in Φ_0, and thus in contact angle, is completely negligible.

3.2 Rough substrates

So far we have outlined the physics of wetting on ideally smooth substrates. In reality, these are never encountered; the surfaces of freshly cleaved mica or graphite single crystals are among the rare exceptions. In granular media, no such surfaces are found. On a well rounded sand grain, the depth of the roughness, δ, is typically one order of magnitude less than the grain radius. Glass beads, which are a common model for sand grains used in experiments, have a roughness on the order of a few microns, or a few percent of the grain size (cf. Fig. 2.7). In any case, the roughness we have to deal with is on the micron scale, and thus large as compared to the extension of the van der Waals tails of the liquid surfaces near three-phase contact lines. Hence we can assume that the concept of a well-defined contact angle can be used wherever the liquid–gas interface meets the solid surface on the slopes and troughs of the roughness.

Based on this assumption, we will now develop a theory of liquid adsorption on the rough surface of a grain. We shall see that under certain circumstances, which are fulfilled in typical experiments, there may be macroscopic amounts of liquid on the grain surfaces *due to their roughness*. In fact, it has been shown in many studies that grooves and troughs, as they are present on rough surfaces, tend to exhibit filling transitions at critical contact angles, which are somewhat reminiscent of capillary condensation [89, 90]. However, these studies have invariably dealt with idealized topographies which are not present on a rough surface. We intend here to go one step further and develop a theory of wetting on rough surfaces which is valid for an as large as possible class of roughness topographies. It should be capable to predict under what circumstances a liquid film with appreciable thickness will be present on the grain surfaces. The presence of such films has significant consequences for the dynamics of wet granular gases, and possibly as well for liquid transport in wet granular piles.

3.2.1 *Presentation of the problem*

Wenzel was the first to report on systematic studies of wetting on randomly rough surfaces [91]. He characterized the roughness by a single parameter, r, which he defined as the ratio of the total substrate area divided by the projected area. Obviously, r $\geq$ 1, and r $=$ 1 corresponds to a perfectly smooth surface. The free energy which is gained per unit area when the rough substrate is covered with a liquid is then given by r$(\gamma_{sg} - \gamma_{sl})$. If this gain is larger than the surface tension of the liquid, γ, we expect a vanishing *macroscopic* contact angle because covering the substrate with the liquid releases more energy than is required for the formation of a free liquid surface of the same (projected) area. More specifically, force balance at the three-phase contact line yields

$$\cos\theta = \frac{\mathrm{r}(\gamma_{sg} - \gamma_{sl})}{\gamma} = \mathrm{r}\cos\theta_Y, \qquad (3.29)$$

where θ (without the index 'Y' referring to Thomas Young) is the macroscopic contact angle seen on a length scale much larger than that of the roughness. When θ_Y is as small as $\theta_W = \arccos(1/\mathrm{r})$, θ vanishes, and the substrate is covered with an 'infinitely' thick liquid film.

We describe the rough solid substrate by a function $f(x, y)$, which shall approximate the actual physical surface at any required precision, but be mathematically smooth, such that ∇f and Δf exist everywhere. The roughness is assumed to be homogeneous and isotropic, i.e., its statistical parameters shall be the same everywhere on the sample, and independent of rotation of the sample about its normal axis. A small amount of liquid deposited on this substrate will make an interface with the surrounding gas, which is described by a second function, $g(x, y)$. The support of g is the wetted area, which we call $\mathcal{W}$. Continuity of the liquid surface assures $g = f$ on the boundary of $\mathcal{W}$, i.e., the projection of the three-phase contact line, henceforth denoted by $\partial\mathcal{W}$ (cf. Fig. 3.10(a)).

Before we proceed, it is instructive to have a look at typical roughness topographies we may encounter. Figure 3.11 shows a collection of data for different surfaces which are relevant for the systems under study. In

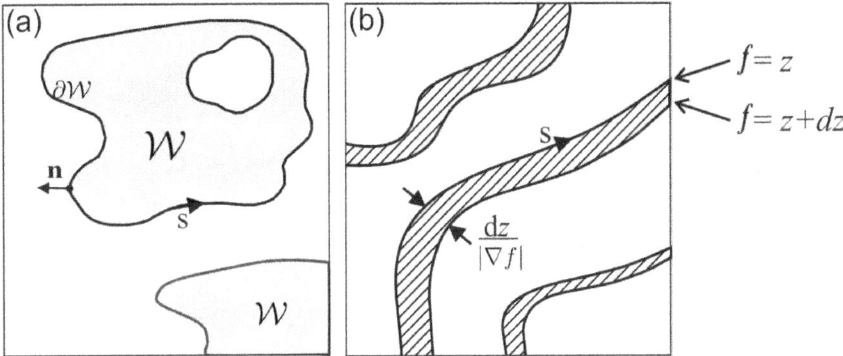

Fig. 3.10 (a) Top view of the sample, showing the wetted areas in grey, the bare substrate in white. The normal vector to ∂W, $\mathbf{n}$, lies in the (x, y)-plane. (b) Two sets of contour lines of f at heights h and $h + dh$. The hatched area between the lines is equal to $p(h)dh$.

Fig. 3.11(a), the surfaces of glass spheres with different degrees of roughness are compiled. The roughness has been introduced by means of chemical etching. Figure 3.11(b) shows a surface such as in Fig. 3.11(a) after only mild etching as it appears with scanning force microscopy.[4] In Fig. 3.11(c), similar scanning force microscopy data are presented of a similar glass sample, but after stronger etching and on a flat substrate. The vertical scale is the same here as the lateral scale. We see that the depth of the roughness is usually smaller than the typical lateral scale of features. This is also true for many natural surfaces, as Fig. 3.11(d) shows. Here we compiled data from etched glass surfaces and from natural rock samples. All data lie well below the dotted line, which represents structures with a slope angle of 0.4. It therefore appears as justified to assume that the amplitude of most natural roughness is much smaller than its dominant lateral length scale. We thus assume that

$$|\nabla f| \ll 1, \tag{3.30}$$

which allows for substantial mathematical simplifications, as we will see shortly. The same shall hold for g. The contact angle with the substrate, θ_Y, then yields the boundary condition

$$|\nabla(g - f)| \approx \tan\theta_Y \approx \theta_Y, \tag{3.31}$$

which is to be fulfilled everywhere on ∂W.

---

[4]For a description of scanning force microscopy, see Section 4.3.3.

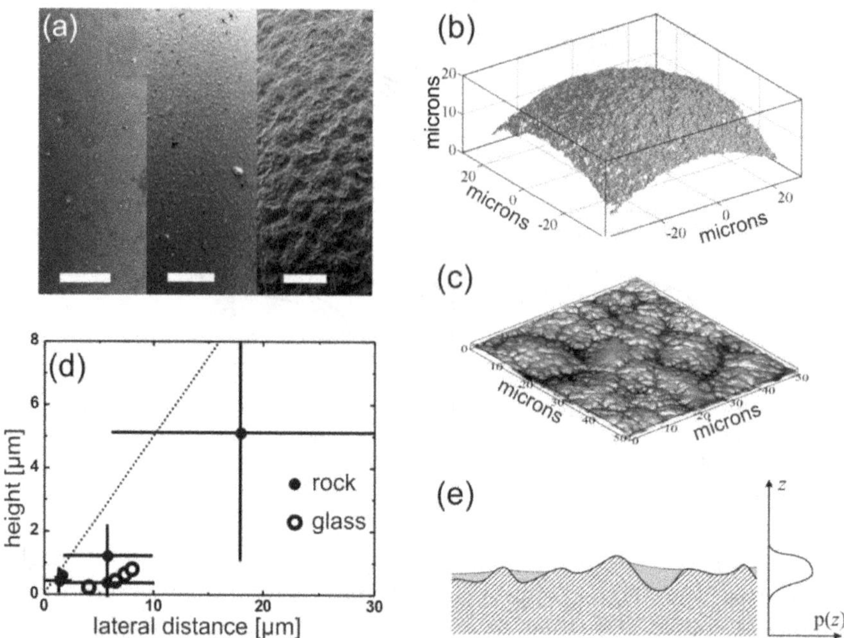

Fig. 3.11   (a) Scanning electron micrographs of etched glass bead surfaces at three different stages of etching (scale bar: 20 microns). (b) Scanning force micrograph of the surface of a glass bead (data provided by Sonja Utermann, Göttingen). (c) Scanning force micrograph of a flat etched glass surface (same etching stage as rightmost image of (a); vertical and lateral scales are identical; courtesy of Matthias Schröter, Göttingen). (d) Typical examples of roughness amplitudes on solid grains of rock (full circles) and etched glass beads (data courtesy of Cornelius Fischer, Bremen). The dotted line represents a slope of 0.4. (e) Sketch of a roughness profile with puddles of wetting liquid adsorbed. On the right-hand side, the height distribution, p(z), of the roughness profile is sketched.

Now we make use of the fact that the typical scale of the roughness, as shown in Fig. 3.11, is in the range of micrometers, or at least several tens of nanometers. This is much larger than the extension of the 'van der Waals tails', h_{vdW}, as estimated in Eq. (3.12). Hence for the discussion of the mesoscopic roughness present on natural surfaces such as most grains, we can assume that the free liquid surface described by $g(x,y)$ and the substrate surface described by $f(x,y)$ intersect at the contact line at an angle of θ_Y. This represents all of the relevant structure of the effective interface potential, and no more interactions need to be taken into account.

Applying then Green's theorem to $(g - f)$ in Eq. (3.31), we obtain

$$\int_{\partial \mathcal{W}} \mathbf{n} \cdot \nabla(g - f) \, \mathrm{d}s = \int_{\mathcal{W}} \Delta(g - f) \, \mathrm{d}^2 \mathbf{x}, \tag{3.32}$$

where s is the distance along $\partial \mathcal{W}$, $\mathbf{n}$ its unit normal vector, and $\mathbf{x} = (x, y)$ (cf. Fig. 3.10). Since $g = f$ on $\partial \mathcal{W}$, $\nabla(g - f)$ is everywhere perpendicular to $\partial \mathcal{W}$. Hence Eq. (3.31) may be written as $\mathbf{n} \cdot \nabla(g - f) \approx \theta_Y$, and Eq. (3.32) can be recast into

$$l \, \theta_Y + \int_{\mathcal{W}} [2\mathsf{H} - \Delta f] \, \mathrm{d}^2 \mathbf{x} = 0, \tag{3.33}$$

in which l denotes the length of $\partial \mathcal{W}$, and $\mathsf{H} \approx \frac{1}{2} \Delta g$ is the mean curvature of the liquid-vapor interface. Note that at equilibrium H is directly related to the gas pressure, Eq. (3.28).

3.2.2 *Introducing useful descriptors for roughness*

Before we can exploit Eq. (3.33), we need to derive a few useful relations. Let $\mathrm{p}(f)$ be the height distribution of $f(\mathbf{x})$, with normalization $\int \mathrm{p}(z) \mathrm{d}z = |\mathcal{S}|$ (total sample area, cf. Fig. 3.11). Then the total area between the contour lines at $f = z$ and at $f = z + \mathrm{d}z$ is given by $\mathrm{p}(z) \, \mathrm{d}z$, which corresponds to the hatched area in Fig. 3.10(b). The average slope on that set, $\sigma_1(z) = \langle |\nabla f| \rangle_z$, is given by

$$\sigma_1 = \frac{\int |\nabla f(z)| \, \mathrm{d}\tau}{\mathrm{p}(z) \mathrm{d}z}, \tag{3.34}$$

where $\mathrm{d}\tau = \mathrm{d}s \mathrm{d}z / |\nabla f(z)|$ is the differential of the hatched area. Hence for the total length of the contour line at height h we obtain

$$L(z) = \int \mathrm{d}s = \sigma_1(z) \mathrm{p}(z). \tag{3.35}$$

In order to express the integral over Δf which appears in Eq. (3.33), we apply Green's theorem again, this time to the area enclosed by a contour line. This yields

$$\int_{\mathcal{C}(z)} \Delta f \, \mathrm{d}^2 \mathbf{x} = \int_{\partial \mathcal{C}} \mathbf{n} \cdot \nabla f \, \mathrm{d}s, \tag{3.36}$$

where $\mathcal{C}(z)$ denotes the set $\{ \mathbf{x} \mid f(\mathbf{x}) \leq z \}$, and $\partial \mathcal{C}$ its boundary, i.e., the contour line itself. Introducing

$$\sigma_2(z) = \langle |\nabla f|^2 \rangle_z, \tag{3.37}$$

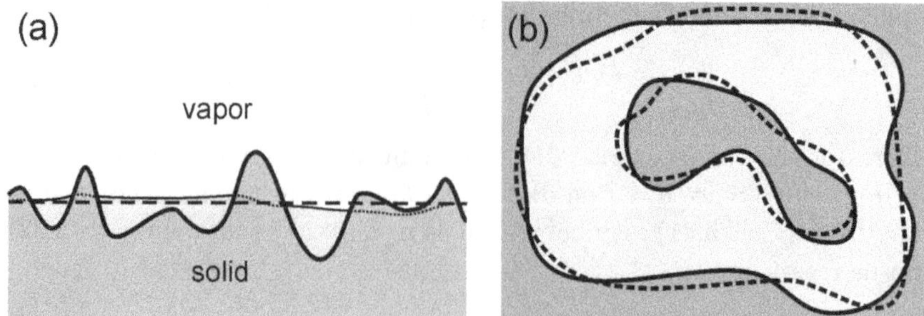

Fig. 3.12 (a) Cross-section of rough substrate (dark grey) with adsorbed liquid (light grey) and liquid–vapor interface (thin solid and dotted curves). The bold dashed horizontal line represents the average height of the three-phase contact lines. (b) Top view. Solid curves: three-phase contact lines encompassing the wetted areas (light grey). Dashed curves: contour lines for the substrate at the height indicated as the dashed line in (a). Dry substrate: dark grey.

we readily see that Eq. (3.36) can be rewritten as

$$\int_{\mathcal{C}(z)} \Delta f \, \mathrm{d}^2\mathbf{x} = \sigma_2(z) \, \mathrm{p}(z). \tag{3.38}$$

In order to fulfill the boundary condition, Eq. (3.31), the vertical position of the three-phase contact line, which may be symbolically written as $f(\partial \mathcal{W})$, will vary along $\partial \mathcal{W}$ about an average value, $\langle f(\partial \mathcal{W}) \rangle$. The projection of the contact line onto the plane will thus approximately follow the contour line at $f(\mathbf{x}) = \langle f(\partial \mathcal{W}) \rangle$, with excursions towards both the outside and the inside of $\mathcal{W}$ (cf. Fig. 3.12). These will in cases represent detours, sometimes shortcuts with respect to $\partial \mathcal{W}$. As a reasonable approximation, we may thus use $l \approx L(\langle f(\partial \mathcal{W}) \rangle)$ for the length of the three-phase contact line. Similarly, we set

$$\int_{\mathcal{W}} \mathrm{d}^2\mathbf{x} \approx \int_{-\infty}^{h} \mathrm{p}(z) \, \mathrm{d}z, \tag{3.39}$$

for the wetted sample area, with $h = \langle f(\partial \mathcal{W}) \rangle$. The integral on the right-hand side is just the cumulative height distribution of the roughness. Inserting these expressions in Eq. (3.33), we obtain

$$2\mathrm{H} \int_{-\infty}^{h} \mathrm{p}(z) \, \mathrm{d}z \approx \left[\frac{\sigma_2(h)}{\sigma_1(h)} - \theta_Y \right] L(h). \tag{3.40}$$

This allows, if $p(z)$, $\sigma_1(z)$, and $\sigma_2(z)$ are known from experimental characterization of the sample, to determine the adsorbed amount of liquid as a function of θ_Y and H.

Up to this point, we have not made any specific assumption about the roughness profile, except its being sufficiently shallow for the approximations made above to hold. Now we shall go one step further, observing that roughness profiles generated by wear, weathering, erosion, or similar processes will invariably have a finite codomain. In other words, the support of $p(z)$ is the interval $[z_-, z_+]\ \forall \mathbf{x}$, where z_- represents the depth of the deepest trough, and z_+ the height of the highest elevation on the sample. This has severe consequences of σ_1 and σ_2, as both must go to zero as $h \to z_\pm$. Since $\sigma_2 \propto |\nabla f|^2$ while $\sigma_1 \propto |\nabla f|$, it is clear that the ratio σ_2/σ_1 goes to zero as well for $h \to z_\pm$.

Aside from these global properties, both σ_1 and σ_2 are expected to be largely featureless, due to the general fact that the processes leading to roughness exhibit only very limited lateral correlation. For any pronounced feature to develop in σ_i, distant places on the sample would have to 'conspire' to contribute to that feature at the same depth. This can happen only for composite surfaces, where the roughness topography may penetrate through a coating or other stratigraphic variation of material properties, but is not of interest for granular systems, as long as the grains are assumed to consist of homogeneous materials.

3.2.3 *The wetting phase diagram*

The generic shape of the function

$$K(h) = \left[\frac{\sigma_2(h)}{\sigma_1(h)} - \theta_Y\right], \tag{3.41}$$

which appears in Eq. (3.40) is sketched in Fig. 3.13, according to the discussion above. Following Eq. (3.40), the film thickness at coexistence ($H = 0$) can be derived from the zeros of K, of which there are either two or none, depending on θ_Y. In the latter case, the contact angle is too large for forming a liquid surface between the spikes and troughs which complies with the boundary condition, Eq. (3.31). If, however, $K(h)$ intersects the h-axis, the slopes of the zeros decide upon the stability of the corresponding solutions. This can be seen by appreciating that K may be interpreted as a deviation from the force balance expressed by Eq. (3.31) [87]. For the left zero, which is marked by an open circle in the figure, a displacement of the three-phase contact line would give rise to an imbalance of wetting forces which drives it further away from the zero. The opposite is true for the right zero, marked by the closed circle. The latter therefore corresponds to the stable solution, and thus to the adsorbed film thickness which will develop.

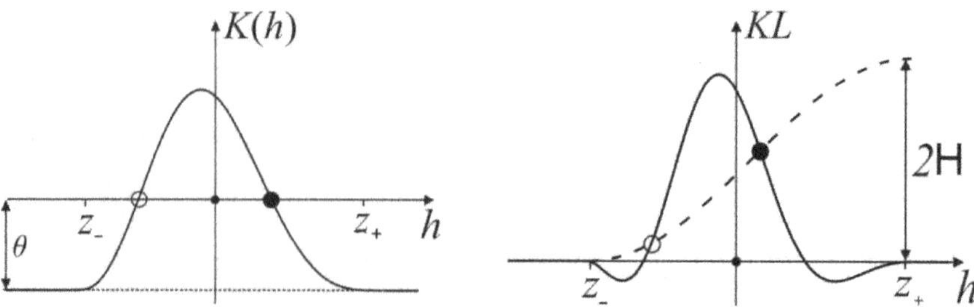

Fig. 3.13   Graphic construction for solving Eq. (3.40). The dashed line represents the
l.h.s. of Eq. (3.40).

The graph of $K(h)$ makes a first contact with the horizontal axis when
θ_Y reaches

$$\theta_f = \max\left(\frac{\sigma_2}{\sigma_1}\right). \qquad (3.42)$$

At this point, the formerly dry substrate is covered with a liquid film of
'thickness' $h_f = \mathrm{argmax}(\sigma_2/\sigma_1)$.

It is important to note that θ_f always lies above θ_W. To see this, we
note that $\mathrm{r} = 1/\cos\theta_W \approx 1 + \frac{1}{2}\theta_W^2$. Furthermore, $\mathrm{r} = \langle\sqrt{1 + |\nabla f|^2}\rangle$. Thus
we have

$$\theta_W^2 \approx \frac{1}{|\mathcal{S}|}\int_{z_-}^{z_+} \sigma_2(h)\mathrm{p}(h)\mathrm{d}h < \max(\sigma_2). \qquad (3.43)$$

On the other hand, $\sigma_2 > \sigma_1^2$, such that

$$\theta_f^2 = \max\left(\frac{\sigma_2}{\sigma_1}\right)^2 > \max(\sigma_2). \qquad (3.44)$$

From Eqs. (3.43) and (3.44), it follows directly that $\theta_f > \theta_W$.

Let us now consider the system off coexistence, again invoking Eq. (3.40)
as the condition determining h. A graphical solution of Eq. (3.40) is
sketched in Fig. 3.13(b). As long as $\theta_Y > 0$, both relevant zeros of $KL(h)$
(representing the solutions of Eq. (3.40) for $\mathrm{H} = 0$) lie well within the interval $[z_-, z_+]$. For $\mathrm{H} > 0$, the closed circle indicates again the stable solution.
Obviously, the two points of intersection will merge when the dashed and
solid curves touch each other only in a single point. This occurs at a certain
curvature $\mathrm{H}_f(\theta)$ of the liquid surface. For $\mathrm{H} > \mathrm{H}_f$, solid and dashed curve
meet only for $h \to z_-$: there is no liquid adsorbed, and the substrate is dry.
Hence the average position of the liquid surface, h, jumps discontinuously
at $\mathrm{H} = \mathrm{H}_f$. As H is reduced below H_f, h increases continuously until at coexistence it reaches a value corresponding to the right zero of KL. Because

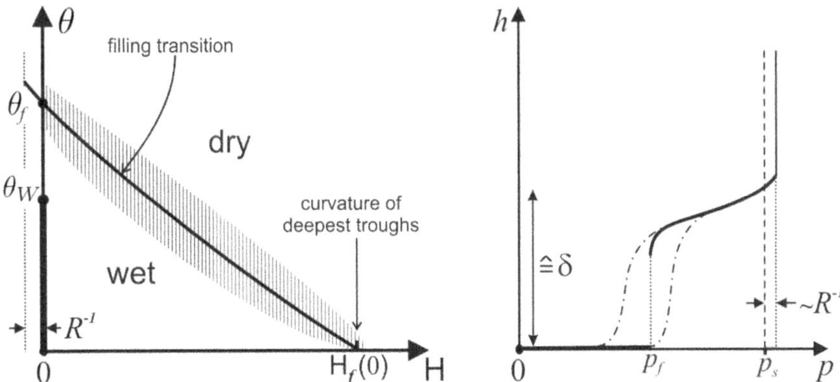

Fig. 3.14 Left: The wetting phase diagram wetting on homogeneous, isotropically rough surfaces. The most prominent feature is the existence of a filling transition line, $H_p(\theta)$, at which the adsorbed average film thickness jumps from zero to a finite value. The hatched area qualitatively indicates the hysteresis to be expected. Right: The liquid coverage as a function of the pressure of the surrounding vapor. Due to the many different solutions to the wetting problem, the 'jump' at the transition line is in fact a hysteresis loop, the width of which corresponds to the distance between the two dash-dotted curves (hatched area in the left panel).

of the phenomenological similarity of the jump in adsorbed film thickness to the prewetting transition encountered in standard wetting scenarios on flat substrates [92], it has been proposed to term this transition 'Wenzel prewetting' [93]. We will here use the more conventional term 'filling', as the troughs of the roughness are filled with liquid at this transition. When the microscopic contact angle is varied, a transition line results, which is shown in the left panel of Fig. 3.14 as the solid curve. Note that from the construction in Fig. 3.13 one may expect that conceptually, the transition line terminates in a critical end point. In the qualitative discussion here, it lies below the abscissa. We will return to this aspect below.

At coexistence, an infinitely thick film will not be able to form even at zero contact angle, due to the finite size of the grain. The finite mean curvature of the grain surface entails a positive Laplace pressure of the liquid-vapor interface. Hence the liquid would drain into any larger droplet, or into a wetting layer covering the container walls of the sample. This ramification is indicated by the dotted vertical line in Fig. 3.14 (left panel), which is at a curvature of $-1/R$ and corresponds to the pressure which would have to be reached to condense a macroscopic layer of liquid on the grain. A typical value of $H_f(0)$ can gained by estimating the maximum curvatures to expect of the roughness topography. If λ is a lateral correlation length of the roughness, we may set $H_f(0) \approx \delta/\lambda^2$. Since λ is certainly very much smaller than R, but only somewhat larger than δ, we find that $H_f(0)$ is significantly farther away from the origin than the dotted line at $1/R$. A typical value of θ_f is the maximum slope of the roughness, which can be up to a few tens of degrees.

We have thus seen that the occurrence of a film of adsorbed liquid with macroscopic thickness (comparable to the amplitude of the roughness) in some range of contact angles well above the Wenzel transition is a very general phenomenon. It is well conceivable that liquid transport takes place very effectively through a film of that thickness. This still depends, however, on the morphology of the film. We have above only discussed an averaged thickness, h, but for lateral transport it is necessary that the patches covered with liquid form a percolated set. For a large class of explicit roughness functions, the location of the percolation transition can be analytically calculated in a similar way as above [93]. It turns out that for zero mean curvature (i.e., at coexistence), the liquid film percolates at a critical microscopic contact angle θ_c lying between θ_f and θ_W. The locus of percolation transitions for finite H forms a line which starts from this point and proceeds roughly parallel to the filling transition line (see left panel of Fig. 3.14). Under most of the conditions where filling occurs, the liquid is also percolated, and well capable of supporting lateral liquid transport.

The dependence on H has an interesting further implication. For typical roughness amplitudes and lateral scales, the mean curvature at which the filling transition occurs can be converted to a vapor pressure by virtue of the Kelvin equation, Eq. (3.28). As one finds, this vapor pressure corresponds to just a few meters above the liquid reservoir under gravity (barometric formula, Eq. (3.1)). Hence we expect that in a thick layer of soil, there should be a transition a few meters above the water table at which the transport properties of the grain surfaces changes more or less abruptly, from good transport below that line to almost no lateral transport above. It is interesting to search for signatures of such a transition in soils and other wet granular piles of sufficient thickness.

3.2.4 *Inadequacy of Gaussian roughness models*

It is important to stress that mathematical models which are conventionally used to describe randomly rough surfaces may spectacularly fail to correctly predict the wetting behaviour of rough surfaces, even qualitatively.

Frequently, one assumes that roughness which has been produced by some intrinsically random process, such as sand blasting (uncorrelated impacts of grains) or weathering, can be reasonably well described through Gaussian roughness models. Hence one approximates the height distribution of the interface profile by a Gaussian and assumes correlations to decay rapidly in all lateral directions. This assumption comes with a number of useful mathematical features in data analysis (such as multivariate analysis), hence its frequent use.

However, it is readily seen that it leads to *qualitatively* wrong predictions when it comes to wetting. It turns out that the quantity $\sigma_1(z)$ defined in Eq. (3.34) is strictly independent of z for Gaussian roughness. As a result, the structure shown in Fig. 3.13 is lost. Instead, one would expect a transition from a completely dry to a completely wet surface (infinite film thickness) at a critical contact angle $\theta = 2\theta_W/\sqrt{\pi} \approx 1.13\ \theta_W$ [94]. Furthermore, there would be no dependence upon H.

It has been demonstrated that this must be in disagreement with the true wetting behaviour [94]. Profilometer data from a sand blasted copper surface have been used as a random interface data set. By numerically distorting the profile, random topographies of different statistical characteristics have been obtained, as apparent in differences in the skewness, Sk, of the height distribution. For variable values of the microscopic contact angle, θ, 'liquid' surfaces with constant mean curvature, H, which make a contact angle θ wherever they meet the substrate, have been obtained numerically. Such computations can be carried out by means of the so-called Surface Evolver, which is a very versatile public domain software provided by Ken Brakke [95]. It approximates liquid surfaces through adaptive triangulation, minimizing total interfacial energy. Starting with a configuration close to liquid–gas coexistence (H $\approx$ 0), H was gradually increased. At some value of H, which depended on θ, the volume of the adsorbed film (i.e., the volume between the constant mean curvature surface and the substrate) made a discontinuous jump to a much smaller value.

Two examples, for two different values of the skewness of the height distribution, are shown in Fig. 3.15. Each of the grey discs corresponds to a jump in adsorbed liquid volume as obtained from the numerical calculation. The diameter of each disk corresponds to the height of the corresponding jump. Note that the jumps were observed upon variation of the mean curvature, not the contact angle. The solid curve represents the filling transition line calculated directly from the used roughness profile, employing the formulae derived in Section 3.2.3. The 'swarm' of jumps grouped next to the

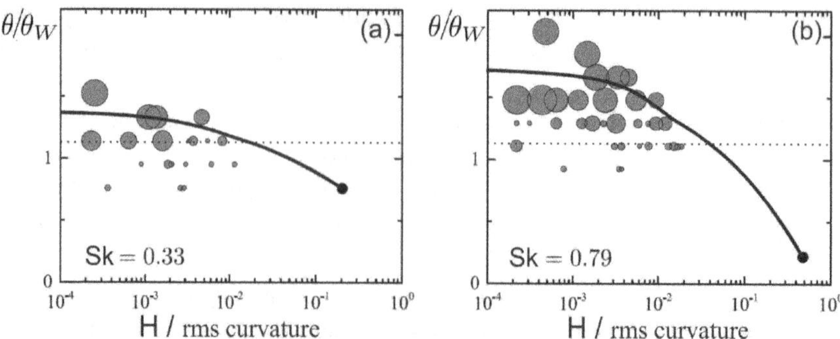

Fig. 3.15   Numerical corroboration [94] of the wetting phase diagram on a randomly rough substrate (Fig. 3.14). Different random substrate topographies were generated by numerically distorting profilometry data from a sand blasted copper surface, thereby obtaining height distributions with different skewness, Sk. Results for two such topographies are shown. Solid curves: filling transition lines calculated from the height distributions (i.e., through σ_1 and σ_2). Black dots are critical end points. Grey circles: loci of filling transitions found upon gradually increasing the mean curvature of the liquid surface. Dashed lines: predictions, assuming Gaussian roughness, of transitions from completely dry to completely wet (infinite film thickness) upon varying contact angle (no transitions to finite thickness upon varying curvature). Redrawn with modifications from [94].

transition lines may be seen as representing the hatched area in the left panel of Fig. 3.14. It should be noted, however, that the numerical study could be carried out only on a very limited sample size for reasons of computation time. For a larger sample, and even more for a real experiment, statistics is expected to be much superior, and the position of the jumps better defined (narrower swarms).

In contrast, the horizontal dotted lines indicate the transition expected for Gaussian roughness. It is a transition from a completely dry surface for $\theta > 2\theta_W/\sqrt{2}$ to an infinitely thick liquid layer (complete wetting) for $\theta < 2\theta_W/\sqrt{2}$. The Gaussian model is conceptionally unable to account for a finite jump in film thickness, leave alone to predict its height. Furthermore, it does not account for any transition upon varying the mean curvature. In other words, it predicts that upon varying the vapour pressure, there should be *no filling transition at all*. It only predicts that a transition occurs at a critical value of the contact angle, when this is varied. This must be considered a serious drawback, as in experimental scenarios, it is usually the vapour pressure (and hence H) which can be varied, not the contact angle, and we see from Fig. 3.15 that these transitions are real.

In conclusion, we have seen that when concerned with wetting properties of rough surfaces, and hence whenever dealing with wet granulates, we should dismiss Gaussian surface models altogether for describing grain surfaces. Other 'classical' descriptors, like the lateral correlation length or the roughness exponent, which is used in self-affine surface models, are not predictive of wetting properties of relevance. In order to describe surface roughness, we should instead use the quantities $\sigma_1(z)$ and $\sigma_2(z)$, which are defined in Eqs. (3.34) and (3.37), and can be easily computed from data obtained with standard profilometry methods.

3.2.5 *Adsorption isotherms on a rough substrate*

We have seen that roughness gives rise to a particular structure in the adsorption isotherms, singling out a liquid film thickness roughly equal to the roughness amplitude. From the construction in Fig. 3.13, we can plot the liquid coverage as a function of pressure for fixed contact angle. The result is sketched in the right panel of Fig. 3.14. It is important to note that h is monotonously increasing with p. As a consequence, two grains which can exchange liquid will always equilibrate such as to approach equal coverage with liquid.

In order to conveniently discuss these aspects, it is useful to describe the wetness of the granulate under study through the parameter

$$w = \frac{\text{amount of liquid wetting a grain}}{\text{grain volume}}. \tag{3.45}$$

For a spherical grain to which a film with average thickness h is adsorbed,

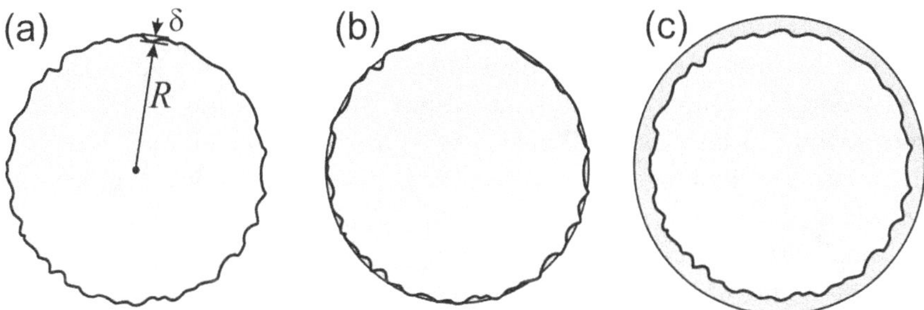

Fig. 3.16   The relation between the physically accessible range of h and the surface roughness. (a) Schematic of grain shape, with δ being a measure of roughness depth. (b) Grain with liquid film at $w \approx w_\delta$. (c) The situation when $h > \delta$. The sphere formed by the free liquid surface is freely mobile with respect to the grain, such that its position is undefined. The pressure is above saturated vapor pressure.

we obtain $w = 4\pi R^2 h / \frac{4}{3}\pi R^3 = 3h/R$. In all cases of relevance, $w \ll 1$. In order to arrive at a refined definition of the roughness parameter δ, we first define a notion of convexity of the wet grain which strongly adheres to the ideas we developed in Chapter 2 (cf. Fig. 2.8). Let $\mathcal{B}_s$ represent the set of points in space which are filled with the solid grain material, and let $\mathcal{B}_l$ represent the set of points in space filled with liquid. We shall then call a grain *convexly wet* if the line connecting any two points out of $\mathcal{B}_l$ lies entirely within the set union $\mathcal{B} = \mathcal{B}_s \cup \mathcal{B}_l$. It is immediately clear that this entails $\mathsf{H} \leq 0$, such that we can assume that whatever resembles a trough or scratch on the grain surface is filled with liquid. We then define the roughness parameter δ as the liquid coverage at which liquid convexity is reached. The corresponding amount of liquid per grain volume is $w_\delta = 3\delta/R$, where R must be considered as an equivalent radius. R can be defined as one half times the so-called Sauter equivalent diameter. The latter is quite commonly used for characterizing granular systems in engineering and the geosciences. It is defined as the diameter of spherical grains which would have the same surface-to-volume ratio as the system under study. For our purposes, and for the sake of clarity, we define R as the radius of a sphere which has the same surface-to-volume ratio as the convex grain body we have defined in Chapter 2. With this definition, we can consider w_δ to be the 'wetness' of the granulate at which isolated grains just become convexly wet.

As illustrated in the right panel of Fig. 3.14, we should for fixed contact angle expect to have, in most cases, either completely dry grains (which we have discussed already above) or a liquid film thickness which is comparable to the depth of the roughness, depending on vapor pressure. The quantity of liquid per grain can be expressed by w. If $h \approx \delta$, we have $w \approx 3\delta/R$. If there is considerably more liquid in the system, the free liquid surface will follow the convex shape of the grains, which according to Eq. (3.20) corresponds to an elevated Laplace pressure. The liquid then tends to drain into any liquid reservoir with lower Laplace pressure which may form anywhere in the system, away from the granulate. Hence for $w > 3\delta/R$ an equal distribution of all liquid among the grain surfaces is not stable, and the liquid will collect spontaneously in some low-pressure reservoir. Furthermore, if two grains can exchange liquid and $w > 3\delta/R$, the film on one grain will acquire liquid at expense of the film on the other grain: the situation is not stable, in contrast to the case where $h \approx \delta$ as discussed above.

3.2.6 *Contact angle hysteresis*

Equation (3.4) defines the equilibrium value for the contact angle, θ_Y. It is based on a microscopic force balance at the three-phase contact line, and has been used in the previous section to elucidate the wetting behavior of rough surfaces. Clearly, θ_Y is characteristic solely for perfect surfaces, and there is a lot more complex physics to be expected as we consider roughness and chemical heterogeneities, which are present on almost all natural solid surfaces. We will now discuss effects of dynamics, i.e., the effect a moving contact line has on the value of the contact angle. This is of particular interest for applications to wet granular materials, as it affects the shape and dynamics of liquid capillary bridges between grains when these are moving relative to one another.

Consider thus the contact line of a liquid droplet on a rough surface, as sketched in Fig. 3.17(a). The contact line, which we see head-on in the sketch and therefore do not resolve, will not be straight in this case, but wiggle about an average position, x, as one moves along the spatial coordinate y, going perpendicular to the plane of the figure. We now (and henceforth in this book) refer to the macroscopic contact angle, θ, seen on scales larger than the scale of the roughness. This is the quantity actually measured in optical contact angle measurements. The shape of the contact

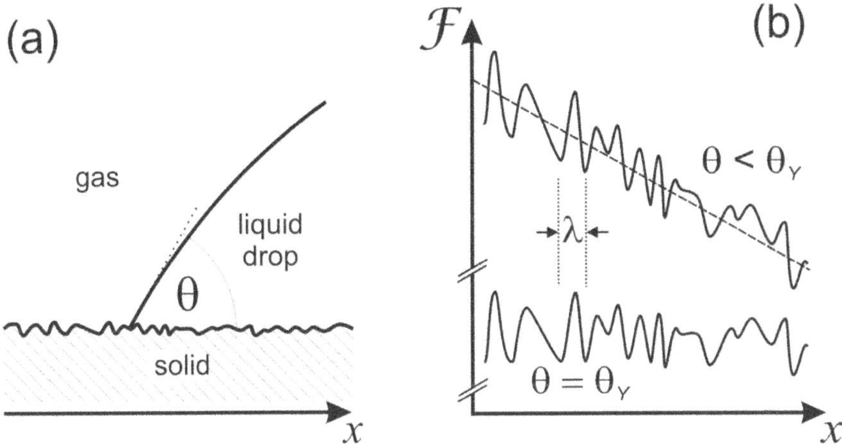

Fig. 3.17 (a) Three-phase contact line on a rough substrate. Microscopically, the contact angle according to Young and Dupré holds everywhere, provided the dominant roughness scale is not in the nanometer range. (b) The 'energy landscape' for $\theta = \theta_Y$ and for $\theta \neq \theta_Y$. The characteristic lateral scale of the roughness is denoted by λ.

line is governed by the minimization of the total free energy of the involved interfaces, in line with the considerations in the previous section. If the average position of the contact line is varied along x, the value of the minimized free energy varies, as sketched in Fig. 3.17(b). If $\theta \approx \theta_Y$, it will vary erratically around a constant value. If, however, θ is smaller than θ_Y, there is a net free energy gain when the contact lines is displaced, such that the free energy landscape acquires a slanted average topography. Its slope, S, is given by

$$\mathsf{S} = \gamma(\cos\theta - \cos\theta_Y) + \eta v C(\theta), \tag{3.46}$$

where the second term comes from the viscous dissipation in the contact line region. This has a complex dependence on the contact angle, which is captured in the function $C(\theta)$, but shall not concern us here any further, as we are just aiming at a very qualitative discussion.

At finite temperature, we may expect the contact line to move at some finite velocity due to the different rates of jumps the contact line performs between neighboring minima. Assuming Boltzmannian jump rates, one readily finds

$$v = v_0[e^{\mathsf{S}\lambda/2k_BT} - e^{-\mathsf{S}\lambda/2k_BT}] = 2v_0 \sinh\frac{\mathsf{S}\lambda}{2k_BT}, \tag{3.47}$$

where λ is the typical lateral length scale of the substrate inhomogeneity, and v_0 some constant. This leads to the result

$$\cos\theta = \cos\theta_Y + \frac{k_BT}{\lambda\gamma}\operatorname{arsinh}\frac{v}{v_0}, \tag{3.48}$$

which is plotted in Fig. 3.18.

Since $\lambda\gamma$ is, for typical roughness scales, much larger than k_BT and v_0 is only a very slow, thermally assisted creeping velocity, the second term represents a rather steep jump at $v = 0$. Hence if a drop is deposited on the substrate, such that the contact line is advancing on the substrate, it will at all reasonable laboratory time scales appear to come to rest at the lower arrow, where the 'advancing' branch terminates. The corresponding contact angle is called the advancing contact angle, θ_{adv}, and is always lather than θ_Y. Similarly, if a major part of a drop is taken off the substrate by means of a pipette, such that the contact line recedes, it will come to rest at the upper arrow, where the 'receding' branch terminates. The corresponding contact angle is called the receding contact angle, θ_{rec}, and is always smaller than θ_Y. The gap between θ_{rec} and θ_{rec} is called the contact angle hysteresis.

We should note that the simple picture developed here is meant to just outline the qualitative features. Several conflicting microscopic theories

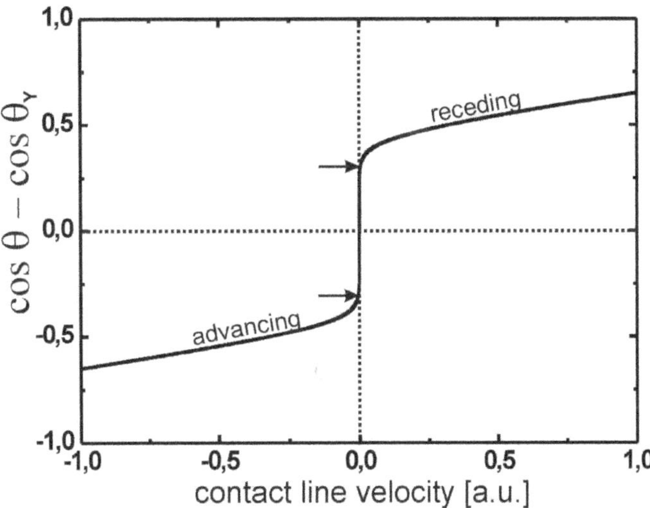

Fig. 3.18 The contact angle hysteresis, as it follows naturally from the simple picture outlined in Fig. 3.17 through thermal activation. The arrows indicate the advancing (lower arrow) and receding (upper arrow) contact angles.

of contact line dynamics and contact angle hysteresis are currently under discussion and strongly debated. For the present book, it is sufficient to keep in mind that there is always contact angle hysteresis on any real (i.e., inhomogeneous) surface. This is certainly of relevance for granular matter physics, since the surfaces of grains are always far from perfect. Even if they were perfect at the beginning of an experiment, the unavoidable wear during the inelastic collisions would result in defects and thus eventually in contact angle hysteresis. One should thus generally keep in mind that contact angle values should be taken with care in the realm of granular matter.

3.3 Conclusions

In this chapter, the basic principles of the physics of wetting have been outlined. A direct connection between molecular interactions, adsorption isotherms, and the contact angle has been established. We have seen that although the contact angle is usually determined (and defined) at liquid–vapor coexistence, its concept pertains as well to situations off coexistence, where strongly curved liquid interfaces are involved. This is of particular importance for wet granular systems, since both the capillary bridges within a wet granular pile and the omnipresent roughness on the grain surfaces necessarily entail more or less strongly curved liquid interfaces.

We have treated the important case of rough solid substrates in an approximative manner, where we have exploited that most natural roughness encountered on grains is on larger scales than the typical range of van der Waals tails. We could therefore use the microscopic contact angle, θ_Y, as a parameter which fully describes the wetting properties of the grain material itself. On that basis, we found that the depth of the roughness on the grains provides a natural scale for the 'wetness' of a granulate. A certain degree of wetness is singled out at which the troughs of the roughness are filled, yielding an almost completely wet grain surface. The grains will then behave in effect like rather smooth grains which are perfectly wetted by the liquid.

We have found that Gaussian surface models, used widely to describe randomly rough surfaces, are conceptionally incapable of accounting for important phenomena of wetting on such surfaces. As an alternative to widely used standard tools, we have developed a set of easily accessible descriptors which allow to characterize roughness in a way which is much more powerful for predicting wetting phenomena.

Another consequence of surface inhomogeneities is the ubiquitously observed contact angle hysteresis, which we have discussed within just a very simple model. It is present on all heterogeneous surfaces, and we should bear in mind that quantitative contact angle values must always be taken with a grain of salt. They should therefore not be given a central role in detailed models of granular matter. Throughout this book, we will try to keep our modelling as independent as possible from specific values of the contact angle. This is not too difficult in most cases, as the contact angle enters in most expressions as its cosine, which varies only slightly in the case of small contact angles. The latter is in fact the most important case for wet granular matter, since liquids with a large contact angle do not easily intrude a granular pile.

Further reading

There have been extensive reviews on the physics of wetting, as far as ideal surfaces are concerned. A wide scope is covered by the comprehensive textbook by De Gennes, Brochard, and Quere, where one finds both static and dynamic aspects of wetting well covered [96]. The same is true for the recent review by Bonn *et al.* [97]. For those who are interested in particular in the interaction mechanisms and corresponding forces, the recent book by Butt and Kappl should be very useful [3].

The wetting properties of (the much more relevant) rough surfaces are comparably sparsely covered in the literature. A classical text on the wetting properties of randomly rough substrates is the early paper by Wenzel [91], which describes the roughness by a single parameter, r. Most of the more recent work is on nanometer scale roughness, where the possible competition between the van der Waals tails of the effective interface potential and the features of the substrate topography are discussed [98, 99]. On the length scales between tens of nanometers and a millimeter, which are most relevant to the roughness found on grains, there is a lot of literature which considers simplified topographic features [89, 100]. In contrast, only few has been published on random roughness in that range of length scales, although this is the most relevant case for naturally occurring surfaces. Two recent papers [93, 101] outline some of the reasons why this is particularly difficult to treat in any depth beyond what we have outlined in this chapter.

Contact angle hysteresis and contact line dynamics are still largely elusive, despite their ubiquity on natural surfaces. This unsatisfactory situation is somewhat 'stabilized' by some very engaged debates about competing models, but one should appreciate that the conceptual problems in treating the boundary conditions near a contact line, as well as the dynamic response of the liquid surface to substrate inhomogeneity, are in fact enormous. A few seminal papers [102–107] should be useful to guide the reader into this important field of current research.

Chapter 4

Capillary Forces

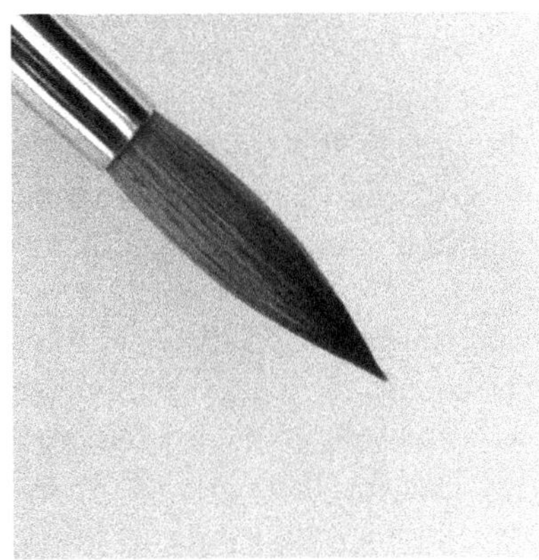

Fig. 4.1 The capillary forces which shape a wet brush, by attracting neighboring hairs into intimate contact, are the same as those which are acting between grains, giving the granulate its stiff consistency.

The most obvious change a dry granulate is subject to when a wetting liquid is added is a transformation from a more or less fluid state to a pasty material. We have seen in the previous section that the formation of liquid capillary bridges between adjacent grains certainly plays an important role here. Theoretical concepts connecting wetting processes to the mechanical properties of wet granulates have emerged around the turn of the millennium [108–115]. One of the simpler questions to be asked is this: is it the (random) network of capillary bridge forces itself which yields the stiffer appearance, or is it just the increase of the friction force between the grains due to their being pushed harder together by the capillary force? More complex questions concern the connectivity of the capillary bridge network, and the transformations of the latter upon shearing the granulate. Once these are sufficiently understood, one may try to understand effects like soil liquefaction, or simply the presence of a yield stress, possibly as dynamic phase transitions in this rather complex non-equilibrium system. The present chapter is devoted to the static properties of an individual capillary bridge extending between two adjacent grains, and the forces which emerge due to its presence. We will start with liquid structures between idealized axi-symmetric surface configurations, and later discuss the effects of possible imperfections.

4.1 Capillary bridge between flat walls

4.1.1 *Extremal surfaces*

When a certain volume of wetting liquid is deposited close to the point of contact between two grains, one observes the immediate formation of a 'liquid bridge' connecting the grain surfaces. Determining the shape of this bridge means to find the liquid surface morphology which minimizes the interfacial free energy of the system, under the constraint of constant liquid volume, V. As far as the interfaces with the solid substrate are concerned, we have learned in the previous chapter that energy balance can be expressed in a fixed equilibrium contact angle, θ, between the liquid and solid surfaces. What then remains is to find a shape of the liquid surface which minimizes its own area, at a given contact angle with the solid, and at a given liquid volume.

It should be noted that gravity can be safely neglected in the discussion of the shape of the capillary bridges. Typical grain sizes are a millimeter or less, such that the size of the bridges is usually in the sub-millimeter regime. This is well below the capillary length of most liquids, which determines the length scale above which gravity plays a significant role in the morphology of liquid structures. The capillary length of a liquid is given by $\sqrt{\gamma/g\varrho}$, where ϱ is the density of the liquid, and g is the acceleration due to gravity. The capillary length for water is about 2.7 mm, much larger than the typical radii of curvature encountered in wet granular systems. More conventionally, this can be expressed in terms of the Bond number,

$$\mathsf{Bo} = \frac{g\varrho l^2}{\gamma}, \tag{4.1}$$

where l is a characteristic length of the liquid structure. For a capillary bridge, one would choose the radius of curvature of the outer meniscus, $l = r_1$. Throughout this chapter, we will assume $\mathsf{Bo} \approx 0$.

Two (assumed spherical) grains constitute an axi-symmetric configuration of solid surfaces. We will therefore restrict our discussion to axi-symmetric liquid shapes, and describe the position of the liquid–gas interface by a function $r(z)$, where z is the axis of symmetry of the system.[1]

---

[1]Note that at large contact angles ($\theta > \pi/2$) the bridge may spontaneously break the

Table 4.1 Relevant properties of a few common liquids. Values are for room temperature where not stated otherwise.

Liquid	Surface tension γ [mN/m]	Viscosity η [mPas]	Density ϱ [g/cm]	Capillary length $\sqrt{\gamma/g\varrho}$ [mm]
Helium (T = 4.2 K)	0.089	0.0033	0.125	0.27
Water	72	1.0	1.0	2.70
n-Nonane	24	0.9	0.79	1.7
n-Dodecane	25.4	1.38	0.75	1.85
Silicone oil (Si AK 5)	19.2	4.6	0.76	1.60
1-Decanol	28.5	15.9	0.831	1.86
Glycerol	64	1000	1.26	2.28

The total area of that interface is given by

$$\int 2\pi r \sqrt{1 + r'^2}\, dz, \tag{4.2}$$

where the prime denotes differentiation with respect to z. If this is to be minimized, the corresponding surface shape can be found through the Euler–Lagrange equation (3.14), this time with the Lagrangian $\mathcal{L} = 2\pi r \sqrt{1 + r'^2}$. After some straightforward manipulation, this leads to

$$rr'' = 1 + r'^2. \tag{4.3}$$

As it is readily checked, this has the simple solution

$$r = r_0 \cosh \frac{z}{r_0}, \tag{4.4}$$

which represents a catenoid, a well-known member of the class of minimal surfaces.

So far, however, we have not taken the constraint of constant volume into account. Accordingly, the surface described by Eq. (4.4) represents a surface the mean curvature of which is zero everywhere. This corresponds to vanishing Laplace pressure, Eq. (3.20), and hence to liquid–vapor coexistence, Eq. (3.28). In order to find the full set of shapes relevant for capillary bridge geometry, we have to add the volume constraint,

$$\int \pi r^2 dz = V. \tag{4.5}$$

Inserting accordingly a modified Lagrangian, $\mathcal{L} = 2\pi r \sqrt{1 + r'^2} + \lambda \pi r^2$ into Eq. (3.14), we obtain

$$rr'' = \left(1 + r'^2\right)\left[1 + \lambda r \sqrt{1 + r'^2}\right], \tag{4.6}$$

axial symmetry. This effect, however, can be neglected in the present discussion since for wet granular matter we only have to discuss liquids which wet the grain surface well (contact angles well below $\pi/2$).

where λ is a Lagrange parameter. It is readily checked that the sphere, $r(z) = \sqrt{r_0^2 - z^2}$, is in fact a solution to Eq. (4.6), as we know from liquid drops. For the Lagrange parameter, we obtain $\lambda = -1/r_0$ in this case. It is immediately clear, however, that this cannot be the only type of solution to Eq. (4.6). A capillary bridge between two flat and parallel walls, for instance, can only be spherical if the distance of the walls is equal to $2r_0 \cos\theta$, and if $\theta > \pi/2$. The most important solutions to Eq. (4.6) are so-called unduloids. These are obtained by rolling an ellipse along the z-axis, tracing the locus of one of its foci, and then rotate the resulting curve around the z-axis.

We will see below, however, that we will not need to make use of these rather complex mathematical objects here. In all cases of practical relevance, the shape of the bridge can be represented as $r(z) = r_0 + f(z)$, with $f \ll r_0$ in the whole range of relevant z [116, 117]. In this case, Eq. (4.6) reduces to

$$r_0 r'' \approx \left(1 + r'^2\right)\left[1 + \lambda r_0 \sqrt{1 + r'^2}\right]. \tag{4.7}$$

If we insert $f(z) = -\sqrt{r_1^2 - z^2}$, representing a toroidal shape, we find

$$\lambda r_1 = 1 - \frac{f}{r_0}, \tag{4.8}$$

such that the toroidal shape is indeed a solution if only f is sufficiently small. For the Lagrange parameter, we then have $\lambda = 1/r_1$.

It is easy to see why the toroidal shape should be very close to solving Eq. (4.6). r_1 will be of the same order of magnitude as f, at least in the relevant case when the contact angle is small. Therefore we have $r_1 \ll r_0$, which means that the mean curvature of the surface is completely dominated by the small radius, r_1. Hence the toroidal surface is rather close to a surface of constant mean curvature.

4.1.2 *Attractive force of a toroidal bridge*

Let us thus assume a toroidal shape for the liquid bridges we discuss here. For the case of a liquid bridge between two parallel flat walls, the corresponding geometry is sketched in Fig. 4.2. The dash-dotted line indicates the cylinder symmetry of the capillary bridge, which is a direct consequence of the homogeneity of pressure. The Laplace pressure is given by $p_L = \gamma(1/r_1 + 1/r_2)$. Since the (dominant) principal curvature, $1/r_1 = 2\cos\theta/s$ is constant, so must be the other principal curvature, $1/r_2$,

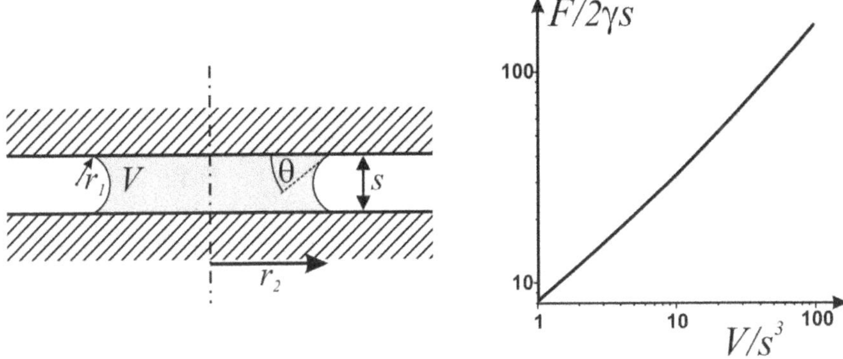

Fig. 4.2   Left: Sketch of a liquid capillary bridge between two flat walls. Right: The attractive force exerted on the walls by the capillary bridge, in normalized quantities, plotted for vanishing contact angle (complete wetting).

in order for the pressure to be constant as well. Hence the capillary bridge must have cylinder symmetry.

When the distance between the walls, s, is increased, volume conservation entails a reduction of r_2, and thus the exposure of solid surface to the gas phase. This is connected to an increase of free energy by $\gamma \cos\theta$, hence there is an attractive force between the two walls. Its magnitude can be easily evaluated by considering a small change in surface separation, ds, and demanding the volume to be conserved:

$$dV = 2\pi r_2 s dr_2 + \pi r_2^2 ds = 0. \tag{4.9}$$

The contribution of each solid surface to the total change in free energy is

$$dE = -4\pi r_2 dr_2 \gamma \cos\theta + 2\pi\gamma r_2 \frac{\left(\frac{\pi}{2} - \theta\right)}{\cos\theta} ds. \tag{4.10}$$

The fraction comes from the evaluation of the area of the outer meniscus along the circumference of the bridge. Combining Eqs. (4.9) and (4.10), we obtain for the force

$$\frac{dE}{ds} = \pi\gamma r_2 \frac{\left(\frac{\pi}{2} - \theta\right)}{\cos\theta} + \frac{\gamma}{r_1}\pi r_2^2. \tag{4.11}$$

It is instructive to write (4.11) as a function of liquid volume, V. Defining a dimensionless volume, $\hat{V} = V/s^3$, we find

$$F = \frac{dE}{ds} = 2\gamma s \left[\frac{\left(\frac{\pi}{2} - \theta\right)}{\cos\theta} \sqrt{\pi\hat{V}} + \hat{V}\cos\theta \right]. \tag{4.12}$$

The first term comes from the circumference, the second from the Laplace pressure acting on the entire cross-section of the bridge. The force according to Eq. (4.12) is plotted for vanishing θ (i.e., for complete wetting) in the right panel of Fig. 4.2. The small contributions of the volume corresponding to the in-bending of the surface at the periphery, with radius r_1, have been neglected in Eq. (4.12). As one intuitively expects, the attractive force increases strongly with the liquid volume of the capillary bridge.

It is interesting to see which of the two terms in Eq. (4.12) is dominant. Both terms are equal when

$$r_2 = s \left(\frac{\pi}{2} - \theta\right)^2 / (\cos^4 \theta) \tag{4.13}$$

as it is readily verified. As long as θ stays below about 60 degrees, the fraction is of order unity. Hence the Laplace pressure term dominates when $r_2 \gg s$. If it is of order s, the circumference term is dominant.

4.2 Capillary bridge between spherical bodies

Let us consider two spherical grains, each of which carries a liquid film of homogeneous thickness, being brought into contact. At the point of contact between the grains, the liquid surface makes a sharp bend, as shown schematically in Fig. 4.3(left). It is clear that this is not an equilibrium situation: a curved liquid surface gives rise to a Laplace pressure, as we have seen above. Where the liquid films meet, this curvature is large and negative, such that there is a substantial under-pressure which sucks the liquid towards the contact region. Equilibrium is reached when the liquid surface has acquired a spatially constant mean curvature, and forms the equilibrium contact angle with the surface of the grains as depicted in the center of Fig. 4.3.

4.2.1 *Formation of the capillary bridge*

In experiments, one observes that the formation of a capillary bridge proceeds largely in two steps. Immediately after the two solid grain surfaces have established mutual contact, there is a capillary bridge with an initial volume, V_i, which is recruited from the liquid squished out of the immediate contact region during the impact. After that, liquid is flowing from the surrounding liquid film into the capillary bridge, increasing its liquid volume to a final value, V_f (cf. Fig. 4.4.). This flow is driven by the negative mean curvature (and thus the negative Laplace pressure) of the free liquid surface of the capillary bridge immediately formed. We will call

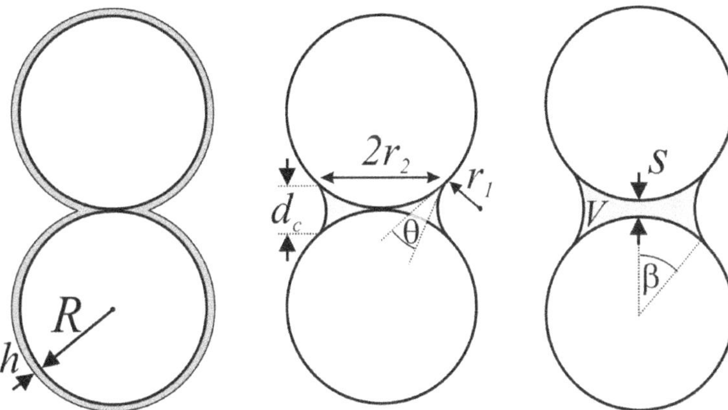

Fig. 4.3   Left: Two spherical grains touching each other, with a wetting layer of thickness h on each of them (thickness exaggerated). Where the liquid film surfaces meet, a sharp bend results which gives rise to a large negative Laplace pressure. Center: Equilibrium is reached when a capillary bridge with constant mean curvature has formed. The contact angle, θ, is indicated. Right: A pendular bridge between two spherical grains which are not in mutual contact. The separation of the grain surfaces is denoted by s. Note that the liquid makes a finite angle θ with the spheres, which at equilibrium is Young's contact angle, θ_Y.

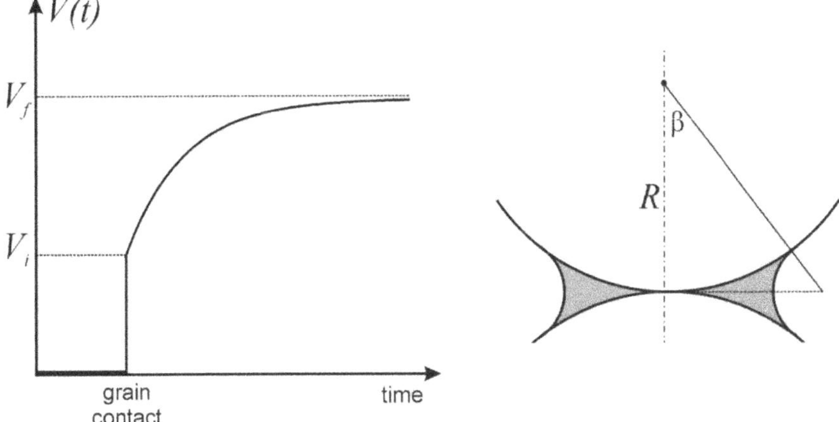

Fig. 4.4   Left: The formation of a capillary bridge in two steps. As the grains establish contact, a bridge of initial volume V_i is formed by squeezing out liquid from the immediate contact region. This is followed by gradual accumulation of liquid via advection of liquid towards the capillary bridge, which proceeds until the bridge has reached its equilibrium (final) volume, V_f. Right: Capillary bridge geometry, with bridge angle β.

this process capillary bridge ripening. Depending on experimental conditions, such as the size and roughness of the grains, as well as the contact angle and the viscosity of the liquid, this can take minutes to hours, or be entirely absent on laboratory time scales.

In order to conveniently discuss these processes, it is of interest how the size of the capillary bridge, as quantified in the bridge angle β defined in Fig. 4.4, depends upon the bridge volume. This is a tedious but straightforward exercise in elementary geometry, which we do not discuss here in detail, as it does not provide any specific insight. We restrict ourselves to presenting a rough outline. First of all, we compute the volume of a cylinder with radius $R \sin \beta$ and height $2R(1 - \cos \beta)$. For unit radius, this yields

$$\text{Cylinder} = 2\pi \sin^2 \beta (1 - \cos \beta). \tag{4.14}$$

From this we subtract two spherical caps (representing involved segments of the spherical grains), each of which has volume

$$\text{Cap} = \frac{2\pi}{3}(1 - \cos \beta) - \frac{\pi}{3} \sin^2 \beta \cos \beta. \tag{4.15}$$

Furthermore, we have to subtract a segment of a torus the surface of which makes an angle θ with the surface of the grains (cf. Fig. 4.3, center). Its volume can be computed using Steiner's theorem, with the result

$$\text{SegTorus} = r_1^2 (\chi - \sin \chi \cos \chi) 2\pi \left(\sin \beta - r_1 \frac{3 \sin \chi - \sin^3 \chi - 3\chi \cos \chi}{3(\chi - \sin \chi \cos \chi)} \right), \tag{4.16}$$

where $\chi = \frac{\pi}{2} - \beta - \theta$, and r_1 is again the radius of curvature of the outer meniscus. The result (for unity R) is then

$$\tilde{V}(\beta, \theta) = \text{Cylinder} - 2\text{Cap} - \text{SegTorus}. \tag{4.17}$$

In general, we have $V = \tilde{V} R^3$. $\tilde{V}$ is plotted for several contact angles (including complete wetting, $\theta = 0$) in Fig. 4.5. Since it becomes clear from the figure that no qualitative changes are to be expected as θ is varied in the region of wetting (θ well below $\pi/2$), we will henceforth almost exclusively discuss the case of complete wetting ($\theta = 0$). In the inset, the same data are presented in a double-logarithmic manner. Clearly, $V \propto \beta^4$ in good approximation.

Using the parameter $w = 3h/R$ defined in Chapter 3 (Eq. (3.45)), we can now consider both β and V as a function of w. We assume that aside from the liquid volume expelled from the immediate contact region, the bridge will as well incorporate the liquid film on the grain surface covered

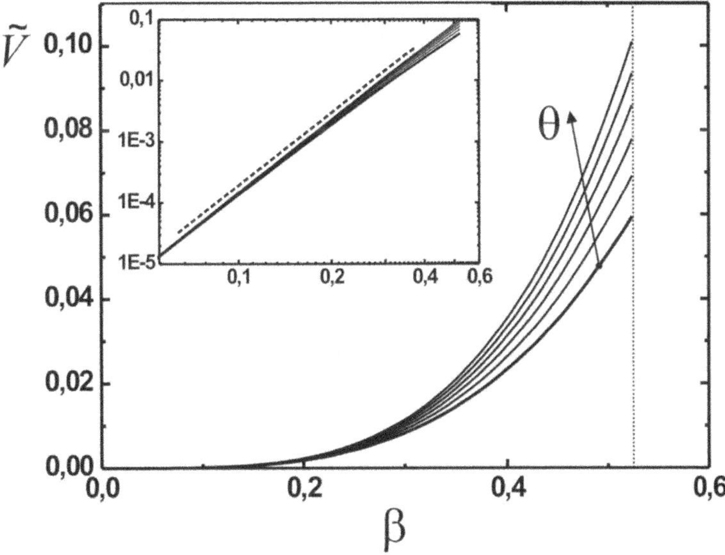

Fig. 4.5 The capillary bridge volume as a function of the bridge angle, β, for five different values of the contact angle. The latter increases from bottom to top, with values $\theta = 0$, 0.2, 0.4, 0.6, 0.8, and 1.0 radians. The vertical dotted line indicates $\beta = \pi/6$, at which neighboring capillary bridges coalesce in a granular pile, such that the discussion of an individual capillary bridge loses its point. The inset shows the same data on log-log scale, showing that the volume scales as the fourth power of β (the slope of the dashed line is equal to four).

by the bridge. Once the geometric relation between β and V is known, the initial bridge volume, V_i, is then implicitly given by

$$V_i = 4\pi R^2(1 - \cos\beta_i)h \tag{4.18}$$

or, in normalized form,

$$\tilde{V}_i = \frac{4}{3}\pi w(1 - \cos\beta_i), \tag{4.19}$$

where $\tilde{V}_i$ is again V_i/R^3. In order to come up with the relation between w and the parameters of the final bridges, β_f and V_f, we need to specify the number k of capillary bridges present on average on each grain surface. We will find in Chapter 6 that $k \approx 6$ accounts well for reasonably densely packed piles. It is readily verified that

$$V_f = w\frac{2}{k}V_{grain} \tag{4.20}$$

or, in dimensionless form,

$$\tilde{V}_f = \frac{8\pi w}{3k}. \tag{4.21}$$

We must reckon indeed with large differences between the volume of the capillary bridges formed immediately after contact and those formed after some time of undisturbed inter-granular contact.

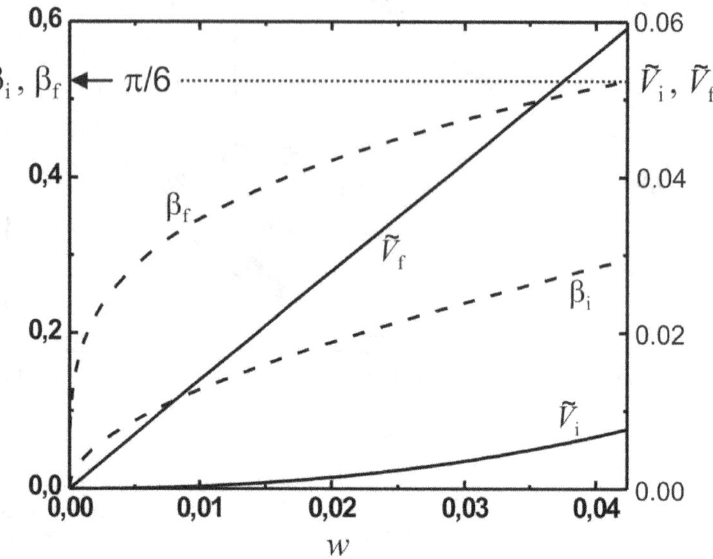

Fig. 4.6   The initial and final capillary bridge volume, $\tilde{V}_i$ and $\tilde{V}_f$ (solid curves, right axis), and the initial and final bridge angle, β_i and β_f (dashed curves, left axis), as functions of the grain wetness, w, for an ideally spherical grain. For the computation of $\tilde{V}_f$ and β_f, a coordination number of $k = 6$ has been assumed. When β reaches $\pi/6$ (dotted line), neighboring bridges in a granular pile will most likely coalesce.

The results obtained for $\tilde{V}_i$ and $\tilde{V}_f$ are presented in Fig. 4.6. The range of the abscissa is chosen such that the maximum value of β_f, indicating the 'size' of the final capillary bridge formed as time tends to infinity, is equal to $\pi/6$. At this value of β, neighboring bridges in a granular pile (which is the only situation where bridges have enough time to ripen) will start to coalesce forming larger liquid clusters (see below, Chapter 6).

In Fig. 4.7, the two volumes V_i and V_f are plotted on logarithmic scale as a function of w, in order to better appreciate the large difference which may occur between both. Formally, we have

$$\frac{V_f}{V_i} \approx \frac{8\pi w}{12 k w^2} = \frac{2}{kw}. \tag{4.22}$$

From the figure, we see that whenever w is appreciable, $\tilde{V}_i \approx 4w^2$ (dotted line) is a quite reasonable (and useful) approximation.

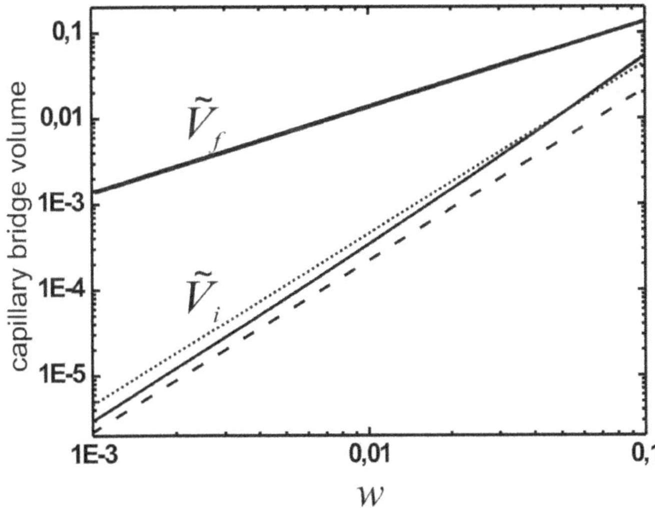

Fig. 4.7 Logarithmic presentation of V_i and V_f as a function of w. While V_f is proportional to w, V_i scales roughly as w^2. The dashed line indicates the minimum volume required to form a singly connected capillary bridge between rough grains if $w = w_\delta$, i.e., if the roughness topography is completely filled with liquid.

If the grain surface is not ideal, as it is always the case in both experiments and applications, the lateral transport properties of the film depend on its morphology. If the film is percolated on the surface (i.e., if a molecule can travel along the grain surface without leaving the liquid phase), transport will be possible. If it is not, the capillary bridge volume will remain at its initial value, if not transport via the gas phase turns out to be substantial. As we have seen in Section 3.2, the morphology of the adsorbed film depends sensitively on the roughness of the surface and on the microscopic contact angle. In fact, it has been shown that capillary bridge ripening strongly depends on the roughness and the microscopic contact angle. With the same set of glass beads as the model grains, bridge ripening was found to take place when the liquid makes a contact angle around zero with the glass surface, but was suppressed if a liquid with a contact angle of about 20 degrees or more was used [118]. The transition between isolated liquid patches on the grains surface and a percolated liquid film can be shown to be generally quite close to the transition line displayed in Fig. 3.14 [93].

4.2.2 *Capillary bridge force*

For our purposes, the toroidal approximation indicated in Fig. 4.3 is sufficiently accurate, because several effects which are beyond experimental control give rise to more severe deviations from quantitative predictions than the toroidal approximation. These are chiefly the ubiquitous roughness of the grains and the contact angle hysteresis. The formulae derived above for the toroidal capillary bridges will therefore form the basis of the subsequent discussion.

Let us calculate the attractive force at zero separation in the toroidal approximation, in the same way as we have done for the capillary bridge between flat walls. When the grains are removed from each other by an infinitesimal amount ds, the contact line will recede, reducing r_2, due to the conservation of the liquid volume:

$$dV = \pi r_2^2 ds + 2\pi r_2 \cdot d_c \cdot dr_2 = 0. \tag{4.23}$$

Consequently, some of the grain surface is exposed to air, which is accompanied by an energy increase of $\gamma \cos\theta$ per unit area. Using the relation

$$r_2^2 = R d_c + \frac{d_c^2}{4} \tag{4.24}$$

which can be read off Fig. 4.3, we readily find

$$F_{cb} = \pi\gamma \frac{\cos\theta}{\cos\beta} \left(2R + \frac{d_c}{2} \right) = 2\pi R\gamma \cos\theta \frac{1 + d_c/4R}{1 - d_c/2R} \tag{4.25}$$

for the capillary bridge force. Since $d_c \ll 4R$, the fraction on the right-hand side is close to unity. We thus can set

$$F_{cb} \approx 2\pi R\gamma \cos\theta \tag{4.26}$$

in good approximation [119].

We have so far neglected the fact that the free liquid surface area also changes when the grains are removed from each other, resulting in an additional term to F_{cb}. As the result of a somewhat lengthy (but straightforward) calculation, it turns out that this contribution is smaller by a factor of approximately r_2/R. Given the approximative character of the treatment we present here, we will neglect this term here.

Remarkably, the result summarized in Eq. (4.26) is independent of the liquid volume, V, of the bridge, in contrast to the above result for the force between two planar walls. It is straightforward to see that this should indeed be so: From the sketch at the center of Fig. 4.3, one can see that $r_1 \propto r_2^2$, or more precisely $r_2 = \sqrt{2Rr_1}$ for complete wetting. Since $r_2 \gg r_1$, we

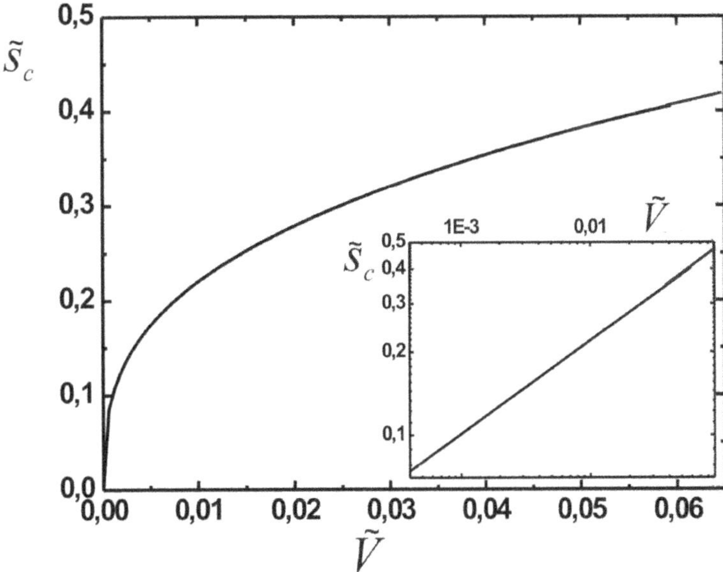

Fig. 4.8 The normalized separation at which rupture of the capillary bridge occurs. As $V \ll R^3$, the first term in Eq. (4.28) is strongly dominant. Consequently, the log-log plot in the inset shows an almost straight line with a slope of 1/3.

have $p_L \approx \gamma/r_1$. The covered area is πr_2^2, such that $F_{cb} = \pi r_2^2 p_L = 2\pi R\gamma$, independent of the size of the bridge. We should note here that F_{cb} has an additional contribution from its perimeter, $2\pi\gamma r_2$, but this is small as compared to the Laplace pressure term as long as $r_2 \ll R$. This is not a serious restriction, as all the above holds only under that condition. We will see below that before the liquid bridge volume becomes large enough to substantially question the validity of Eq. (4.26), the bridges coalesce and form more complex geometric structures. The notion of individual capillary bridges then loses its justification.

An excellent and comprehensive presentation of the physics of capillary bridges between spherical bodies has been given by Willett *et al.* [119]. This paper provides a large amount of particularly high quality experimental data and demonstrates excellent agreement with state-of-the-art theory. The main facts are as follows. At zero distance, the attractive force exerted by the capillary bridge is given by the expression in Eq. (4.26). For the variation of the force at finite separation,

$$F_{cb} = \frac{2\pi R\gamma \cos\theta}{1 + 1.05\hat{s} + 2.5\hat{s}^2} \tag{4.27}$$

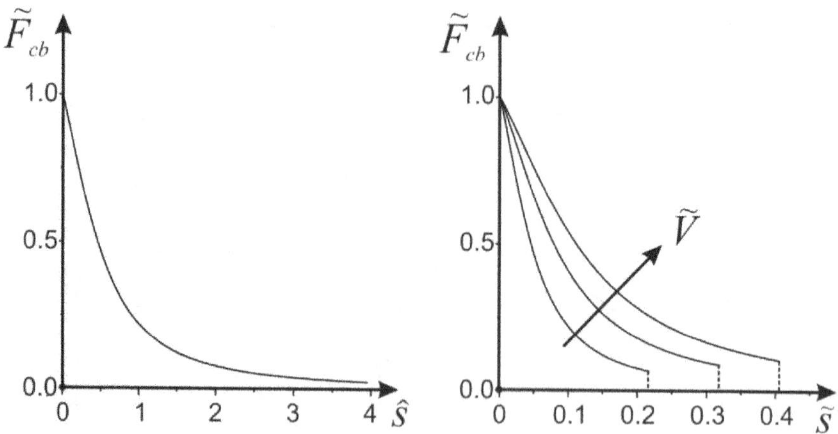

Fig. 4.9 Force characteristic of a capillary bridge (Eq. (4.27)), evaluated for $\theta = 0$. The plotted quantity is $\tilde{F}_{cb} = F_{cb}/2\pi R\gamma$. Right panel: force characteristic at dimensionless liquid volumes $\tilde{V} = 0.01$, 0.03, and 0.06.

is shown to be an excellent approximation, with $\hat{s} := s\,\sqrt{R/V}$. This characteristic is plotted in two normalized versions in Fig. 4.9.

For the critical separation s_c at which the bridge pinches off, the authors find

$$\tilde{s}_c = (1 + \theta/2)\left(\tilde{V}^{1/3} + 0.1\tilde{V}^{2/3}\right),\tag{4.28}$$

where $\tilde{s}_c := s_c/R$ and again $\tilde{V} := V/R^3$. This is valid for equally sized spheres, with only minute deviations for nonequal ones. In that latter case, Derjaguin's approximation is found to hold, which consists in averaging the curvatures,

$$\frac{1}{R} := \frac{1}{2}\left(\frac{1}{R_1} + \frac{1}{R_2}\right),\tag{4.29}$$

where R_1 and R_2 are the radii of the two spheres in contact. In other words, if $z(x,y)$ is the distance between the two surfaces measured at a lateral position (x,y) relative to the contact point, $(0,0)$, all that is relevant is the mean curvature of the surface described by the graph of z, or $\mathsf{H} = \frac{1}{2}\Delta z$ (cf. Eq. (3.19)). This gives rise to important commonalities in wet granular matter, since $z(x,y)$ will to leading order vary quadratically away from (almost) any contact point. More precisely, for any two convex bodies this holds everywhere except for a set of zero measure. We will come back to this aspect later when we discuss capillary forces between non-spherical grains, and in the context of the yield stress of wet granular piles.

4.2.3 *The interaction potential between wet grains*

Let us construct the full interaction potential, both from capillary bridge force and elasticity, between two grains, which we assume to be spherical with radius R as before. In order to obtain the capillary bridge potential, U_{cb}, we integrate Eq. (4.27). Using

$$\int_0^s \frac{dx}{ax^2 + bx + c} = \frac{2}{\sqrt{4ac - b^2}} \arctan \frac{2as + b}{\sqrt{4ac - b^2}}, \qquad (4.30)$$

we readily find

$$U_{cb} = 4.216 R^2 \gamma \cos\theta \sqrt{\tilde{V}} \ \arctan(1.678\hat{s} + 0.352). \qquad (4.31)$$

The elastic potential given in Eq. (2.11) can be written as

$$U_{el} = \begin{cases} \frac{4ER^3}{15\sqrt{2}(1-\nu^2)}(-\tilde{s})^{5/2} & \text{if } \tilde{s} < 0 \\ \\ 0 & \text{else} \end{cases}. \qquad (4.32)$$

Note that we do not need to limit the capillary force potential to positive values of the separation, since the spheres effectively overlap when the grains are deformed in the region of contact, and the wetting forces are determined from the geometry well outside the contact region. The total potential can then be written as

$$\frac{U_{tot}}{\gamma R^2} = 0.189\mathsf{E}(-\tilde{s})^{5/2} + 4.216 \cos\theta \sqrt{\tilde{V}} \arctan 1.678 \frac{\tilde{s}}{\sqrt{\tilde{V}}} + 0.352, \quad (4.33)$$

where the first term is taken to be zero for positive $\tilde{s}$. We have introduced here what we may choose to call the elasticity parameter,

$$\mathsf{E} = \frac{YR}{\gamma(1 - \nu^2)}. \qquad (4.34)$$

Even for very soft grains materials, this is a very large number. The elastic modulus of a soft rubber is in the MPa range, such that E is of order 10^3 at the very least. Hence the repulsive branch of the potential is extremely steep as compared to the attractive branch. This reflects the obvious fact that the capillary forces will not noticeably deform the grains.

For a wet granular pile, it is of substantial interest to calculate the internal pressure generated by the capillary forces. If we were to increase the radius of each grain by an amount dR, the sample would expand, and the relative change in volume would be $dV/V = 3dR/R$. If we instead introduce a separation $s \ll R$ between adjacent grains at each point of contact, the overall result on the volume of the sample is the same, with

$dR = s/2$. The change in energy which would be associated with such expansion is $dE = N\frac{k}{2}F_{cb}s$, where N is the number of spherical grains, and k is the number of contacts on each grain. The 2 in the denominator accounts for the fact each contact belongs to two grains. We then obtain for the pressure

$$p_{\text{int}} = \frac{dE}{dV} = \frac{kF_{cb}}{3}R\frac{N}{V} = \frac{kF_{cb}\phi}{4\pi R^2}, \tag{4.35}$$

where we have used the identity $\phi = NV_g/V$, and $V_g = 4\pi R^3/3$ for the grain volume. Using Eq. (4.26) for F_{cb} we arrive at

$$p_{\text{int}} = \frac{\gamma}{2R}k\phi\cos\theta. \tag{4.36}$$

Close to the maximum packing density, k is around 6 for spheres, as mentioned above.

The attractive capillary force leads to some finite equilibrium deformation, which is found by balancing the capillary bridge force with the Hertz elastic response of the grains. We obtain

$$(\tilde{s}_{eq})^{3/2} = 3\sqrt{2}\pi\mathsf{E}^{-1}\cos\theta. \tag{4.37}$$

The second derivative of the elastic potential, taken at $s = s_{eq}$, can then be seen as a spring constant which governs the elastic response of the system to minute perturbations around the equilibrium deformation. We readily find

$$D_{el} = 1.234\gamma\mathsf{E}^{2/3}(\cos\theta)^{1/3}. \tag{4.38}$$

Since the load on each contact point is given by the capillary bridge force, D_{el} should be the same for all inter-granular contacts in a pile of equally sized spherical grains.

Moeller and Bonn have shown experimentally that this is in fact the case. By means of a standard rheometer, they measured the elastic response of wet granular piles at extremely small amplitudes. As expected, they observed an elastic restoring force which scaled just in the way suggested by Eq. (4.38) [120].

It suggests itself to compute the characteristic resonance frequency of a grain, which is given by $2\pi f_{\text{res}} = \sqrt{D_{el}/m}$. Taking into account that the effective restoring force in one direction is governed by two (approximately) opposing contact points (boundary of the first Brillouin zone), we find using above results

$$f_{\text{res}} = 0.122 \ \cos^{1/6}\theta \ \gamma^{1/6}E^{1/3}\varrho_s^{-1/2}R^{-7/6}, \tag{4.39}$$

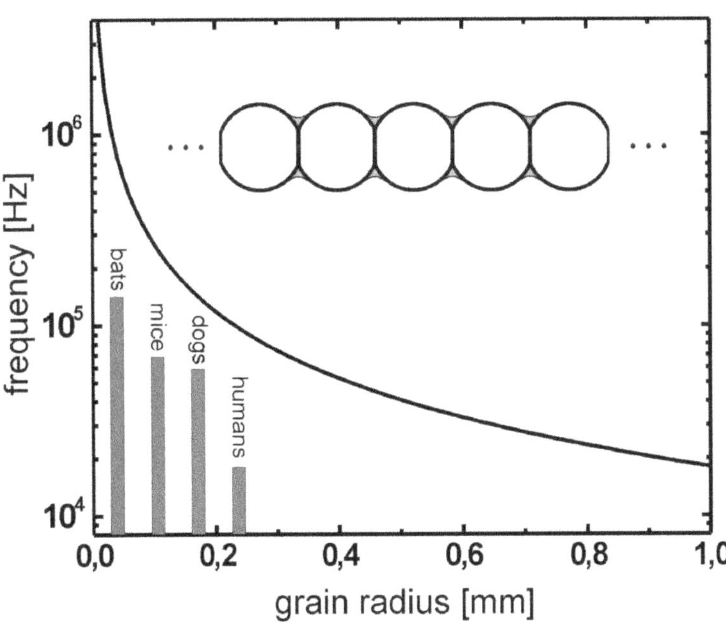

Fig. 4.10   The 'resonance frequency' of a wet granular pile as a function of the radius of the (assumed spherical) grains. The grey bars indicate the audible frequency range for a few common species. While for human ears the 'ringing' of sand is clearly out of reach, it may be well audible for bats, and also for mice and dogs if the sand is sufficiently coarse.

where $(\cos\theta)^{1/6}$ can be considered very close to unity. This can be evaluated, e.g., for quartz sand grains ($E \approx 6 \times 10^{10}$ Pa, $\varrho \approx 2900$ kg/m^3) and water as the liquid ($\gamma = 0.072$ N/m). The result is plotted in Fig. 4.10 as a function of the radius, R. In a granular pile, it should be seen as the position of the peak in the phonon density of states.[2] To be precise, what is plotted is the frequency of the zone boundary phonons of an infinite linear chain of spherical grains, held together by capillary bridges (see inset). Clearly, the frequency is well below the elastic vibration frequencies of the grains alone, which would be in the higher Megahertz regime. For large grains, they even come close to the range accessible to the human ear. It is interesting to note that for many animals, this 'ringing' of wet sand should be well audible. The audible frequency ranges are indicated in the figure for bats, mice, and dogs. We may imagine the experience of walk on a

---

[2]The acoustic density of modes of a granular system has been demonstrated experimentally [121].

beach has interesting acoustic facets to such animals, which are completely eclipsed to our ears.

4.2.4 *The hysteretic nature of the capillary bridge force*

Aside from the independence of the attractive capillary bridge force of the liquid volume of the capillary bridge, the other striking feature of the attractive force is its *hysteretic* nature: as the grains are separated from each other, the bridge ruptures at a certain critical distance $s_c(V)$, which does depend on the liquid volume (as opposed to the capillary bridge force, which does not). As however, the grains are approaching each other again, there will be no force until their surfaces come into contact ($s = 0$), such that a capillary bridge can reappear. Hence a finite amount of energy, E_{cb}, is lost in each impact, which corresponds to the area under the solid curve displayed in Fig. 4.9.

Using the expression in (4.27), we can directly calculate this energy loss, which is given by $E_{cb} = \int F_{cb} ds$, where the integral extends from zero to s_c. Using Eq. (4.30) we obtain

$$E_{cb} = 4.21 \ R^2 \gamma \cos\theta \sqrt{\tilde{V}} \ \arctan(1.678 \hat{s}_c + 0.352). \qquad (4.40)$$

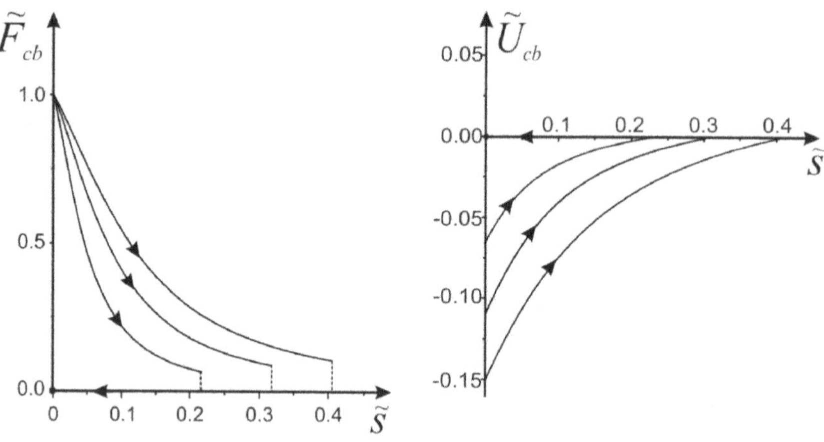

Fig. 4.11   Left: Force characteristic of a capillary bridge (cf. right panel of Fig. 4.9). There are two different branches, one with finite force for retraction, one with zero force for approach.   Right: Pseudo-potential representation of the force.   The approaching branch is identical with the $\tilde{s}$-axis. The retraction branch is normalized such that $\tilde{U}_{cb} \to$ 0 for $\tilde{s} \to 0$.

Since

$$\hat{s}_c = \left(1 + \frac{\theta}{2}\right)\left(\tilde{V}^{-\frac{1}{6}} + 0.1\tilde{V}^{+\frac{1}{6}}\right) \tag{4.41}$$

(note the difference between $\hat{s}$ and $\tilde{s}$, Eq. (4.28)!) the argument of the arctan never comes close to zero, such that it contributes almost no variation to E_{cb}. We thus obtain as a very good approximation

$$\tilde{E}_{cb} = \frac{E_{cb}}{\gamma R^2} = 4.37 \ \cos\theta\sqrt{\tilde{V}}. \tag{4.42}$$

Hence we find that the capillary bridge energy, E_{cb}, varies as the square root of the liquid volume of the capillary bridge. This is sketched in Fig. 4.12.

This is in contrast to the dominant dissipation mechanism in dry granular systems, where one generally describes the dissipative character of the grain interaction through a restitution coefficient, ε. The latter is defined as the relative momentum after an impact divided by the relative momentum before the impact [12], such that $\varepsilon < 1$. It only weakly depends on the initial impact energy, and is assumed constant in most studies. Hence in dry granulates, a *certain fraction* of the energy is lost in each impact, while

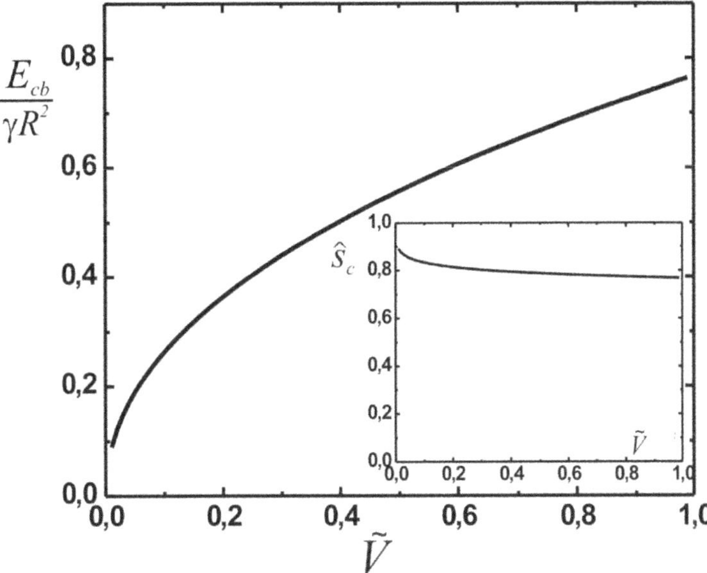

Fig. 4.12 The energy lost upon rupture of a capillary bridge, normalized with respect to γR^2, as a function of $\tilde{V} = V/R^3$. Inset: the normalized, non-dimensionalized separation, $\hat{s}$, varies only slightly in the range of interest.

in the wet case, it is a *fixed amount* of energy which is dissipated. Although this may appear as a minor difference at first glance, we will see below that it gives rise to very substantial changes in phenomenology. These will the topic of Chapter 5.

4.3　Capillary bridge between irregular grains

4.3.1　*Effects of grain shape*

We have seen above that there may be strong qualitative differences in the capillary force characteristics for different shapes of the grain surfaces. For parallel flat walls, we found an marked dependence of the force upon the liquid volume of the bridge, while for two spherical grains in mutual contact, the force was largely independent of the volume. One might therefore anticipate that the shape of the grains has a strong influence on the qualitative force characteristics in general. Since the grains of a granular system have all different shapes in most cases, we might expect to be in need of an independent formula for the force for each grain, or even for each pairing of two contacting grains. Fortunately, this is not the case. We will instead find that the setting of two parallel flat walls is singular in some sense, and does not represent the generic behavior we will encounter in general for wet grain contacts.

Following the discussion in Section 2.1.4, let us assume for the moment that aside from a certain roughness, the grains are in general convex bodies. The superimposed roughness will be treated further below. A region of contact between two convex grains is characterized by a well-defined point of contact, as sketched in Fig. 4.13. For the discussion to follow, the tangential plane of this contact point shall be the (x, y)-plane, and we consider the vertical distance of the grains surfaces, $d(x, y)$, measured perpendicular to the (x, y)-plane. It is important to observe that for almost all possible contacts (i.e., all except a set of zero measure) between two grains, $d(x, y)$ varies to first order quadratically with the distance from the contact point.

Recalling the discussion of the capillary bridge force between two spheres (Section 4.2.2), we realize that the independence of that force of the liquid volume of the bridge was due to the fact that the Laplace pressure varied quadratically with the lateral extension of the bridge. Clearly, this is the case here as well: at least in the relevant case of small θ, the Laplace pressure will scale as $1/d(x, y)$, and thus quadratically with the distance from the contact point, (x_c, y_c). Hence we find that as far as the capillary bridge

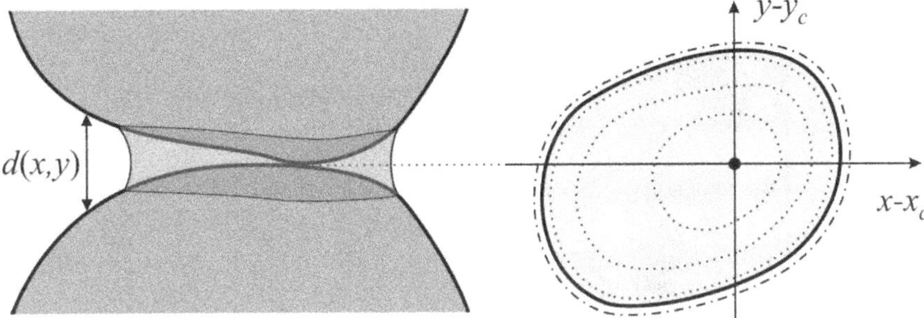

Fig. 4.13   Capillary bridge between two non-spherical grains. For small contact angles, the contact line follows a contour line of the separation of the grain surfaces, where this separation is about twice the radius of curvature of the outer liquid surface.

is concerned, a contact between two convex bodies is in good approximation equivalent to the contact between two spheres. Relation (4.29) for contacts of spheres of different radii (Derjaguin's approximation) is just a special case of this result. The effect of the non-spherical shape is that the radius R entering Eq. (4.26) is related to the local curvature of the surfaces via $R = 4/\Delta d(x_c, y_c)$. In a real system involving convex grains and many contacts, the capillary bridge force will thus vary statistically around a value corresponding to a contact between two spheres of radius R, where R is the half average diameter of the grains. The width of that variation will correspond to the deviation of the grain shapes from the sphere.

4.3.2   *Effects of grain surface roughness*

A quite noticeable consequence of the imperfection of the grain surface is that the attractive force, F_{cb}, is not generally as independent of the liquid volume as suggested by Eq. (4.26). For very small liquid content, it is intuitively clear that the crevices in the grain surface must first be 'filled' before a bridge can form at all. In the case of a single rough surface, this filling process is described by the phase diagram developed in Chapter 3 (Fig. 3.14). For two rough surfaces in contact, the process will be distinctly different. At zero compressive force, there will be three points of atomic contact between the two grains, at any finite force there will be many more [122]. On top of that, there will be many 'almost-contacts', where the separation of the two solid surfaces is only a few nanometers. When liquid is added, it will preferentially accumulate at these points of contact or 'near-contact', forming nanometer scale liquid bridges [123]. As liquid

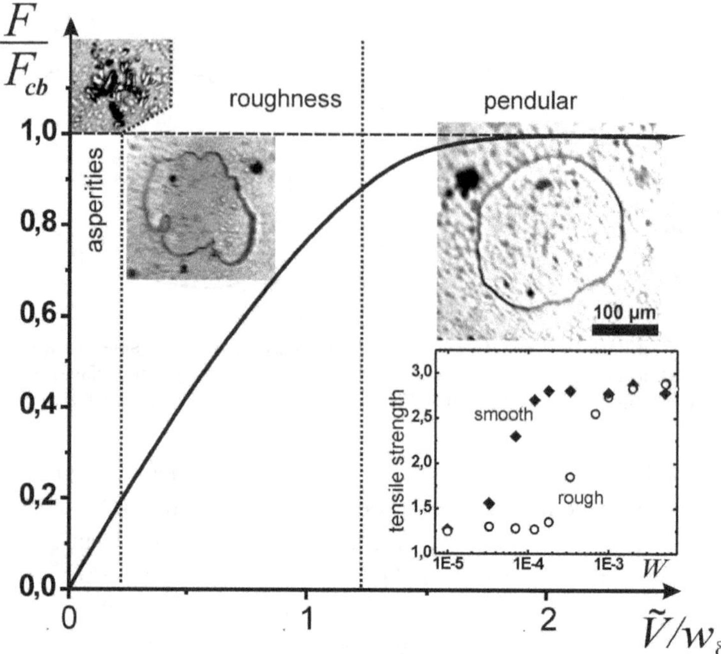

Fig. 4.14   The capillary force at small volumes of the capillary bridge, where roughness comes into play. The solid curve sketches its qualitative behavior. Optical micrographs corresponding to the different regimes are positioned accordingly (taken from [126]. Scale bar applies to all images). The exact position of the dotted lines schematically separating the regimes depends on the characteristics of the roughness. At the smallest volumes, liquid accumulates at individual asperities. At intermediate volumes, the space between the two rough surfaces is filled with liquid over a certain area, but the shape of the latter is far from a circular disk. At sufficient liquid content, a fully developed capillary bridge emerges. This corresponds to what we will later call the pendular regime (in accordance with current nomenclature, see Chapter 6). The lower inset shows the variation of the tensile strength of a wet granular pile for rough and smooth grains in the wetness range under discussion.

keeps accumulating, these grow and coalesce into extended wet patches. Their total area is given by $V/2\delta$, as δ is by definition the amount of liquid which can hide within the roughness of one grain. Since it provides as well an estimate for the typical average separation of the solid surfaces, the Laplace pressure in these patches will be roughly $-2\gamma\delta\cos\theta$. Following the discussion in Section 4.1.2, we assume that the Laplace pressure term dominates over the capillary force acting along the perimeter of the wetted patches. The attractive force is then approximately given by the (negative)

Laplace pressure times the total area of the patches. This yields

$$F_{\text{rough}} = \frac{V}{2\delta}\frac{2\gamma\cos\theta}{\delta} = \gamma R\cos\theta\frac{\tilde{V}}{\tilde{\delta}^2} = \tilde{F}_{cb}\frac{\tilde{V}}{2\pi\tilde{\delta}}. \qquad (4.43)$$

We see that the force is expected to grow linearly with the added liquid volume, in contrast to the behavior of a fully developed capillary bridge. This result has been obtained by Halsey and Levine [123], who speculated in addition on another regime at even smaller W, assuming self-affine roughness. While the latter is difficult to identify due to the scarcity of truly self-affine roughness on real surfaces, the linear regime has been established experimentally by a number of authors [43, 124, 125].

The linear regime is expected to cross over to the constant force predicted by Eq. (4.26) when sufficient liquid has accumulated in the contact region to form a single connected capillary bridge. This takes place when $F_{\text{rough}} = F_{cb}$ (cf. Eq. (4.26)). For the corresponding liquid volume, we directly obtain

$$\tilde{V}_c = 2\pi\tilde{\delta}^2 \qquad (4.44)$$

with $\tilde{\delta} = \delta/R$, which can as well be written as $\tilde{V}_c = \frac{2}{3}\pi w_\delta^2$. This is plotted as the dotted line in Fig. 4.7, which lies below V_i for all values of the wetness w. We thus can conclude that if the grains are sufficiently wet for the troughs of their surface roughness to be filled with liquid, which should be the case close to liquid-vapor saturation, a fully developed capillary bridge will form immediately after two grains have come into contact.

4.3.3 *Small scale capillary bridges*

We should still discuss what happens if we reduce the liquid content further, such that there is not even enough liquid to fill the roughness topography between the grains in the region of contact. In this case, we may expect that there are many nanometer-scale contacts emerging, each between just tiny asperities on the surfaces of the grains [123, 127]. That this is indeed the case can be directly demonstrated by means of a scanning force microscope (SFM), which has become one of the most widely used instruments in surface characterization. In an SFM, a tiny tip, with a typical radius of curvature of a few nanometers, is attached to a an elastic silicon cantilever produced by standard lithography techniques. This tip is scanned over the sample of interest, and the deflection of the cantilever is traced by means of a laser beam reflected from its upper surface. More precisely, the standard mode of operation consists in vertically displacing the cantilever mount by means of a piezo-actuator while scanning the sample, such as to keep the cantilever deflection angle constant. The so-adjusted height of the cantilever mount, as a function of the position on the sample, represents then the sample topography.

It is clear that the geometry of the contact between the tip and the sample closely resembles the geometry of the contact of an asperity on a grain with the opposite grain surface. Since the SFM allows for the measurement of minute forces, it is possible to capture the formation and rupture of capillary bridges between the tip and the sample [128–130]. In order to do this in a reliable manner, it is advantageous to use a different mode of operation, usually called 'tapping mode'. In this mode, the cantilever is oscillated at its resonance frequency (which is typically a few 100 kHz) and slowly approached to the sample until a change in oscillation amplitude is detected. While the tip is scanned over the sample, the vertical position of the cantilever mount is continuously adjusted such that the amplitude of the cantilever oscillation remains constant. This does not only allow to determine the sample topography from the cantilever mount vertical position: the relative phase shift between the cantilever oscillation and its drive is extremely sensitive to subtle variations in the sample properties. This can be exploited to detect the presence of extremely small capillary bridges between the sample and the tip when a liquid is present in the system under study.

The formation of nanometer-size capillary bridges between the tip and the sample can be deliberately induced by exposing the substrate to high humidity air. When the tip is approached in tapping mode to a flat, clean surface in sufficiently humid atmosphere, characteristic features appear in the dependence of the amplitude and phase of the cantilever oscillation on the average tip-to-sample distance, d_0. The right-hand panel of Fig. 4.15 shows the phase of oscillation as a function of d_0 (corresponding to the vertical cantilever mount position). Depending on conditions, either a curve like the dashed curve or like the solid curve was observed. In the latter, one clearly sees a hysteresis loop between approaching and retracting.

When the humidity is varied, one finds that the system switches from a characteristic corresponding to the dashed curve in Fig. 4.15 (no jump) to a characteristic like the solid curve (with hysteretic jump). It suggests itself to attribute this humidity-dependent jump to the transient formation of capillary bridges. In fact, a detailed numerical simulation on the basis of the hysteretic capillary forces outlined above is capable of reproducing precisely the characteristics as shown in Fig. 4.15 [128, 131].

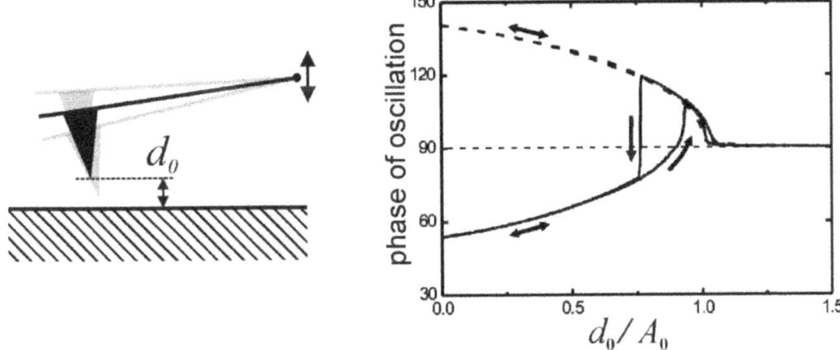

Fig. 4.15   Left: Sketch of the setup. The cantilever is being oscillated at its resonance frequency of 280 kHz. The free amplitude of vibration, A_0, was about 40 nm. The radius of curvature of the tip was about 20 nm. Right: Relative phase of oscillation of the cantilever of the scanning force microscope with respect to the driving piezo, for variable average distances, d_0. There either is a clear hysteresis and change in sign, or just a monotonous increase towards the substrate. At a critical value of the free amplitude A_0, the system jumps from one behavior to the other.

Figure 4.16 shows a 'mode diagram' of the system in the plane spanned by the relative humidity and the free cantilever amplitude, A_0. The significant shift to larger amplitudes at higher humidity is obvious. In addition, we see that there is a transition as well at very low humidity, when there can be no capillary bridge. This is because the mechanical interaction between the solid sample, the cantilever, and the driving device induce a hysteretic jump already by themselves. Hence it is indispensable to perform quantitative simulations of the system in order to clearly identify the capillary bridge effects. The fact that the experimental results are well described by numerical simulations of the experiment assuming capillary forces [128] strongly suggests that we can use the concept of wetting even for the smallest liquid features, which may occur at the scale of surface roughness.

4.4   Force networks

4.4.1   *Frustrated wet force networks*

Since we will be interested in the mechanical properties of wet granular piles, it is worthwhile to contemplate the degrees of freedom such pile has. If we restrict the discussion to grains which are spherical but rough, we can model a pile of them with the concept of cogwheels in intimate contact with

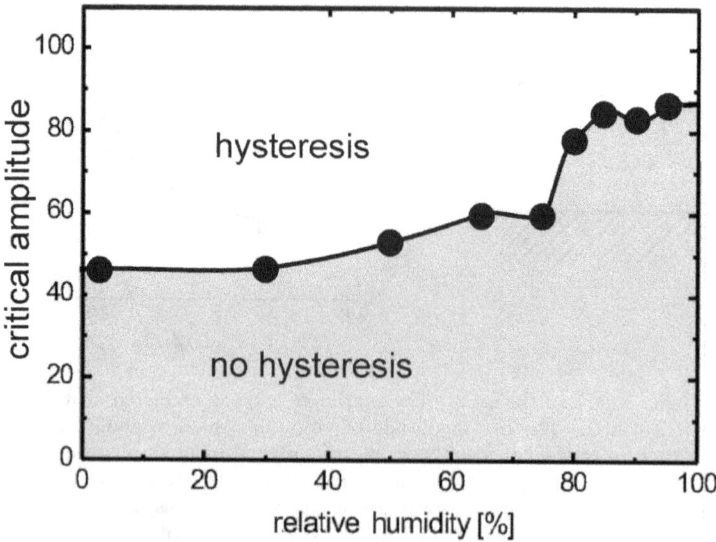

Fig. 4.16   The critical free amplitude, A_0, for the occurrence of the hysteresis loop. The dependence on the humidity of the air is obvious, thus corroborating its connection to the intermittent formation of capillary bridges between the tip and the substrate [128].

each other. The attractive capillary forces press the contacts together, such that the grain surfaces will not move tangentially with respect to each other unless the tangential stress exceeds the friction force. This is sketched in Fig. 4.17 for two different geometries. They differ in the number of steps one has to make, going from one grain to a contacting next one, to return to the original grain. This number will be either odd (a) or even (b). It is immediately clear that in the odd case, the rotation of the cogwheels is frustrated, such that their angles will be fixed. In the even case, however, they are free to rotate. At first glance, one may thus conclude that only piles which have odd 'orbits' (closed self-avoiding paths going from one grain to the next) are mechanically frustrated. In piles with only even 'orbits', grains can rotate, and one may expect that they can then as well be sheared and deformed.

This is, however, not the case. It turns out that the conditions for shearing the pile are in fact considerably stricter than for grain rotation. Consider namely a single even orbit. For the sake of simplicity, we stay in two dimensions, as in the sketch. If the orbit is deformed, the sum of the angles between the contact points on each grain must be constant (if

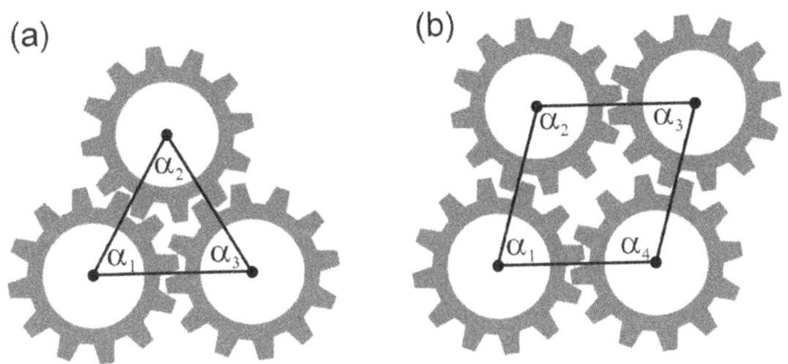

Fig. 4.17 A cogwheel model for illustrating mechanical frustration in granular piles. If there is friction between grains (and there always is), grains can rotate freely only under certain geometric constraints of their arrangement. (a) An uneven number of grains in a closed 'loop'; not rotation is possible without exerting force against friction. (b) An even number of grains in a closed 'loop'; collective rotation is possible.

friction is strong), hence the sum of their variations must vanish,

$$\sum \partial \alpha_n = 0. \tag{4.45}$$

It is straightforward to show that the angular positions of the grains, or cogwheels, impose

$$\sum \partial \alpha_e - \sum \partial \alpha_o = 0, \tag{4.46}$$

where the sum on the left-hand side extends only over even, the one on the right-hand side only over odd indices. This structure derives from the reversal of the direction of rotation as one goes from one grain to the next. Since we have

$$\sum \partial \alpha_n = \sum \partial \alpha_e + \sum \partial \alpha_o, \tag{4.47}$$

it immediately follows that the sums over the odd and over the even indices must both vanish. Hence we have one constraint more than for the structure without the cogwheels. Hence the geometry in Fig. 4.17(b) cannot be deformed (i.e., the angles α_n cannot be changed) without overcoming friction (or, in the sketch, breaking cogs) although the grains are free to rotate. We shall call this a mechanically frustrated network. It turns out that in order not to be frustrated, a granular pile (in two dimensions) would have to have only orbits of six or more steps. It is not possible to build any stable structure under this constraint (i.e., it would be hypostatic).

Hence we conclude that all reasonably stable granular piles are mechanically frustrated in the sense that any deformation needs to overcome tangential friction forces at the grain contacts, even in cases where grains are free to rotate with their center positions fixed.

4.4.2 *Self-assembled granular walkers: Ratcheting*

We have now a widely developed toolbox for understanding the forces between grains under dry and wet conditions. In the case of spherical (i.e., model) grains, most quantities of interest can be calculated quantitatively by means of the simple formulae we have developed. We will end this chapter with discussing a simple but impressive experiment which allows to appreciate the interplay of capillary, inertial, and friction forces for a granular system consisting of a few wet spherical grains.

The experimental setup is as sketched on the left side of Fig. 4.18. A glass vial (for optical access) containing glass spheres of different sizes, with a few volume percent of wetting liquid added is oscillated vertically, by means of either a loudspeaker or a professional electromagnetic shaker. One frequently observes that clusters of glass beads, containing small and large ones, glued together and to the glass wall by the wetting liquid, leave the sample and climb up the container wall. At a typical vibration frequency of 100 Hz, the climbing velocity is a few millimeters per second. It turns out that it is always a large bead which leads the cluster, with the smaller ones trailing behind.

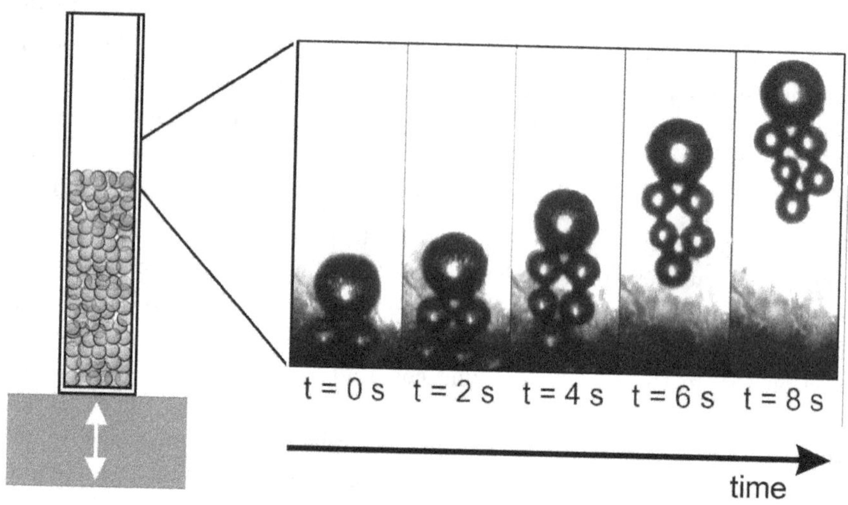

Fig. 4.18   A container with set spherical beads of different size (poly-disperse sample) is agitated by applying a sinusoidal vertical motion, with a frequency of about 100 Hz. A cluster of glass beads of different sizes, glued together by a wetting liquid, has spontaneously formed and climbs up the container wall (image reproduced with permission from Institute of Physics [132]).

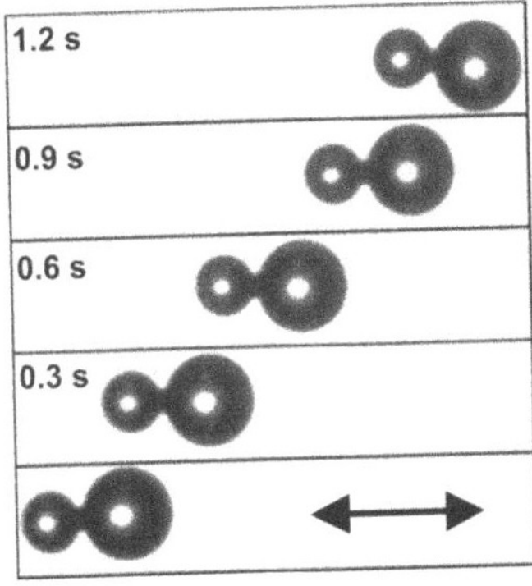

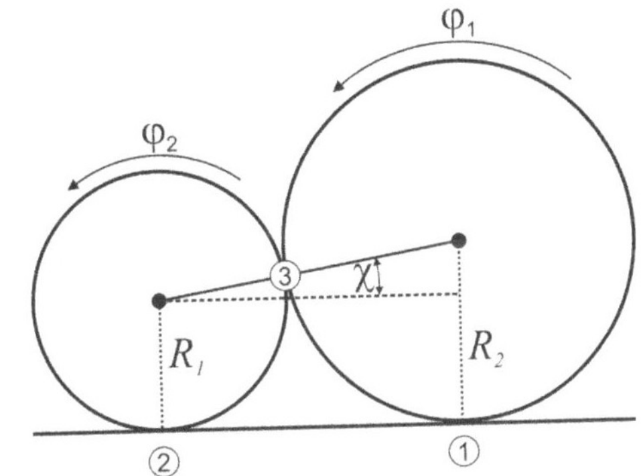

Fig. 4.19 A pair of glass beads of different size, held together by a capillary bridge, on a horizontal substrate which is oscillated within its plane (image reproduced with permission from Institute of Physics [132]). Left: Time lapse images showing the directed motion. The arrow indicates the direction of shaking. Right: Geometry of the system and quantities for its description.

It is easy to show that gravity plays no significant role in this scenario, by repeating the experiment on a horizontal substrate which is shaken horizontally. Figure 4.19 shows a top view of this setup, this time with a deliberately assembled 'walker', consisting of just one large and one small bead. Again, the large bead is leading and the small one trailing. In what follows, we will outline a simple theory which accounts for these observations, and which gives an idea of the ubiquity of such effects in wet granular matter.

Consider a walker composed of two spherical beads with radii R_1 and R_2 and masses m_1 and m_2, which are held together and attached to a flat horizontal substrate by liquid capillary bridges. The walker is subject to an external driving force which is directed parallel to the substrate surface. When the walker is aligned with the direction of the excitation, the plane defined by the three contact points is the plane of symmetry of the system, and all forces lie within that plane. The beads can rotate and slide on the substrate and on each other. The angular velocity of sphere i about an axis perpendicular to the symmetry plane is denoted by $\dot{\varphi}_i$, where the dot indicates differentiation with respect to time. Its sliding velocity on the substrate is denoted by v_i. The sliding velocity of bead 1 on bead 2 is thus $v_3 = R_1\dot{\varphi}_1 + R_2\dot{\varphi}_2$. Assuming that the spheres remain in hard mechanical contact with each other and with the substrate, the velocity of the center of mass of both beads with respect to the substrate is described by a single variable along the shaking direction, $v_x = v_1 - R_1\dot{\varphi}_1 = v_2 - R_2\dot{\varphi}_2$. The acceleration of the whole walker with respect to the substrate is then $\dot{v}_x$.

If the external acceleration of the substrate is $a(t) = a_0 \cos \omega t$ in the reference frame of the laboratory, the driving force consists of the inertial forces acting on the beads. Adhesive forces act on each contact point and are normal to the respective contact plane. Given that their directions are fixed, they can be represented in a compact form by using a three-component vector notation, where each component of the vector represents the magnitude of one of the forces. In this notation, the contact forces at rest, which are due to the capillary bridges, can be written as

$$\mathbf{F}^0 = \begin{pmatrix} F_1 \\ F_2 \\ F_3 \end{pmatrix} = 4\pi\gamma\bar{R} \begin{pmatrix} 1 + \sin\chi \\ 1 - \sin\chi \\ \frac{1}{2}\cos^2\chi \end{pmatrix} \tag{4.48}$$

with the averaged radius $\bar{R} = \frac{1}{2}(R_1 + R_2)$. This is readily verified from the sketch in Fig. 4.19. The asymmetry of the system is characterized by $\sin\chi = (R_1 - R_2)/(R_1 + R_2)$, and the prefactor $4\pi\gamma\bar{R}$ sets the characteristic force scale of the system. At rest ($a_0 = 0$), all forces are normal to the contact planes.

When there is an external driving ($a_0 \neq 0$), there will be force components acting tangential on the contacts. These are balanced by friction and will be denoted by $\mathbf{F}^{\|}$. At the same time, the normal forces will change, and we write

$$\mathbf{F} = \mathbf{F}^0 + \Delta\mathbf{F}. \tag{4.49}$$

The balance of all forces (both tangential and normal) in the rest frame of the walker leads to system of equations which contains sums of force components and is thus linear. Hence there will be a linear relationship between $\Delta\mathbf{F}$ and the set of tangential forces, $\mathbf{F}^{\|}$, which can be written in the form

$$\Delta\mathbf{F} = \mathbf{M}\mathbf{F}^{\|}, \tag{4.50}$$

where $\mathbf{M}$ is a 3×3 matrix expressing the geometry of the walker.

If the externally applied oscillation frequency is small as compared to the internal vibrational eigenmodes of the system, we may adopt a quasi-stationary treatment, considering the acceleration constant at every single point of time. As the typical eigenfrequencies are at least tens of kilohertz (cf. Fig. 4.10), these assumptions are well justified at excitation frequencies at a few hundred Hertz. In this quasi-static limit, a constant angular velocity $\dot{\varphi}_i$ for both beads means that the overall torque due to the friction forces applied on bead i must vanish. As a result, we have

$$F_1^{\|} = -F_3^{\|} = F_2^{\|}. \tag{4.51}$$

Similarly, a constant locomotion velocity entails

$$F_1^{||} + F_2^{||} = -(m_1 + m_2)a. \tag{4.52}$$

After normalizing all of the forces by the capillary bridge force, Eqs. (4.51) and (4.52) lead to

$$\frac{\mathbf{F}^{||}}{4\pi\gamma\bar{R}} = \frac{\Gamma}{2}\begin{pmatrix} -1 \\ -1 \\ 1 \end{pmatrix}, \tag{4.53}$$

where $\Gamma = a(m_1 + m_2)/4\pi\gamma\bar{R}$. The normal contact force can then be written as

$$\frac{\mathbf{F}}{4\pi\gamma\bar{R}} = \frac{\mathbf{F}^0}{4\pi\gamma\bar{R}} + \frac{\Gamma}{2}\mathbf{M}\begin{pmatrix} -1 \\ -1 \\ 1 \end{pmatrix}. \tag{4.54}$$

Finally, the friction law on each contact point relates the friction force, $F_i^{||}$, to the normal contact force, F_i, and the sliding velocity, v_i. This relation is an odd function of the sliding velocity, such that

$$F_i^{||}(F_i, -v_i) = -F_i^{||}(F_i, v_i). \tag{4.55}$$

According to Eq. (4.53), $F_i^{||}$ is as well an odd function of the dimensionless driving force, Γ, such that

$$F_i^{||}(F_i(-\Gamma), -v_i(-\Gamma)) = -F_i^{||}(F_i(\Gamma), v_i(\Gamma)). \tag{4.56}$$

For the behavior of the walker, it is of interest whether the v_i are odd functions of Γ. If they are, the motion of the walker would cancel out in the time average. Let us assume that this is the case. Then we have, knowing that $F_i(v_i)$ is odd,

$$F_i^{||}(F_i(-\Gamma), -v_i(-\Gamma)) = -F_i^{||}(F_i(-\Gamma), v_i(\Gamma)). \tag{4.57}$$

Comparing Eqs. (4.57) and (4.56), we see that this requires

$$F_i(\Gamma) = F_i(-\Gamma) \tag{4.58}$$

which is at variance with Eq. (4.54). This shows that the tangential velocities cannot be odd functions of Γ.

A compilation of experimental results is presented in Fig. 4.20. The left panel shows the walker velocity as a function of the peak acceleration, Γ_0. As predicted, there is no significant dependence on frequency (which did not show up in the theory). The simple force law used in the derivation

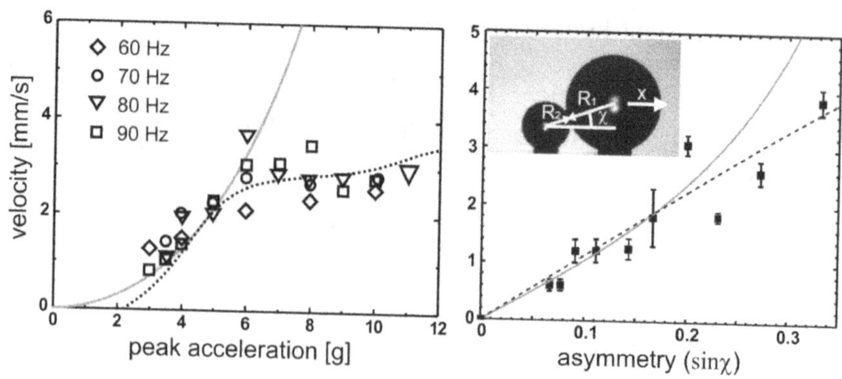

Fig. 4.20 Left: Velocity of the granular walker as a function of peak acceleration. Right: Velocity as a function of the asymmetry, $\sin\chi$. Figure taken with permission from [132].

yields the solid curve. The dotted curve, which provides a better fit to the data is obtained with a more realistic force law,

$$F^{||} = \mu F^{\perp}\mathrm{sgn}(v^{||}) + \eta R v^{||}, \tag{4.59}$$

where $v^{||}$ is the tangential relative velocity of the surfaces in contact (cf. Eq. (2.18)). The fact that the walker moves even if the average external acceleration vanishes (if $\Gamma = \Gamma_0 \cos\omega t$) thus turns out to be a very robust property, which does not depend on the special characteristics of the friction forces. Even with a completely linear friction force law, like

$$F^{||} \propto v^{||} F^{\perp} \tag{4.60}$$

would lead to a ratcheting effect, and thus to locomotion, if only the two spheres are of different size.

On the basis of the robustness of this locomotion effect, it is well conceivable that within the complex cohesive force network a wet granular pile represents, many such processes can take place. Given the fact that, e.g., the soil covering the earth is always subject to seismicity from tectonic processes, or even just traffic, rearrangements within the soil among its grains could then over time accumulate to substantial transport effects.

4.5 Conclusions

In this chapter, we have learned about a number of important properties of capillary forces between adjacent grains. First of all, it turned out that the force exerted by a capillary bridge between grains in contact does, to first order approximation, not depend upon the volume of liquid constituting the bridge. It only depends on the (suitably averaged) radii of curvature in the vicinity of the contact and the surface tension of the liquid. For not so well wetting liquids, it is important to take its dependence on the contact angle into account: it is proportional to $\cos\theta$.

The second important property of the capillary force is its hysteretic (non-conservative) nature, which gives rise to a certain amount of energy which is dissipated during formation and rupture of a capillary bridge in an agitated granular medium. This sets a natural energy scale, which is to be compared to the granular temperature. The interplay between these two quantities provides the basis of the phase transitions to be discussed in the subsequent chapters. The rupture energy scales in good approximation as the square root of the amount of liquid.

We have furthermore seen that the interplay of more than two wet grain contacts with their friction and capillary forces leads quite generically to ratcheting effects, which may give rise to macroscopic movements when the system is periodically agitated. Since periodic agitation is very relevant not only in experimental settings but also in applications and geosystems (seismicity), this must be considered as an important candidate for gradual macroscopic deformations, like creep or demixing (see also [39]).

Further reading

Although we have seen that the mathematics of minimal surfaces is of limited importance for wet granular matter due to the inherent imperfections of the system, it is worthwhile to point out some literature to those who are interested in this beautiful field of knowledge [116, 133]. For the practical and physically relevant aspects of capillary bridges, the reader is referred to the physical papers and books on that matter [3, 119, 134, 135]. In particular, the book by Butt and Kappl [3] provides most of what one needs to know about capillary interactions. For those who are interested in such phenomena on ultra-small scales, the paper by Farshchi-Tabrizi *et al.* and the literature it guides to is interesting to read [136] as well.

Chapter 5

Wet Granular Gases

Fig. 5.1 Granulation is a technique widely used in industry to clot fine powders into larger granules by means of adhesive capillary forces. Inspired from such concepts, the artist David Huycke creates artwork exhibiting similar structure.[1]

Now that we have acquired the tools to deal with wetting forces between granular particles, let us tackle the problem of a dilute gas of wet grains. Such systems are of utmost practical relevance in industrial granulation processes, where fine powders, which would give rise to intolerable dust if handled as received, are converted into a coarser granular material by means of agglomeration [138]. In large reactors, the powder is wetted with a spray of a suitable liquid, and kept floating in air for a while, such that the fine wet powder particles stick together by virtue of the adsorbed liquid, thus forming larger grains. This is a standard procedure, e.g., in pharmaceutical and food industry, but the process is interesting as well from a physical science point of view. First of all, it is an aggregation process which involves not only diffusive transport of the initial particles, as in numerous classical studies, but also advective transport in a fluid, which may be turbulent. Similar situations are encountered in clouds, where small water droplets or ice crystals gradually aggregate to rain drops, hail, or snow flakes. In fact, one of the main reasons for the limited success of

---

[1] *Fractal Piece*, 2007 (Copyright David Huycke); patinated silver 925/1000, Ø16 cm × 15 cm [137].

143

long-term weather and climate forecasts is due to the lack of understanding the mechanisms of precipitation, i.e., of cohesive aggregation with advection in clouds.

In any case, it is to be expected that the statistical properties of the resulting agglomerate particles are different from those of objects generated in diffusion-limited aggregation (DLA [139]), as advection follows scaling laws different from those pertaining to diffusion. For instance, if the agglomerates are fractal (and in case of granular aggregation we will indeed find they usually are), we may expect their fractal dimension to be distinctly different from those found in DLA processes. Second, we will see that the well-defined energy scale set by capillary bridge forces renders the collective behavior of wet grains fundamentally different from dry granular gases, giving rise to non-equilibrium phase transitions. Wet granular gases thus represent easily accessible model systems which may help to understand the physics of collective phenomena far from thermal equilibrium.

As far as liquid capillary bridges between adjacent grains are concerned, we have so far only been concerned with static (or quasi-static) aspects. For a thorough understanding of the dynamic properties of wet granular gases, however, we have to discuss what happens in the vicinity of a wet grain contact immediately after this contact has been established, and to follow its dynamics as the collision proceeds.

5.1 Dynamical aspects of capillary bridges

5.1.1 *Short time dynamics of a capillary bridge*

When two grains collide, we can identify three phases of the collision process, separated by four points on the time axis, t_1 through t_4, as illustrated in Fig. 5.2. Imagining again two ideally spherical grains, t_1 is defined as the time when the free liquid–gas interfaces of the liquid films adsorbed on the grains come into contact. t_2 is the time at which contact between the two solid bodies of the grains is established, which is then maintained until t_3. Around $t = t_2$, the lateral size of the capillary bridge has reached the angle β_i, which was introduced in Chapter 4. t_4 is the time when the capillary bridge ruptures, such that no more contact of any kind remains between the two wet grains.

Let us estimate the characteristic time and length scales of the processes involved. A typical radius of the contact line of the capillary bridge is of order $\sqrt{Rh}$, which indicates the boundary of the region where the liquid film

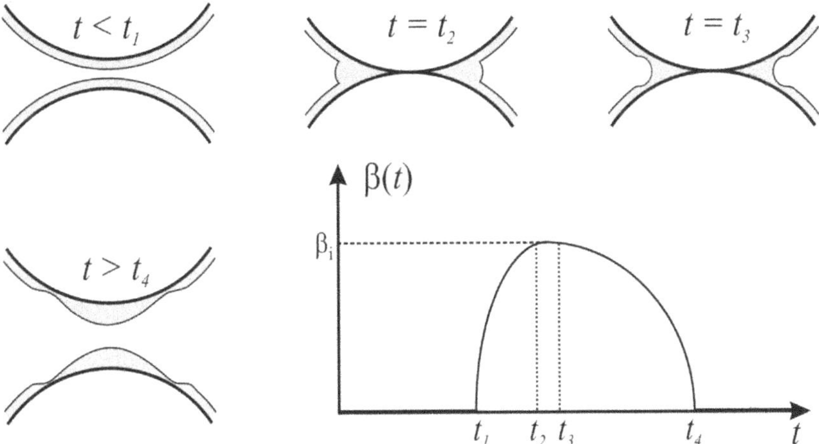

Fig. 5.2 Schematic representation of thin liquid film dynamics in the temporal vicinity of an inter-grain contact. As there is no time for 'ripening' of the bridge, its size will never go beyond β_i (cf. Chapter 4).

takes part in the capillary bridge formation process. If the grains approach each other at a velocity v, we have $t_2 - t_1 = h/v$, which directly yields a characteristic contact line velocity, $v\sqrt{R/h}$, in the approach phase ($t < t_2$). This is to be compared with the capillary velocity of the liquid, γ/η. The ratio of these quantities is the capillary number of the liquid bridge, which we define as

$$\mathsf{Cb} = \frac{v\eta}{\gamma}\sqrt{\frac{R}{h}} = \mathsf{Ca}\sqrt{\frac{R}{h}}, \tag{5.1}$$

thus introducing the standard capillary number, $\mathsf{Ca} = v\eta/\gamma$.

If $\mathsf{Cb} > 1$, the contact line will be moved by 'brute force' due to the inertia of the approaching grains and the volume conservation of the (presumed incompressible) liquid. The same will be the case analogously when the grains are removed from each other after the loss of solid contact. Hence in this case we have $t_4 - t_3 \approx t_2 - t_1$, and there will be neither an extended capillary bridge nor the corresponding hysteresis in the interaction force.

The time of real contact, $t_3 - t_2$, is governed by the elastic response of the grains. It varies only weakly with impact velocity, and is of the same order as the time it takes for a sound wave to travel back and forth through a grain. This is roughly $4R/c$, where c is the velocity of sound in the grain material. This yields the estimate

$$\frac{t_3 - t_2}{t_2 - t_1} \approx \frac{v}{c}\frac{4R}{h}. \tag{5.2}$$

Another time scale of relevance concerns the equilibration of the liquid surface forming the toroidal meniscus after $t = t_2$ (transition from $t = t_2$ to $t = t_3$ in Fig. 5.2). From the dispersion relation of damped capillary waves, $\omega = \gamma q / \eta$, where q is the wave number [140], we can immediately derive the relaxation time. We obtain $\tau \approx h\eta/\gamma$, where we have used that the 'vertical' distance of the contact lines is of order h. From this we get

$$\frac{\tau}{t_2 - t_1} \approx \mathsf{Ca}, \qquad (5.3)$$

such that we can consider the relaxation of the free liquid surface as quasi-instantaneous if $\mathsf{Ca} \ll 1$. In this case, the configuration of the liquid surface will stay close to its equilibrium shape also when the contact line recedes again, even more so since this process takes longer than the approach phase, as s_c is always larger than h. We conclude that if Cb is not too large (such that $\mathsf{Ca} \ll 1$), the model of the hysteretic capillary bridge formation and rupture is expected to provide a reasonable description of the interaction between wet grains. Furthermore, we see that hysteresis plays a significant role whenever $\mathsf{Cb} \leq 1$.

As we are dealing with dynamical processes here, we have to take into account both surface tension and viscous forces. From Chapter 4, where only static forces were explicitly considered, we have (cf. Eq. (4.42))

$$E_{cb} \approx 4.37\gamma R^2 \cos\theta \sqrt{\tilde{V}}, \qquad (5.4)$$

for the energy loss due to the capillary force exerted by the bridge, and due to its hysteretic nature. For the liquid volume of the capillary bridge, we must use $\tilde{V}_i \approx 4w^2$, since we are dealing with short time dynamics here.

The influence of the viscosity of the liquid on the impact kinetics has been studied for a long time, but is still a matter of research. This is mainly because as the impacting bodies come into contact at finite velocity, the viscous force diverges, such that a separation of scales is not readily achievable. Furthermore, the liquid may temporarily undergo transitions into other phases, due to the enormous pressures which may be temporarily reached at the contact point. For our purposes, however, it is sufficient to use the results which have been obtained under milder conditions, such as those encountered in agglomeration processes [110, 141–143]. Ennis *et al.* [110] have discussed this in detail and found that the viscous force is in good approximation given by

$$F_{\text{visc}} \approx \frac{3\pi}{2}R\gamma\frac{\mathsf{Ca}}{\tilde{s}}. \qquad (5.5)$$

As mentioned, this diverges as $\tilde{s} \to 0$. However, this divergence is not of practical relevance in our case. When the velocities are such that no permanent deformation occurs on the grain surfaces, which we assume for our systems, the roughness amplitude can be considered as a lower bound for s. Assuming the velocity to be roughly constant during the impact process, the total energy loss due to viscosity is obtained as

$$E_{\text{visc}} \approx \int_{\delta}^{s_c} F_{\text{visc}} ds \approx \frac{3\pi}{2} R^2 \gamma \mathsf{Ca} \ln\left(\frac{\tilde{s}_c}{\tilde{\delta}}\right). \tag{5.6}$$

Since the velocity will necessarily decrease during the impact, this should be considered as an upper bound. The total energy loss incurred during the wet grain impact is therefore, combining Eqs. (5.4) and (5.6),

$$\Delta E \approx \gamma R^2 \left(8.74 w \cos\theta + \frac{3\pi}{2} \mathsf{Ca} \ln\frac{s_c}{\delta}\right), \tag{5.7}$$

where the first term represents Eq. (5.4), the latter Eq. (5.6). Here we have used again $\tilde{V} \approx 4w^2$ (cf. Fig. 4.7).

Next we observe that in the regime of interest, effects of liquid capillary bridges should significantly affect the system, but at the same time not completely inhibit any relative motion of the grains. It is thus characterized by a granular temperature which is of the same order as the energy loss connected to the formation and rupture of capillary bridges. In order to obtain some idea about the corresponding typical velocities, we can therefore set

$$T_g \approx \frac{mv^2}{2} = \frac{1}{2} v^2 \frac{4\pi}{3} R^3 \varrho_s \approx \Delta E. \tag{5.8}$$

For the discussion which follows, it is useful to introduce a 'transition length',

$$l_t = \frac{\eta^2}{\gamma \varrho_l}, \tag{5.9}$$

which is specific the liquid used in the experiment. It is identical to the inverse of the wave number above which capillary waves on the surface of the liquid are overdamped. Together with a few other useful quantities, it is listed for some liquids in Table 5.1.

Combination of the Eqs. (5.8) and (5.7) leads to a quadratic equation in Ca,

$$\mathsf{Ca}^2 \approx \frac{3}{2\pi} \tilde{R}_t \left(8.74 w \cos\theta + \frac{3\pi}{2} \mathsf{Ca} \ln\frac{s_c}{\delta}\right), \tag{5.10}$$

Table 5.1   Relevant properties of a few common liquids.

Liquid	Density	Viscosity	Capillary velocity	Transition length
	ϱ_l	η	γ/η	$l_t = \frac{\eta^2}{\gamma \varrho_l}$
	[g/cm^3]	[mPa s]	[m/s]	
Helium (T = 4.2 K, boiling point)	0.125	0.0033	27	(1.0 nm)
Water	1.0	1.0	72	14 nm
n-Nonane	0.72	1.0	27	52 nm
n-Dodecane	0.78	1.35	18	96 nm
1-Decanol	0.83	13.8	1.79	9.3 μm
Silicone oil (Carl Roth, M 5)	0.93	4.6	4.2	1.2 μm
Silicone oil (Carl Roth, M 50)	0.97	48	0.43	115 μm
Glycerol	1.26	1470	0.44	2.7 mm
Silicone oil (unspecified [144])	0.97	970	0.21	4.8 cm
Silicone oil (unspecified [144])	0.98	4900	0.0043	1.16 m

Notes: Values are for room temperature where not otherwise stated.

where we have introduced the dimensionless 'transition radius'

$$\tilde{R}_t = \frac{\varrho_s R}{\varrho_l l_t} \tag{5.11}$$

which only depends on properties of the materials used. With Eq. (5.1), we can write $\mathsf{Ca}^2 = \mathsf{Cb}^2 w/3$, and therefore

$$\mathsf{Cb}^2 = \frac{9}{2\pi \tilde{R}_t}\left(8.74\cos\theta + \frac{\pi\sqrt{3}}{2\sqrt{w}}\mathsf{Cb}\ln\frac{s_c}{\delta}\right). \tag{5.12}$$

When we know the grain radius in units of R_c and the wetness, w, of the system, which also determines s_c, Eq. (5.10) (or (5.12), respectively) tells us at which velocities we should expect interesting effects in the system due to the dissipative character of the capillary bridges.

First of all, it is of interest to see under what circumstances the concept of hysteretic capillary bridges holds, and when it is expected to break down. The latter is the case when Cb comes of order unity, or larger. The three-phase contact line is then moved so rapidly along the grain surface that a capillary bridge has no chance to form. The corresponding border in the plane spanned by the grain size and the wetness can be easily obtained setting Cb equal to one in Eq. (5.12). This yields

$$\tilde{R}_t = 12.5\cos\theta + 3.90\frac{\ln(s_c/\delta)}{\sqrt{w}}. \tag{5.13}$$

We still need to evaluate s_c/δ which appears under the logarithm. We have from Eq. (4.28) that $\tilde{s}_c \approx \tilde{V}_i^{1/3}$ for small θ (in what follows, we will evaluate all quantities for vanishing contact angle). We readily find, using $\tilde{V}_i \approx 4w^2$ again,

$$\frac{s_c}{\delta} \approx 4^{1/3}\frac{R}{\delta}w^{2/3} = \frac{3h}{\delta}4^{1/3}w^{-1/3}, \tag{5.14}$$

and accordingly

$$\ln\left(\frac{s_c}{\delta}\right) \approx 1.56 + \ln\left(\frac{h}{\delta}\right) - \frac{1}{3}\ln w. \tag{5.15}$$

Note that since in a wet granular gas we can expect h to be of the same order as δ (cf. Chapter 3), the second term can be considered as small, and will be neglected in the further evaluation. We thus have the boundary

$$\tilde{R}_t = 12.5 + \frac{6.08 - 1.30\ln w}{\sqrt{w}}, \tag{5.16}$$

which is shown as the left of the two solid curves in Fig. 5.3. It provides the boundary between regimes **II** and **III**.

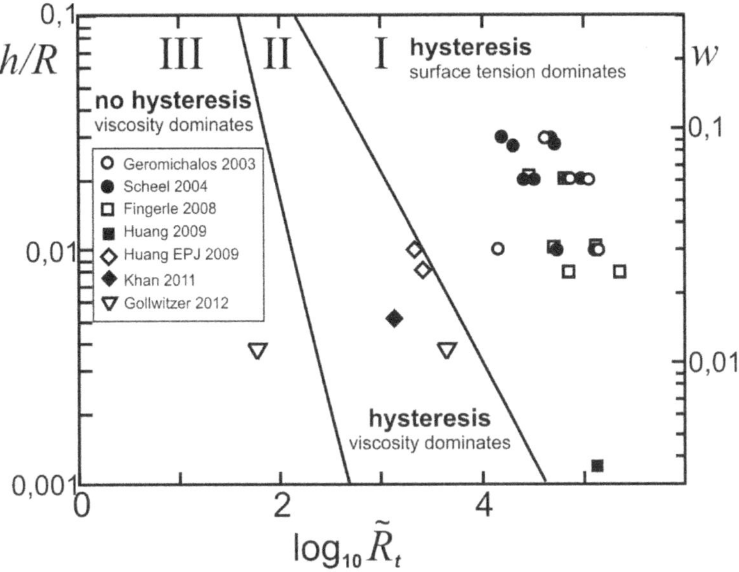

Fig. 5.3 The boundary between the different regimes of energy dissipation in a wet granular gas. Note that $h/R = w/3$. The material parameter which determines the diagram is the crossover radius, R_c. The two solid curves divide the plane into three regimes, which are described in the text. The left curve corresponds to $\mathsf{Cb} = 1$, the right curve to $E_{cb} = E_{\mathrm{visc}}$. Open circles are taken from [115], closed circles from [145], open squares from [146], closed squares from [147], open diamonds from [148], closed diamonds from [132], and open triangles from [149].

Next we would like to know whether the energy loss in a certain system is dominated by capillary forces or rather by viscous forces. We thus compute the locus of points at which E_{visc} and E_{cb} are equal. We solve the quadratic equation (5.10) to obtain Ca as a function of R and w, insert it into Eq. (5.6), and set the result equal to the expression for E_{cb}, Eq. (5.4). This leads to the simple result

$$\tilde{R}_t \approx \frac{2.80(\ln \frac{s_c}{\delta})^2}{w}. \tag{5.17}$$

With the help of Eq. (5.15) this provides an expression for R containing only w as a variable. We find

$$\tilde{R}_t \approx \frac{(2.61 - 0.56 \ln w)^2}{w}, \tag{5.18}$$

which is shown as the right solid curve in Fig. 5.3. It indicates the boundary between regimes **I** and **II**.[2]

The symbols inserted in the figure represent the conditions of various experiments which have been reported with wet granular gases. We see that almost all of them are well within the regime of hysteretic capillary bridges, and of those again the overwhelming majority in the regime of negligible viscous forces. A number of other experiments with a different focus, which used viscous oils, are so far within regime **III** that they do not show up in the figure.

It is of interest to see what are the important experimental parameters which determine the dissipation regime. In Fig. 5.4, the transition length, l_t, is plotted as a function of the viscosity for the liquids listed in Table 5.1. We see that since the viscosity is not only the most strongly varying of the quantities entering l_t, but also enters quadratically, there is a strong correlation between the viscosity and l_t. Furthermore, the range of grain radii is limited as well, as Fig. 1.4 in the introduction (Chapter 1) has already shown. Hence it is mainly the viscosity of the liquid which determines $\tilde{R}_t$, and thereby the position of the system in Fig. 5.3.

5.1.2   *The effective restitution coefficient*

In the discussion above, we have only reluctantly referred to the contact angle θ, which was introduced in Chapter 3. The contact angle is an equilibrium concept and should be applied to dynamic situations only with

---

[2]Experiments reported with wet Newton's cradles [150] would lie well outside (to the left) of region **II** displayed in Fig. 5.3. The same is true for other studies using very viscous liquids [144, 151, 152].

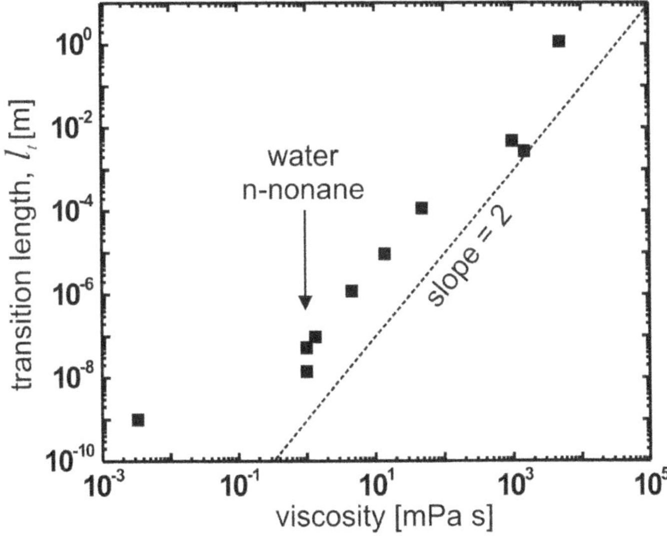

Fig. 5.4 The transition length, l_t, of several liquids vs. the respective viscosity, η. Since of the parameters entering l_t the viscosity is the most strongly varying, and enters quadratically, there is a strong correlation between l_t and η.

utmost care. When the contact line is moving, strong deviations in θ from its equilibrium value are observed, as discussed at the end of Chapter 3. The physics underlying this deviation, which is due to a combination of liquid viscosity and surface heterogeneity, is still not fully understood and a topic of lively debate. We will see, however, that the dynamic behavior of the contact angle is not a great problem for many of the aspects discussed here.

Figure 5.5 shows schematic representations of the force as a function of grain surface separation, in the three regimes introduced above. In the quasi-static case (**I**), the force follows precisely the characteristic developed in Chapter 4 (cf. Fig. 4.9). The pinch-off distance, s_c, is much larger than the distance at which the liquid films make first contact, $s \approx 2h$. As viscous forces become appreciable (**II**), a repulsive (i.e., negative) dip develops in the approaching branch, and a corresponding overshooting in the receding branch. The former represents the force needed to deform the liquid film. The latter has two possible contributions. One is the viscous force again, which is now needed to deform the capillary bridge. The other is the decrease of the contact angle when the contact line recedes, which makes the liquid appear as better wetting the grain surface. Both effects together

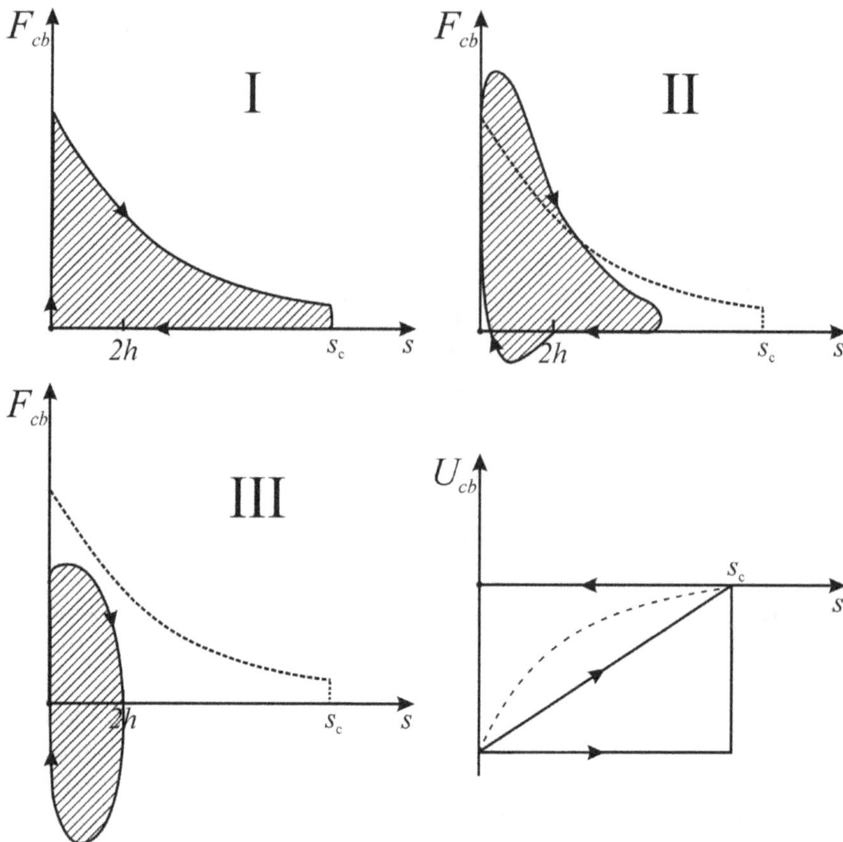

Fig. 5.5   Schematics of the force characteristics in the three different regimes outlined in Fig. 5.3. The hatched area represents the energy lost in the complete impact process. Bottom right: simplified effective interaction pseudo-potentials which can be used to model the hysteretic capillary force acting between adjacent grains. The dashed line represents the quasi-static form, cf. Fig. 4.11. The roman numerals **I**, **II**, and **III** correspond to the regimes indicated in Fig. 5.3.

lead to a reduced rupture distance, since the liquid cannot flow rapidly enough to form the 'equilibrium' bridge shape when the surfaces of the grains recede rapidly from each other. The dashed line indicates the quasi-static characteristic for comparison. If the impact process is so fast that no capillary bridge forms at all (**III**), the characteristic acquires a mirror symmetry with respect to the separation axis, and does not anymore extend beyond $s = 2h$.

The energy losses incurred during the impact process in the three cases sketched in Fig. 5.5 can be summarized as follows. In the case of dominant capillary forces (regime **I**), we have (assuming complete wetting)

$$\frac{\Delta E}{\gamma R^2} = \frac{E_{cb}}{\gamma R^2} \approx 8.74 w. \tag{5.19}$$

When viscosity dominates, but capillary bridges are well developed (regime **II**), we find using $\tilde{s}_c \approx \tilde{V}_i^{1/3}$ and $\tilde{V}_i \approx 4w^2$

$$\frac{\Delta E}{\gamma R^2} = \frac{E_{\mathrm{visc}}}{\gamma R^2} \approx \frac{3\pi}{2}\,\mathsf{Ca}\left(\frac{2}{3}\ln w + 0.46 - \ln \tilde{\delta}\right). \tag{5.20}$$

When finally the impacts are so fast that capillary bridges cannot develop (regime **III**), the energy loss can be obtained from Eq. (5.6) by setting $s_c = 2h$. This yields

$$\frac{\Delta E}{\gamma R^2} = \frac{E_{\mathrm{visc}}(s_c = 2h)}{\gamma R^2} \approx \frac{3\pi}{2}\,\mathsf{Ca}\ln\left(\frac{2h}{\delta}\right) = \frac{3\pi}{2}\,\mathsf{Ca}(\ln w - 0.41 - \ln \tilde{\delta}). \tag{5.21}$$

A convenient way to depict the qualitative difference of dry and wet granular gases is by means of the restitution coefficient, ε, which has been introduced in Chapter 2. Although it is commonly used for the description of dry granular systems, it may be formally evaluated as well for an impact between two wet grains, by just dividing the final and initial momenta, p_f and p_i, respectively, by one another. For general energy loss ΔE, one can write $\Delta E = (1 - \varepsilon^2)E_i$, where E_i is the energy before impact ('initial'). This yields

$$\varepsilon = \sqrt{1 - \frac{\Delta E}{E_i}}. \tag{5.22}$$

In dry granular systems, the restitution coefficient, $\varepsilon < 1$, is usually taken as a constant number characterizing the granulate under investigation. This will henceforth be referred to as the *standard model* of granular physics, and is shown in Fig. 5.6 as dash-dotted line.

In more refined models, the restitution coefficient becomes a function of the initial kinetic energy, E_i, of the collision partners. Quite generally, one finds that ε is a monotonously *decreasing* function of E_i. A well-known theoretical model introduced to account for this dependence considers the grains as viscoelastic media [12]. In this framework, one finds

$$\varepsilon = 1 - C_{ve}E_i^{1/10}, \tag{5.23}$$

where C_{ve} is a constant which is determined from the viscoelastic properties of the grain material. This will be called the *viscoelastic model* below and is represented by the dashed curve in Fig. 5.6. We see that the assumption

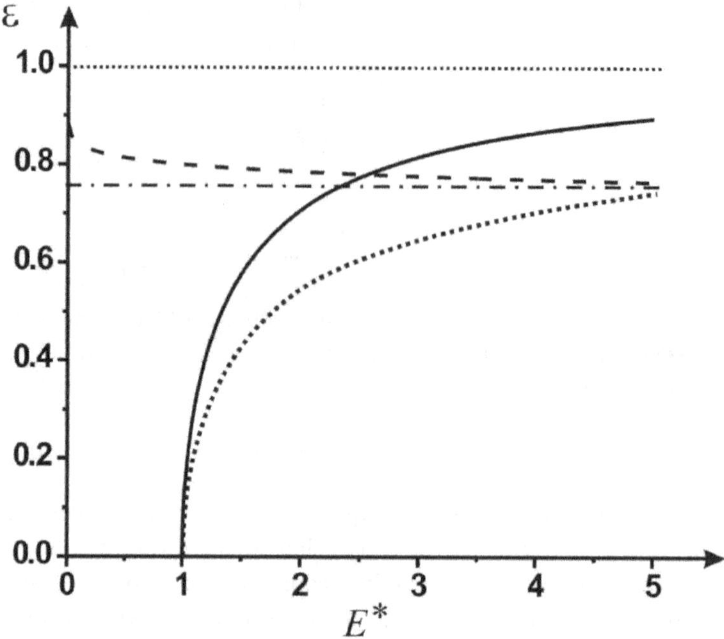

Fig. 5.6 The restitution coefficient as a function of impact energy for three different systems. Dashed: viscoelastic model, which exhibits an algebraic variation with impact energy, $E^* = E_i/E_{cb}$. Dash-dotted: a frequently used approximation for dry systems (constant ε). Solid: effective restitution coefficient for a wet granular system in regime **I**. Dotted: effective restitution coefficient for a wet granular system in regimes **II** and **III**.

of a constant coefficient of restitution still appears quite well justified as long as the limit of vanishing impact energy can be disregarded.

A qualitatively different situation arises when the grains are covered with a liquid film. Let us first consider the case of a fully developed hysteretic capillary bridge with only capillary forces involved. This will be referred to as the *capillary model*. In this case (**I** in Fig. 5.3), the energy lost in a collision is $\Delta E \approx E_{cb}$, which is independent of velocity, and we obtain

$$\varepsilon = \sqrt{1 - \frac{E_{cb}}{E_i}}.$$

(5.24)

This is a strongly *increasing* function of E_i, as opposed to the behavior encountered in the viscoelastic model.

In Fig. 5.6, the effective restitution coefficient for regime **I** is indicated as solid curve. It exhibits a zero at $E_i = \Delta E$. For $E_i < \Delta E$, the particles remain glued together by the liquid after the impact. In principle, there is a zero ($\varepsilon = 0$) also for the viscoelastic model, but this occurs at enormously high energy ($\approx 10^4$ in our example), where the validity of the model is more than questionable.[3] The zero of ε, with positive slope, can be considered a particularity of the capillary model.

It should be noted in passing that systems of grains with a possibility of sticking together are also attracting considerable interest in the context of structure formation in the universe [153–157]. They are referred to as 'sticky dust' or 'sticky gas' in this context, and play a significant role in models of the formation of accretion disks and planetesimals. This demonstrates the ubiquity of the systems we are discussing here, and the significance of investigating their properties from a statistical mechanics point of view.

Let us now turn to regime **II in Fig. 5.3**, where capillary bridges are formed, but viscous forces dominate. For the energy loss, we now have to refer to Eq. (5.20). This is not anymore independent of velocity. Nevertheless, we will readily see that the qualitative behavior of ε is the same. Note that E_{visc} depends linearly, E_i quadratically on Ca. After some straightforward manipulation, we obtain

$$\varepsilon = \sqrt{1 - \sqrt{\frac{C_{\mathrm{visc}}}{E_i}}}, \tag{5.25}$$

where

$$C_{\mathrm{visc}}(R, w) = \frac{27\pi}{8}\tilde{R}_t\left(\frac{2}{3}\ln w + 0.46 - \ln\tilde{\delta}\right)^2 \gamma R^2, \tag{5.26}$$

depends on system parameters, but not on the velocity. This behavior is indicated as the dotted curve in Fig. 5.6, which differs from the solid curve mainly far away from the important range around $E^* \approx 1$. A glance at Eq. (5.21) reveals that even in regime **III**, where no capillary bridges will develop, we encounter a similar behavior of ε, since $\Delta E \propto$ Ca here as well. The main difference is that the rupture distance s_c is negligible in this regime, such that we are left with just a characteristic energy scale, but no extra characteristic length scale of the interaction [158, 159].

5.2   Free cooling and clustering

Let us now turn to the impact of liquid on the collective behavior in granular many-body systems. While there is no significant inter-particle attraction

---

in dry granular matter, the addition of a liquid introduces substantial attractive forces. By analogy with molecular systems, we might then expect that wet granular systems will show condensation, while dry systems do not. However, as we have seen already in Chapter 2, there is clustering even in a completely dry granulate. Hence the truth is close, but a little more intricate: there is clustering in both cases, but the phenomenology is distinctly different. Similar to what we have discussed in Section 2.2, it is the comparison of the two cases which is most revealing about the underlying physical mechanisms and their interplay.

Numerical simulation is a powerful tool to compare the clustering behavior of dry and wet granular gases. A standard setting for studying these processes is to observe the free cooling behavior. One starts with a granular gas, with a high initial 'granular temperature', while the grains are assumed to be distributed evenly in space. Rotational degrees of freedom have been found to exhibit some correlation with the translational motion [160]. This is, however, believed to be of minor importance for the clustering behavior, and rotational degrees of freedom are neglected in the overwhelming majority of simulations published to date [12].

Granular media can be numerically simulated along the lines of standard molecular dynamics simulation techniques. We disregard for a moment the rotational degrees of freedom of the grains. For N particles with N center of mass coordinates, there are then $3N$ Newtonian equations to be solved, which can be done by partitioning the time axis in steps of suitably small size. This type of simulation is commonly called 'time-driven', and can be numerically quite demanding because of the many floating point computations to be executed.

In a dry granular gas, simulations are much easier to implement, since the particles are force free between collisions, and the solutions to the Newtonian equations consist of straight lines in space-time. It is then sufficient to determine at which point in space-time the next collision will occur, and to apply suitable update rules at these points. The only challenge here is to determine which collision is the next to occur.[4] In this way the simulation proceeds from one collision event straight to the next one. Therefore this scheme is commonly called 'event driven'. Concerning computational effort, is very economic as compared to the time driven scheme.

Let us now get back to Fig. 5.5 and consider the bottom right panel. It shows the pseudo-potential U_{cb} of the capillary force which is meant to

---

[4]This is most effectively done, for instance, by means of the so-called heap algorithm, which can be found in the standard literature of numerical recipes [161].

model the actual force characteristic of the capillary bridge. Three different types of curves are depicted. The dashed line indicates the 'exact' form, Eq. (4.31). The solid lines represent two possible approximations. One is a 'triangular' potential well, characterized by a constant force acting all along the retracting path from $s = 0$ to $s = s_c$. The other is a rectangular well, which also extends over the interval $[0, s_c]$. Both approximative types of potential wells and the 'exact' potential have in common that they represent an energy loss of the same $\Delta E = E_{cb}$, and both extend over the same interval in s. It is now of great importance whether the collective properties of the granular gas depend on the shape of the pseudo-potential, or whether ΔE and s_c are the only relevant physical parameters. If the latter is the case, we could, for instance, use the rectangular potential in simulations. This has the particular merit of enabling event-driven simulations, since (just as for the dry case) the force vanishes almost everywhere on the time axis (i.e., everywhere except a set of zero measure). Forces appear only whenever one of the inter-granular separations becomes equal to s_c, and yield then a force pulse of δ-function type.

We anticipate at this point the very favorable outcome that the shape of the pseudo-potential is in fact almost irrelevant, in particular in dilute systems. We will come back to this again later, when the universal properties of wet granular gases become obvious in experiments and simulations of phase transitions. In Fig. 5.7, the motion of two colliding grains is sketched for the rectangular pseudo-potential. In the upper panel (a), $E_i > \Delta E$, while in the lower panel (b), $E_i < \Delta E$. The trajectories are piecewise straight, reflecting the fact that the forces vanish almost everywhere on the time axis.

It is readily appreciated that clustering must occur in both the wet and dry case, but through completely different mechanisms. In the dry case, grains in areas of high grain density undergo particularly frequent impacts. As a certain fraction of energy is lost in each impact, dense regions cool faster, making it even more difficult for particles to leave these areas, which further increases their density. An immediate result is the rapid formation of dense clusters, as shown in Fig. 5.8(a) [27], even though there is no attractive force. As time proceeds, the granular temperature decays algebraically to zero, according to Haff's law [23, 27, 164] (cf. Chapter 2). At the same time, the grains remain free to move, even if they do so at smaller and smaller velocity. We are thus tempted to say that the dry granulate *condenses in momentum space* rather than in real space.

In fact, the more realistic viscoelastic model [12] shows this even more

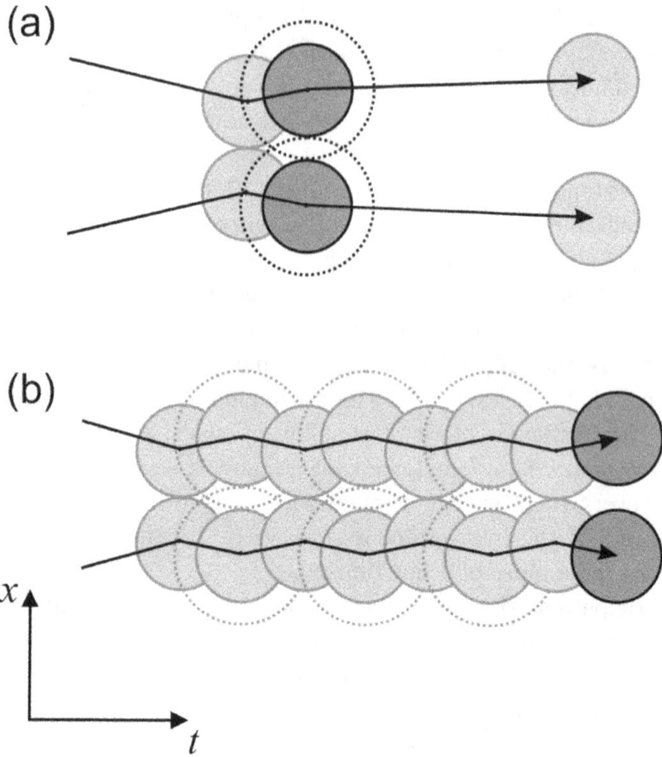

Fig. 5.7 Collision dynamics in an event driven simulation of a wet granular gas, sketched in the plane spanned by the distance of the grains, and time. The grey disks indicate the grains, the dashed circles (radium $R + s_c$) indicate the rupture distance. (a) $E_i > E_{cb}$; the grains are elastically reflected at hard core contact, but abruptly reduce the velocity of withdrawal as they cross the rupture distance, where their kinetic energy is reduced by the amount E_{cb}. (b) $E_i < E_{cb}$; the grains are reflected not only at hard core contact, but also back towards each other as they reach the distance indicated by the dashed line. A bound state results.

clearly than the ideal dry granular gas. As the bottom row of Fig. 5.8 shows, a gas of viscoelastic grains clusters only transiently. Initially, clusters form (d) as in the standard model (b). As time proceeds, however, the density of particles smoothes out again over space (e), becoming more or less homogeneous again as time tends to infinity. This is because there are two competing effects in the viscoelastic model (dashed line in Fig. 5.6). On the one hand, the dissipative nature of the collisions tends to enhance the clustering. On the other hand, the restitution coefficient approaches unity

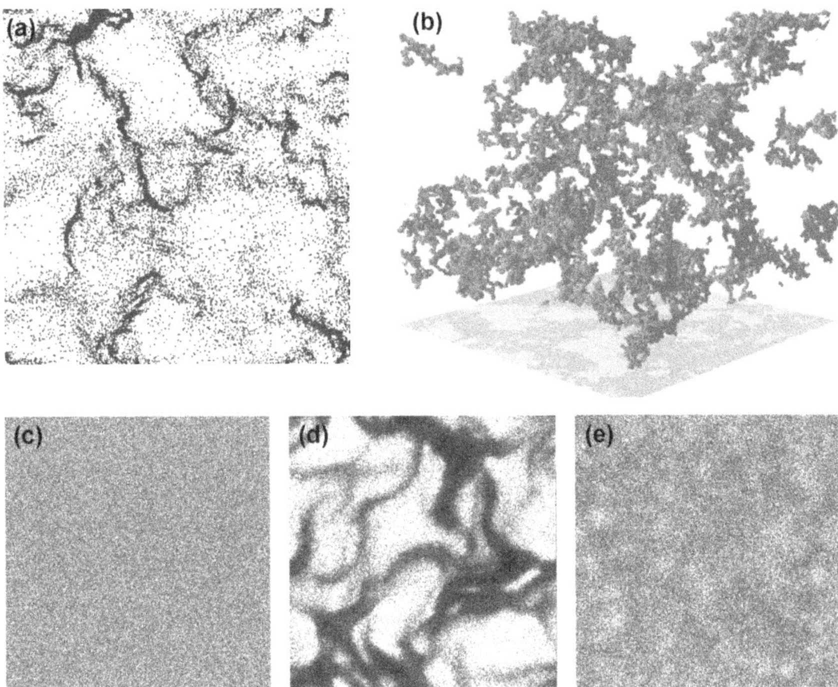

Fig. 5.8   Clustering phenomena in (a) a dry [27] and (b) a wet [162, 163] granular gas (reprinted Fig. 3 of [27], copyright (1993) American Physical Society, and Fig. 2 of [162], copyright (1993, 2009) American Physical Society). The bottom row shows simulations of the clustering behavior of a dry granular gas in the viscoelastic model. (c) early stage; (d) intermediate stage, the dissipative character of the collisions leads to clustering; (e) late stage. As the restitution coefficient approaches unity for small energies, entropy redistributes the grains homogeneously in space. Clustering is rendered a transient phenomenon in this model (images reproduced from Fig. 26.2 of [12] by permission of Oxford University Press).

as the relative velocities become smaller[5] (Fig. 5.6). It therefore approaches the elastic case, in which entropy tends to distribute the grains evenly in space. In the viscoelastic model, the scaling of ε with the impact energy E_i is such that for small energies (i.e., for late times) the second mechanism wins.

In marked contrast, wet grains will cluster by directly adhering to each other due to the wetting forces exerted by capillary bridges. As soon as the kinetic energy of the grains comes below E_{cb}, they will stick to each other, and a growing cluster forms. When this has occurred, the temperature

---

[5]This is quite generally the case for compact grains.

within the cluster drops much more slowly, provided capillary bridge rupture is the only dissipation mechanism, as we assume here for the sake of clarity. These clusters are stabilized by cohesion to mutual distances below s_c. We may rephrase this scenario by saying that the (ideal) wet granulate *condenses in real space* rather than in momentum space. We will revisit this distinction further below. For dissipative forces, such as in regimes **II** and **III**, it is clear that the extension of the system in momentum space will as well continue to shrink as time proceeds, even after cluster formation is completed.

5.2.1　*Granular temperature during free cooling*

Let us investigate the decay of the granular temperature from a given initial value T_g^0 which we assume to be large as compared to ΔE [163]. Then we can be sure that at least in the early stage of free cooling, almost no collision will result in a bound pair of particles, and the amount ΔE is dissipated in virtually every collision. If particles collide with frequency f_{coll}, the average loss per unit time is given by

$$\frac{3}{2}\frac{dT_g}{dt} = \frac{1}{2}f_{\text{coll}}\Delta E. \tag{5.27}$$

The factor $1/2$ takes into account that two particles are involved in each collision. For the collision frequency, we have the well-established result

$$f_{\text{coll}} = 8\pi R^2 g_c n\sqrt{\frac{T_g}{\pi m}}, \tag{5.28}$$

g_c is the pair-correlation function at contact, for which we can again use the Carnahan–Starling form, Eq. (2.8). Since the density is assumed to be constant, the combination of Eqs. (5.27) and (5.28) yields

$$\frac{dT_g}{dt} \propto -\sqrt{T_g}, \tag{5.29}$$

which is solved by $T_g(t) \propto (t - t_0)^2$. Inserting the pre-factors and the initial value $T_g(0) = T_g^0$, one obtains an analytical form of the decay of the temperature

$$T_g(t) = T_g^0(1 - t/t_0)^2 \tag{5.30}$$

with the characteristic time scale

$$t_0 = \frac{\sqrt{9\pi m T_g^0}}{8\pi R^2 g_c \Delta E}. \tag{5.31}$$

Note (5.30) is formally different from Haff's law, Eq. (2.13), such that $T_g(t)$ has a zero at $t = t_0$. In this simplified model, the assumption that *every* collision causes an energy loss ΔE gives rise to a time scale at which all energy is dissipated. Although this assumption does not hold for the later stages of cooling (bonds do not break anymore if the relative kinetic energy is too small), the time scale t_0 has a clear physical relevance. It sets the time after which the temperature is comparable to the bond breaking energy ΔE, and after which persistent clusters will form.

In order to achieve a deeper insight into the processes which are taking place, it is instructive to have a closer look at simulations. The transition from the initial 'algebraic' cooling phase to massive clustering is best appreciated if one considers how the remaining energy is partitioned among the different degrees of freedom of the system. This can be seen in Fig. 5.9, which shows the fraction of kinetic energy stored in the translational motion of single grains, the rotational degrees of freedom of the forming clusters, and the internal energy stored in the relative motion of grains bound to each other by 'capillary bridges'. The time axis is normalized with respect

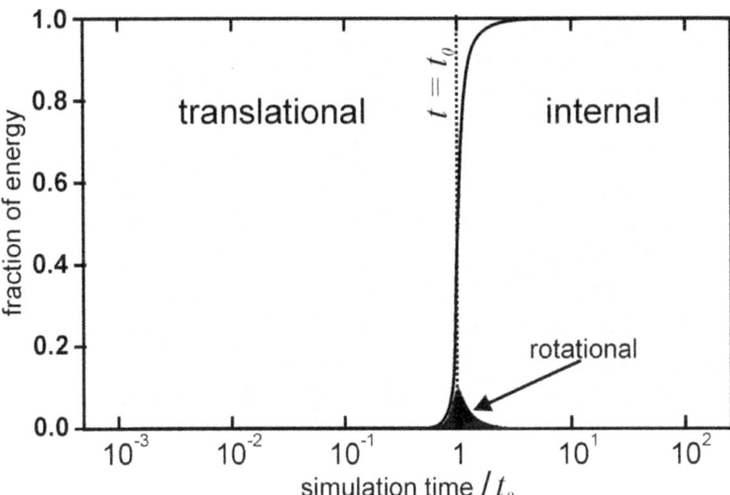

Fig. 5.9 Partitioning of energy among the various degrees of freedom of the system during free cooling, as a function of the time elapsed since the beginning of the simulation. While initially almost all energy is stored in the kinetic energy of the individual particles (translational), all remaining energy is stored in the internal degrees of freedom within the clusters in the final stage. Only in a short intermediate phase there is a significant amount of rotational energy, when the size of the clusters and hence their moments of inertia are small. Redrawn with modifications after [163].

to t_0. We clearly observe a sharp transition around $t = t_0$, at which almost all translational energy is converted into capillary bridge motion. There is some temporary energy storage in the rotation of clusters, but this is obviously a minor effect.

The transition coincides with the breakdown of the temperature decay described by Eq. (5.30). It is natural to ask how the temperature will continue to decay after this transition. Since the probability to break a capillary bridge becomes small for $t > t_0$, the system can be considered more or less Hamiltonian, such that equilibrium statistics is applicable. The probability that a degree of freedom accumulates enough energy to break the bond at the next trial is then proportional to $e^{-\Delta E/T_g}$. We may hence expect that the temperature decays according to

$$\frac{dT_g}{dt} \propto e^{-\Delta E/T_g}, \tag{5.32}$$

which can be solved by separation of variables. It is straightforward to show [163] that this leads asymptotically to a decay according to

$$\frac{T_g}{\Delta E} \propto \frac{1}{\ln(t)}. \tag{5.33}$$

This is much slower than the initial decay, Eq. (2.13), due to the strongly reduced probability to break any further bonds.

5.2.2   *Morphology of the emerging clusters*

The morphology of the emerging clusters can be well described by means of their fractal dimension. This is a generalized concept of dimensionality which refers to the way the volume of an object scales with its size. The most obvious way to measure it is by investigating the minimum number B of cubic boxes of edge length λ which are needed to entirely cover the object, as a function of λ. For an arbitrarily curved line in space, we will find $B \propto \lambda^{-1}$, asymptotically for small λ. For an arbitrarily curved, but infinitely thin sheet, we will obtain $B \propto \lambda^{-2}$, and for any solid object, we have $B \propto \lambda^{-3}$, always asymptotically as λ tends to zero. If for an object one finds $B \propto \lambda^{-D_f}$ with some non-integer D_f, the object is called fractal, with fractal dimension D_f. This method is, quite obviously, called the box counting method.

Since in the simulation we have all particle coordinates for all times, it is straightforward to determine the fractal dimension of the clusters which emerge during the cooling process. Figure 5.10 shows B as a function of

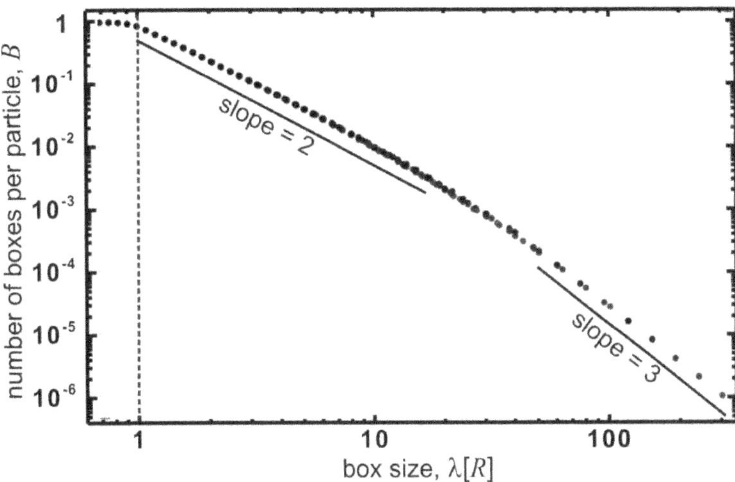

Fig. 5.10 Determination of the fractal dimension of a freely cooling wet granulate by means of the box counting method. Overlay of several data sets obtained for different times elapsed since the start of the simulation. Redrawn with modifications after [163].

box size, λ, where the latter is measured in units of the grain diameter. B has been normalized with respect to the total number of grains in the system. In the log–log plot, the fractal dimension simply shows up as the slope. For large box sizes, we see that B scales as λ^{-3}, as the clusters appear as voluminous objects on large scale. However, as λ is reduced to about 10 grain diameters, the slope crosses over to a value very close to two. Only as we make the boxes smaller than a single grain, we always need one box per grain. This is because we chose not to cover the grains completely, but to cover only the set of their centers of mass. The crossover from $D_f = 2$ to $D_f = 3$ above 10 grain diameters can be attributed to finite size effects. Hence we see that the clusters are neither equivalent to single isolated particles ($D_f = 0$) nor long threads of grains ($D_f = 1$), but some other, non-trivial structure with D_f close to two.

One may criticize here that observing a well-defined slope over just an order of magnitude in box size may not be very compelling yet. It is therefore instructive to determine the fractal dimension as well with a complementary method. This can be done by plotting the radius of gyration of each cluster vs. its mass. The radius of gyration, R_g, is defined as the radius of a spherical shell of the same mass which would have the same moment of inertia. It is easily computed by summing up the squares of

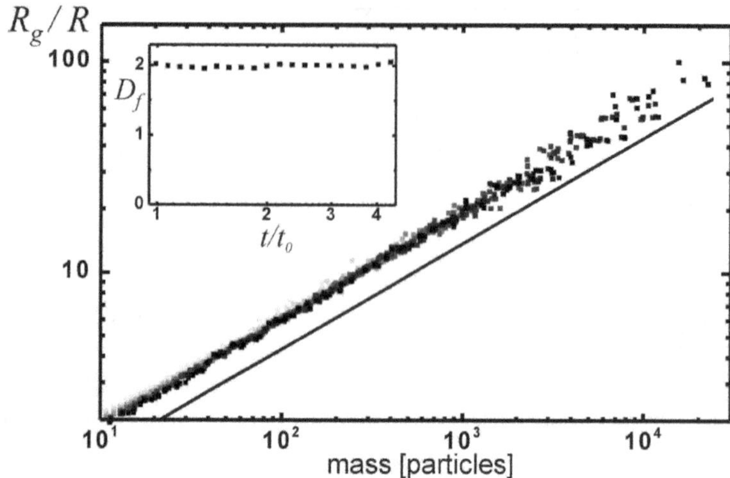

Fig. 5.11 Determination of the fractal dimension of a freely cooling wet granulate by means of the radius of gyration, R_g, of the clusters. Inset: fractal dimension as a function of simulation time. Redrawn with modifications after [163].

distances of all particles belonging to a cluster from the center of mass of that cluster. The result is shown in Fig. 5.11 for all clusters found in the simulation, for several different times. Different shades of grey indicate different simulation times. As we can see, the data points again fall nicely on a single line, and the slope of this line hints to the fractal dimension of the clusters (although the relation is more complex than above). Again we find a fractal dimension very close to two. In the inset, this is resolved for different simulation times, ranging from the early stage of cluster formation to the very late stages. Obviously, $D_f = 2$ is a very robust feature for the system at all times.

We should compare this to the structure of clusters obtained from diffusion-limited aggregation (DLA), which is considered as the paradigmatic cluster formation process in aggregation. In this process, the particles stick together as well, similar to the wet granular gas, but their transport towards the growing cluster is entirely diffusive [139]. Figure 5.12 shows results for the fractal dimension of the emerging clusters, compiled from the literature. There is obviously good agreement between the different approaches. For three dimensions, a fractal dimension of about 5/2 is obtained. Our result for the wet granular gas lies distinctly below these data, indicating a much more open, filigrane structure.

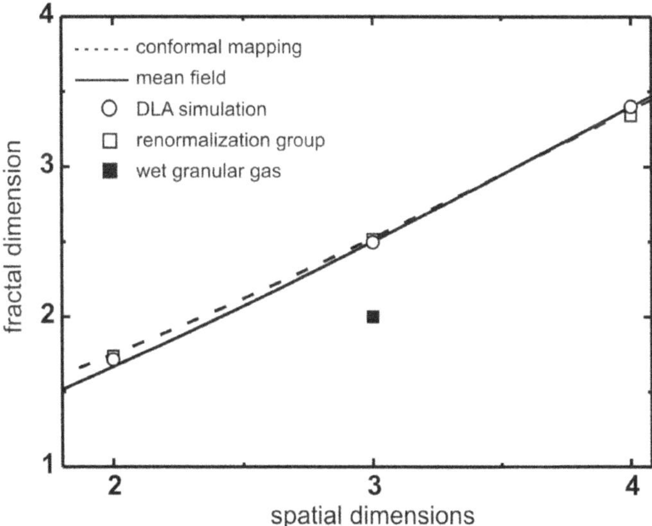

Fig. 5.12   Fractal dimensions of clusters formed in diffusion limited aggregation (DLA), for variable spatial dimensions and obtained from different methods. Solid curve, mean-field theory [165, 139]; dashed curve, conformal mapping [166]; open squares, renormalization group [167]; open circles, computer simulations [168]. The solid square indicates the result from free cooling of wet granular gases, which is distinctly different.

5.2.3   *A formal distinction between dry and wet cooling*

We are now in shape to return to the above-mentioned notion of condensing in either momentum or real space. In order to put these notions on a solid basis, we consider the volume our system occupies in real space and in momentum space. First of all, we have seen above that the volume in real space occupied by the wet grains, V_x^{wet}, has only dimension $D_f \approx 2$. Then clearly

$$\lim_{t \to \infty} \left(\frac{V_x^{\text{wet}}}{V_x^{\text{dry}}} \right) = 0. \tag{5.34}$$

This states that the space occupied by the wet granular gas is, at infinite times, vanishingly small as compared to the space occupied by its dry counterpart. The latter fills space much more evenly, in particular if the dependence of the restitution coefficient on impact energy is taken into account (cf. Fig. 5.8, bottom row). In momentum space, things turn out to be quite the opposite. To define the volume of a granular gas in momentum space, let V_p be the volume of the smallest convex region containing the momenta of all grains. As the momentum scales as the square root of granular temperature, but momenta are three-dimensional, we have $V_p \propto T_g^{3/2}$. From Eq. (5.33), and Haff's law for the dry case, we then have

$$\lim_{t \to \infty} \left(\frac{V_p^{\text{dry}}}{V_p^{\text{wet}}} \right) \propto \lim_{t \to \infty} \left(\frac{\ln t}{t} \right)^{3/2} = 0, \qquad (5.35)$$

which states that the momentum space occupied by the dry granular gas will be vanishingly small as compared to the momentum space occupied by its wet counterpart. Equations (5.34) and (5.35) thus provide a formal way of comparing the condensation behavior of the dry and the wet granular gas, essentially stating what we had said already above: dry granular gases condense in momentum space, wet granular gases condense in real space.

These statements are true, of course, only for the idealized cases. In reality, we are always faced with a mixture of both dissipation mechanisms, with the pertinent impact on the mode of condensation. The formulation given here shall just serve for pointing out the essence of the behavior connected to the two dissipation mechanisms we have discussed.

5.3 Granular liquid–gas coexistence

We have seen that a wet granular gas undergoes real space condensation as time proceeds. It is then natural to ask whether such a system may as well exhibit phase separation, similar to regular cohesive systems (like van der Waals gases), into a dense (liquid) and a dilute (gaseous) phase.

5.3.1 *Equation of state of wet granular gases*

When phase separation is observed, it suggests itself to try to set up an equation of state which can account for this observation. We thus seek an equation of the form

$$p = n T_g Z(n), \qquad (5.36)$$

where p is the pressure of the granular gas, n the number density of its grains, and the function Z is commonly called the compressibility factor. For an ideal gas $Z = 1$. In order to derive Z for our system, we start by considering the dry grains alone, i.e., a gas of hard spheres. It is useful to discuss this in terms of the volume fraction, ϕ, which has a maximum value, ϕ_j (jamming limit, see Chapter 2). Towards this limit, the free space the grains have for moving goes to zero, such that we would expect the pressure to diverge. Z should therefore be roughly the inverse of the free

volume which is left for the grains to move. A straightforward idea is then
to set

$$Z_{\text{dry}} \approx \frac{1}{1 - \frac{\phi}{\phi_j}}. \tag{5.37}$$

We will mention better expressions below, but (5.37) is sufficient for the
qualitative discussion we start with.

If we introduce an attractive force due to the capillary bridges, the
pressure is reduced accordingly. This reduction is equal to the compressive
stress exerted by the capillary bridges, and can be easily estimated. From
Eq. (4.35) we have, by analogy,

$$\frac{dE}{dV} = n\frac{k}{2}F_{cb}R. \tag{5.38}$$

With the approximation $E_{cb} = F_{cb}s_c$, this can be written as

$$\frac{dE}{dV} = n\frac{kE_{cb}}{2\tilde{s}_c}, \tag{5.39}$$

and thus

$$p = nT_g Z_{\text{dry}}(n) - \frac{kn}{2\tilde{s}_c}E_{cb}. \tag{5.40}$$

The expressions for Z in Eqs. (5.37) and (5.40) are sketched in Fig. 5.13
for $k = 6$, $\tilde{s}_c = 0.1$, and $T_g = E_{cb}$. For large ϕ, there are capillary bridges
between all neighboring grains, and we are on the lower branch expressed
by Eq. (5.40). As ϕ is reduced until the average separation between neighboring grains becomes larger than s_c, the system crosses over to the upper
branch, as indicated qualitatively by the dotted line. As expected, the attractive force introduces a structure similar to that known from van der
Waals theory of real gases, where a non-monotone dependence of the pressure on the density indicates phase separation.

If we want to come up with an equation of state which is quantitatively
correct, we are faced with a very complex problem. Fortunately, the equation of state for a gas of hard spheres has been worked out before, such
that we can resort to a considerable body of sound results. In contrast,
including the effect of formation and rupture of capillary bridges requires
careful bookkeeping procedures, in particular when it comes to evaluating
multi-particle correlations and the histories of bridge rupture events. In
principle, the path to be taken is straightforward, but lengthy and tedious,
such that a detailed discussion clearly goes beyond the scope of the present
book. We will therefore just report on the results here, and refer to the

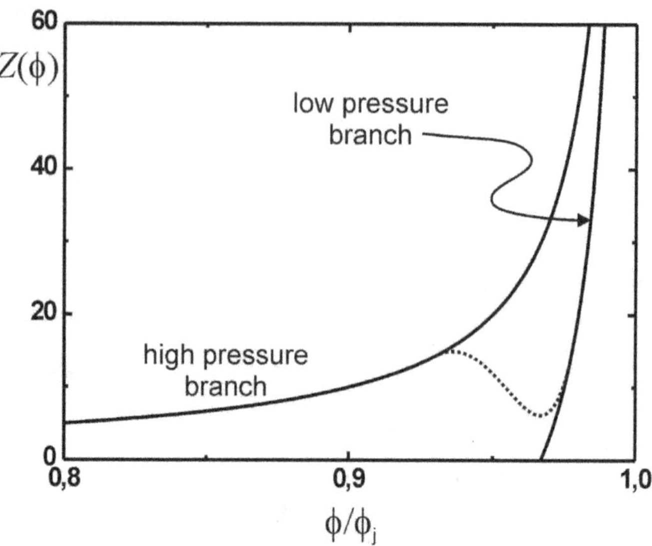

Fig. 5.13 Sketch of the compressibility factor for a real gas, with both attractive and short-range repulsive terms in the interaction potential. ϕ_j represents here the maximum density accessible for the system. The upper solid line includes only excluded-volume effects (Eq. (5.37)). It represents the equation of state for high temperature, when binding energies are small as compared to thermal energies. The lower solid line represents low temperatures, when particles are bound to each other, leading to a substantially reduced pressure (Eq. (5.40)). At intermediate temperature, the system crosses over from one behavior to the other, when the system becomes dense enough for the particles to feel the attraction of their neighbors. This may lead to non-monotone behavior of the pressure, and thus to phase separation.

literature for details [37]. The final result for the equation of state (Eq. (74) in [37]) is

$$p = n_f T_g (1 + 2^{D-1} \phi g_c(\phi)) - \frac{k(\phi) n_f E_{cb}}{D \tilde{s}_c}, \tag{5.41}$$

where $g_c(\phi)$ is again the density correlation function at contact, and n_f is a reduced density, which does not contain the internal degrees of freedom of clusters. This equation is difficult to evaluate in three dimensions $(D = 3)$, because of the complexity of the expressions for g_c, k, and n_f at higher densities. In two dimensions, however, the evaluation is feasible, and comparison can be made with ample literature on the dynamics of gases of hard disks.

Figure 5.14 shows a plot of Eq. (5.41) for $D = 2$ and a temperature large as compared to E_{cb} (dashed curve) and for $T_g = 0.2 E_{cb}$ (solid curve)

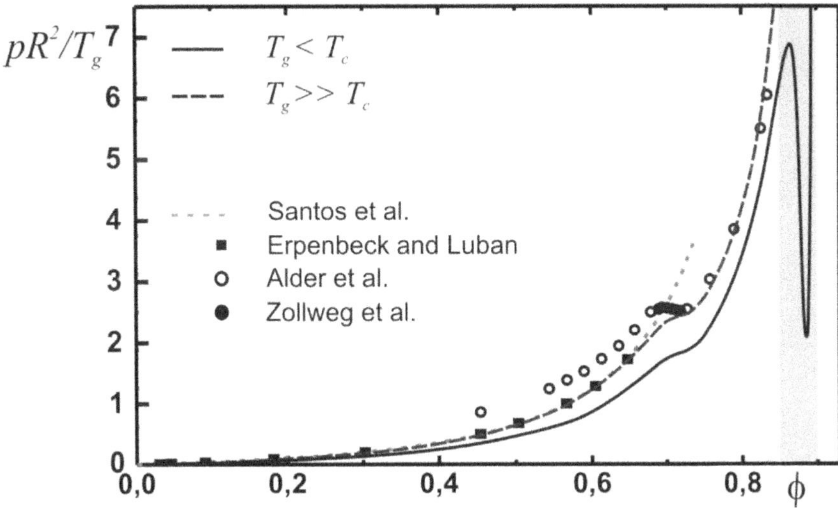

Fig. 5.14 Equation of state of a wet granular gas in two dimensions [37] (solid curve). The density of closest packing here is $\phi_j = \pi/2\sqrt{3} = 0.907$. The anomaly at $\phi \approx 0.7$ indicates two-dimensional melting. The other data are taken from various studies of two-dimensional gases without attraction. The pronounced dip exhibited by the solid curve within the grey shaded region is indicative of the attractive effect of the capillary bridges. Dotted curve, Santos *et al.* [169]; squares, Erpenbeck and Luban [170]; open circles, Alder *et al.* [171]; closed circles, Zollweg *et al.* [172] (redrawn with modifications from [173]).

[37]. We clearly see the development of the non-monotone behavior in the solid curve, corresponding to a granular temperature below the bridge rupture energy. Along with these, we include data from various simulations and analytical results for gases of hard disks. As the dashed curve shows, Eq. (5.41) successfully describes this case as well, including the anomaly at $\phi \approx 0.7$, which indicates the two-dimensional melting transition.

Let us now have a closer look at the non-monotone region at higher volume fraction, and for temperatures around E_{cb}. Figure 5.15 shows a closeup on the region around the critical volume fraction, ϕ_c. We see that the pressure curve exhibits a saddle point when $T_g = 0.274E_{cb}$, which is thereby identified as the critical temperature. At lower temperature, the curve exhibits a maximum and a minimum. The loci of these extrema are indicated by the dashed line, which represents the binodal line of the system. The inset shows this binodal line as a function of temperature. Obviously, the system behaves very similarly to a standard van der Waals gas, at least qualitatively.

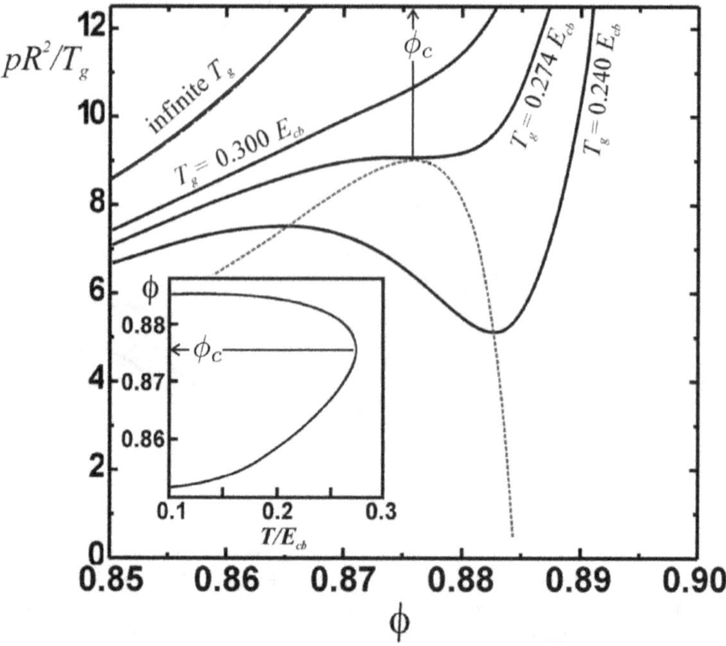

Fig. 5.15 Closeup of the critical region, evaluated at three different granular temperatures around the critical point at $T_c = 0.274E_{cb}$ [173]. The dotted curve represents the binodal of the system. The range of the abscissa corresponds to the grey shaded region in Fig. 5.14 (redrawn with modifications from [173, 37]).

It is of interest to discuss the dependence of the critical density and critical granular temperature as functions of the rupture distance, s_c, for the two-dimensional case. This is shown in Fig. 5.16 for rupture distances up to $\tilde{s}_c = 0.2$. This corresponds roughly to the full range of liquid bridge volumes V_i covered in Chapter 4 (cf. Fig. 4.8). We see that the critical density stays very close to the maximum density, ϕ_j. Furthermore, the dependence of ϕ_c on s_c is almost linear, and thus scales roughly as the average separation between neighboring grains. This accords with the idea that the critical density must be in a range where the number of capillary bridges in the system strongly increases with increasing density. For the critical temperature, we find a very mild variation with the rupture distance when scaled with respect to E_{cb}. Over the range shown, $T_c \approx E_{cb}/4$ is a reasonable approximation.

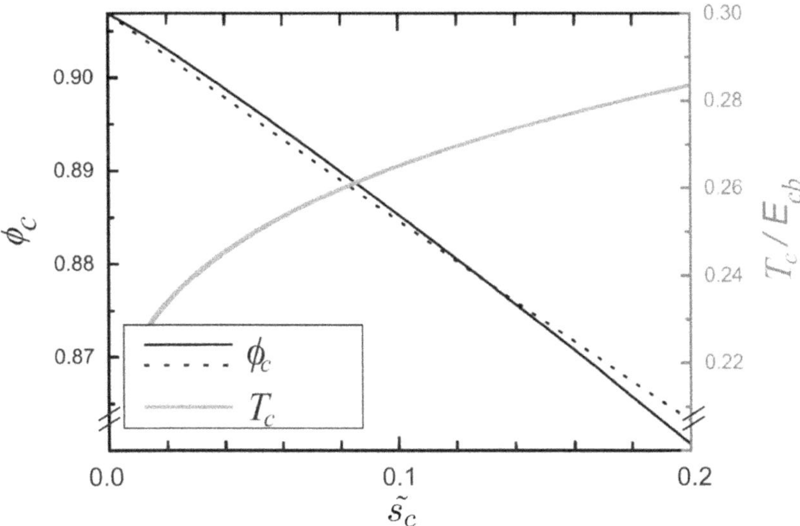

Fig. 5.16 The critical density, ϕ_c, and the critical temperature, T_c, in units of E_{cb} as functions of the normalized separation, $\tilde{s}_c$ [37]. The dashed line is a linear approximation of the black solid line, which describes the data quite well. The critical temperature (grey) remains close to $E_{cb}/4$ in the entire region (note that the zero of both vertical axes is suppressed. Redrawn with modifications from [173]).

5.3.2 *Experimental verification of the critical point*

It suggests itself to test these predictions by means of an experiment. We shall try to agitate a sample of a wet granulate strongly enough for the average kinetic energy of the grains to become of order E_{cb} and look for indications of phase separation. As opposed to a usual gas prepared in some equilibrium state, we have now to deal with dissipative interactions between the grains. Hence we must continuously supply energy to the system in order to prevent it from cooling down due to collisions, as described by the wet granulate analogue of Haff's law, Eq. (5.30). As typical energies correspond to velocities on the order of a few centimeters per second, this is easily done by just shaking the sample cell, e.g., in a sinusoidal fashion. The vertical position of the platform then has the form $z(t) = a_0 \cos \omega t$ with $\omega = 2\pi f_{\mathrm{sh}}$ representing the shaker frequency. The strength of the agitation can then be quantified by the maximum acceleration, which we normalize with respect to the acceleration due to gravity,

$$\Gamma = \frac{4\pi^2 a_0 f_{\mathrm{sh}}^2}{g}, \tag{5.42}$$

just as we did in Section 4.4.2. Alternatively, we can associate a characteristic energy with the vibration, setting

$$E_{\text{vib}} = \frac{1}{2}m(2\pi f_{\text{sh}}a_0)^2 = 2\pi^2 m a_0^2 f_{\text{sh}}^2. \tag{5.43}$$

The dissipative character of the sample has direct implications for the choice of the sample cell geometry. It should be sufficiently small, or narrow, for the grains to be in close contact with the container walls, such that the buildup of appreciable temperature differences is effectively suppressed. A top view of a suitable setup is shown in the left panel of Fig. 5.17. The circular platform and the annular membrane belong to an electromagnetic shaker which is sketched in cross-section in the right panel of the figure. It is capable of vertically translating the platform in a sinusoidal motion with accelerations which can be much larger than the acceleration due to gravity. On the platform we have mounted a steel tube with an inner diameter of a bit more than a centimeter, and about 10 cm long. This contains the granulate, which is effectively 'heated' by the nearby walls once the shaker is turned on.

There is a straightforward way to observe phase separation in the system, even if there is no optical access. As the grains, we used stainless steel

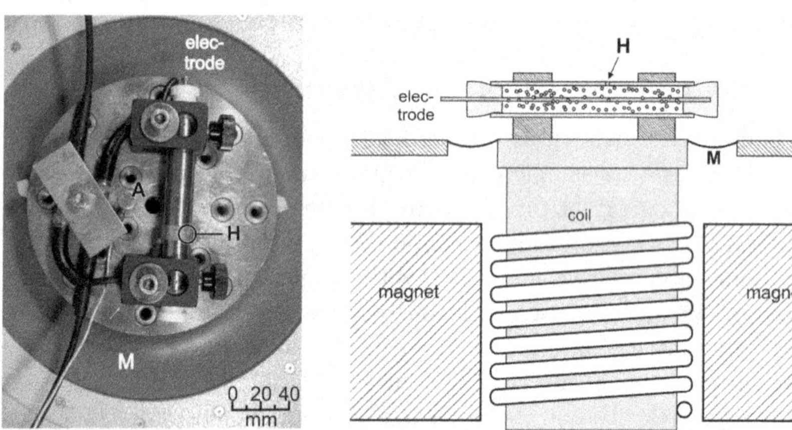

Fig. 5.17 Experimental setup for observing liquid–gas phase separation in a wet granular sample. The granulate consists of about 10^5 steel balls with a diameter of one half millimeter each. Left: top view of the setup; the steel tube containing the sample (with small hole, H, for filling in controlled amounts of liquid) is mounted on a vibration stage (accelerometer A). Right: schematic cross-section of the setup. The vibration stage is driven vertically by an electro-magnet (electro-magnetic shaker). It is connected to the casing with a rubber membrane, M (visible in left and right panel). The resistivity of the sample is measured between the center electrode and the conductive steel tube [173].

balls, as used for ball bearings. Since the material of the grains is a metal, one might naively expect the ohmic resistance of a pile of steel balls to be small. However, neighboring balls make contact with each other only in a very tiny region, and the contact is not of high quality because of the presence of some layers of metal oxide near the surface. Consequently, the resistance of the pile is actually very large, on the order of many Megaohms. This changes drastically when a little bit of conductive liquid is added. This can be done in a controlled way through the little hole in the side wall of the cell by means of a syringe. As already discussed, the liquid forms capillary bridges around the points of contact, thereby dramatically increasing their conductivity, and thus the conductivity of the sample. This can be measured between the steel wall of the cell and the inner electrode rod (cf. Fig. 5.17), yielding a suitable average over the length of the cell.

When the shaker is set in motion and the granulate is strongly agitated, such that the kinetic energy of a steel ball is typically well above the energy required to rupture the liquid bridges on its surface, the latter will pinch off and form in rapid sequence. If, however, we choose the density of the granular filling such that there is ample space for the grains to move, but the average distance between adjacent grain surfaces is still smaller than the rupture distance, s_c, we should expect that there will at all times be a percolated network of capillary bridges throughout the sample, and the electric conductivity will be high. If we lower the strength of external vibration, such that T_g comes of order E_{cb}, we would expect the steel balls to cluster into larger aggregates. As their density is larger than the overall density in the sample, it is clear that the distance between adjacent clusters is now typically much larger than s_c. Hence the network of conducting paths is destroyed, and the sample resistance increases.

This is indeed the case, as can be appreciated from Fig. 5.18. The strength of the external (sinusoidal) agitation is slowly decreased here, starting from values much higher than those required to bestow an average kinetic energy of order E_{cb} to each steel ball. The experiment is performed at three different frequencies. In each run one observes a distinct maximum of the resistivity of the sample at a certain strength of the agitation. If no liquid is added, the resistivity of the sample is too high to obtain reasonable data. If the amount of liquid and the overall density of the steel balls are not properly adjusted, the maximum vanishes.

A compilation of a larger set of data is shown in Fig. 5.19 as a function of agitation frequency. Each data point corresponds to a maximum in the resistivity. The straight line corresponds to a constant ratio Γ/f_{sh},

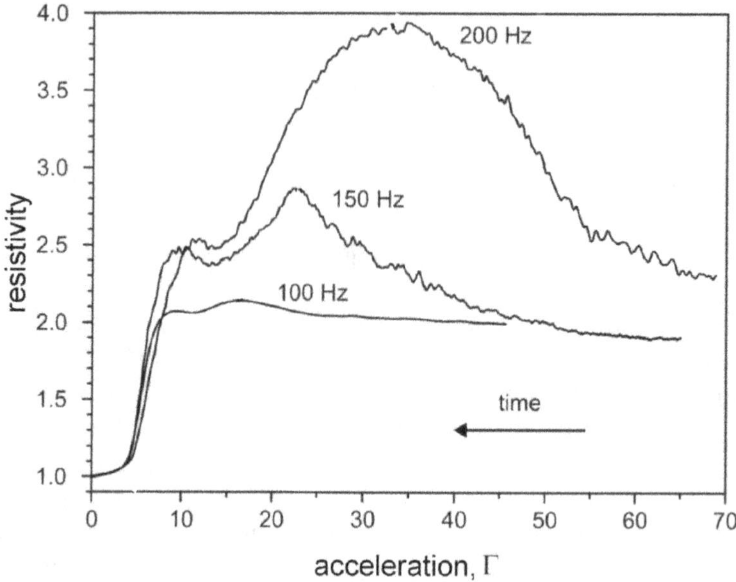

Fig. 5.18   Resistivity of the sample as a function of acceleration, Γ, scaled with respect to its value in the still (solid) state. The steep edge at small acceleration indicates the fluidization of the pile. While the granulate is in the gaseous state, capillary bridges are formed and ruptured independently of the capillary force, which is then much weaker than the kinetic forces. At the high density chosen, this creates a percolated network of bridges which gives rise to a low resistivity. As the gas is cooled down (note the arrow of time), the gas condenses, forming disjoint clusters (resistivity increases). At even lower temperature, the clusters join, such that resistance goes down again [173].

or constant $a_0 f_{\text{sh}}$, according to Eq. (5.42). Following Eq. (5.43), this corresponds to a fixed granular temperature. As a detailed analysis shows, this temperature corresponds to about one quarter of the rupture energy, $T_g \approx 0.25 E_{cb}$, which is well in line with the prediction from the equation of state we discussed in the previous section (cf. Figs. 5.15 and 5.16).

5.3.3   *Non-equilibrium phase separation*

So far, T_g has been considered to be more or less constant within the system under study. In the experiment, the container was accordingly chosen small, in order to prevent large spatial granular temperature differences from building up within the sample. Next we will investigate what happens in larger containers. We will find that the temperature is not anymore spatially constant, in strong contrast to equilibrium systems.

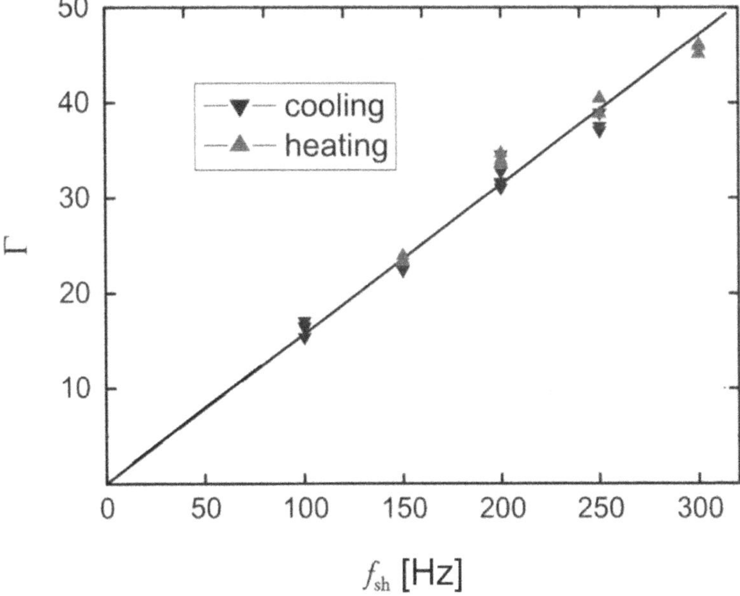

Fig. 5.19 The acceleration, Γ, at which the peak displayed in Fig. 5.18 occurs, for both rising and falling temperature ramp. The linear relationship indicates a characteristic energy at which the peak occurs [173].

This can be demonstrated, again, in a simple experiment [146, 148, 174]. A Petri dish with the lid closed is mounted on an electromagnetic vertical shaker like the one shown in Fig. 5.20. The Petri dish is filled about half way with a granulate consisting of, e.g., small glass beads (balottini) serving as roughly spherical (and therefore well defined) grains. Some liquid is added to provide the hysteretic attractive force. When the shaker is turned on and the amplitude of the (sinusoidal) vertical motion is gradually increased, the grains start to move with respect to each other, resembling the molecular motion in a regular fluid. It is found that this transition from the resting (solid) state to a fluid state takes place at a certain critical peak acceleration of the container, Γ_c, irrespective of the frequency chosen. This is indicated in Fig. 5.21 by the open squares.

It is instructive to plot Γ_c as a function of the total liquid content of the sample, W, which we define as

$$W = \frac{\text{total volume of liquid}}{\text{total sample volume}}. \tag{5.44}$$

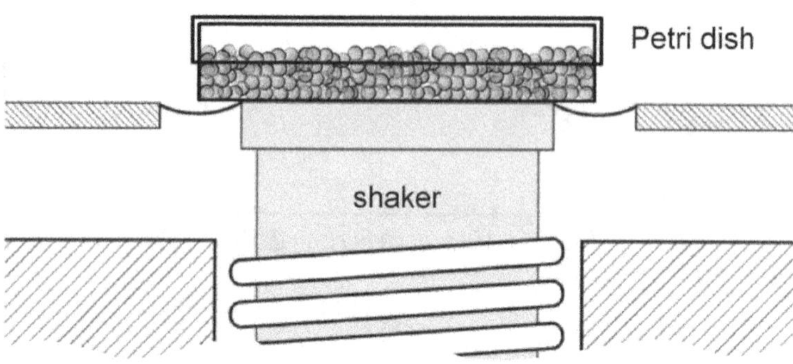

Fig. 5.20 Setup for agitating a larger sample, with optical access. A Petri dish is mounted on the electromagnetic shaker sketched in Fig. 5.17, illuminated by a strobe lamp which can be operated at the shaker frequency, f_{sh}. Images are taken with a short shutter time camera.

It is readily seen that

$$W = w\phi, \tag{5.45}$$

if ϕ is the average volume fraction of the sample. In terms of W, one obtains the result shown in Fig 5.22. As opposed to dry granulates, which fluidize at an amplitude only slightly larger than the acceleration due to gravity (corresponding to $\Gamma_c \approx 1.2$ [34], arrow in Fig. 5.22), fluidization sets in only at dimensionless accelerations as high as about 1.5, and seems to be independent of the liquid content in the wide range investigated. This is reminiscent of the above mentioned independence of the attractive capillary force, F_{cb}, of the liquid volume contained in a capillary bridge, in particular since acceleration is directly linked to force through the particle mass. In fact, a thorough investigation reveals that Γ_c depends upon the grain diameter just in the way one would expect balancing the capillary bridge force (Eq. (4.26)) with acceleration [145].

So far we conclude that there is a transition from the quiescent solid

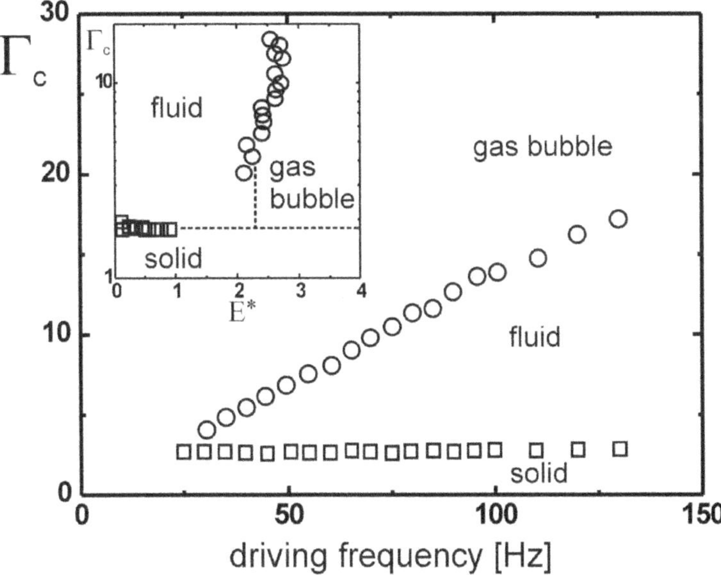

Fig. 5.21 Experimentally determined phase diagram, plotted in the plane spanned by the driving frequency and the acceleration. The sample consists of glass beads, with an average diameter of 850 μm. The liquid is a low viscosity silicon oil. The setup shown in Fig. 5.20 has been used, the shaking frequency is 80 Hz. The diameter of the dish is 14 cm. Inset: same as main panel, but with the abscissa rescaled for the agitation energy, E^*.

state to a fluidized state of about the same density, which is governed by the balance between the forces imparted on the grains by the agitation and the capillary forces attracting the grains to each other. We may call such a transition *force driven*.

As Fig. 5.21 shows, there is yet another transition as well, which is represented by the open circles. It indicates the appearance of a low density 'gas bubble', which coexists with the fluidized dense phase. Figure 5.23(a) shows a typical snapshot, taken with the camera as sketched in Fig. 5.20. Since the glass beads, which are serving as the grains here, show up as little white spots due to the reflections from the strobe lamp, we clearly recognize the low density phase as the darker disk in the center. Hence we have a coexistence between a gas (low density) phase and a liquid (mobile high density) phase. As one appreciates from Fig. 5.21, the acceleration at which this occurs is proportional to the driving frequency. As already mentioned in the previous section, this corresponds to a well defined energy

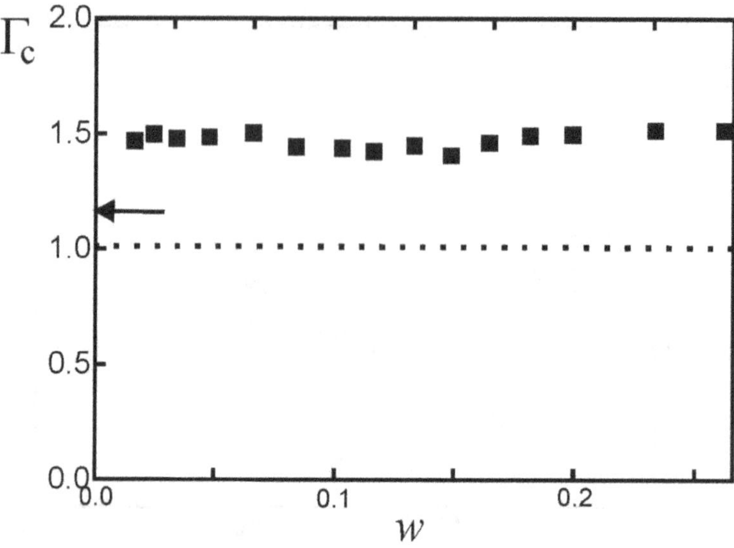

Fig. 5.22   The critical acceleration, Γ_c, for the onset of fluidization. There is no significant dependence upon the liquid content, and Γ_c lies well above the established value for dry granular matter, which is indicated by the arrow (taken from [174]).

at which the bubble appears. In the inset of Fig. 5.21, the data of the main panel are replotted as a function of a normalized driving energy,

$$E^* = \frac{E_{\text{vib}}}{E_{cb}} = \frac{2\pi^2 m a^2 f_{\text{sh}}^2}{E_{cb}}. \tag{5.46}$$

The data points representing the appearance of the gas bubble roughly follow a vertical line. In the parlance we introduced for the above mentioned transition from the solid to the fluid phase, we may call the appearance of fluid–gas coexistence an *energy driven* transition. At first glance, what we seem to observe here is again the fluid–gas coexistence predicted by the equation of state, which we corroborated above (at least as far as the critical region was concerned) by means of the described resistivity measurements (cf. Fig. 5.17). As the temperature is raised to the boiling point, part of the fluid phase 'evaporates' forming bubbles, and at globally constrained density, liquid–gas coexistence is a commonly observed phenomenon.

This is, however, not the case, as a measurement of the typical velocities of the glass beads reveals. Grain velocities can be estimated by simply correlating consecutive camera frames taken with the setup in Fig. 5.20. The result is shown in Fig. 5.23(b), where the bright dots representing the beads have been color shaded according to their kinetic energy. Red corresponds

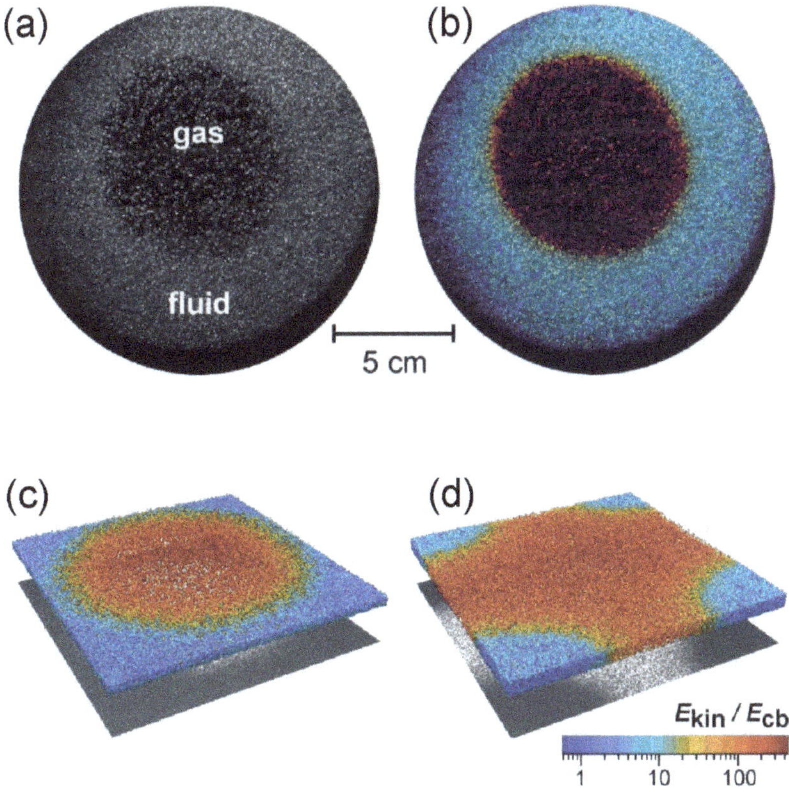

Fig. 5.23   (a) Top view of a sample of wetted glass spheres in a petri dish under vertical vibration. The phase separation into a dilute 'gaseous' phase (center) and a dense 'liquid' phase (outer regions) is clearly visible. (b) same image as (a), but with the bright spots corresponding to the grains colored according to their kinetic energy (see scale bar under (d)). (c) simulation of the experimental situation shown in (a), but in a square shaped container with periodic boundary conditions. Coloration as in (b). (d) Same as (c), but at a stage where the 'liquid–gas' interface has reached the boundary of the simulation box. The liquid 'drops' which form at each corner acquire a circular shape, pointing directly to the presence of a surface tension (adapted with modifications from [146]).

to high, blue to small kinetic energies (for colours, refer to [146]). The color scale spans more than two orders of magnitude in kinetic energy (see color scale bar at bottom of figure). This is in harsh contrast to common phase transitions in thermodynamic systems, where the temperature is always spatially constant after equilibration (equipartition theorem). The fact that the temperature of the 'gas phase' is more than two orders of magnitude higher than that of the 'fluid phase' shows that we are dealing here indeed with a phase transition *far from thermal equilibrium*. In particular, this is sharply distinct from the discussion of the equation of state, which assumed the granular temperature to be a spatially constant system parameter. The steel tube hosting our sample in the experiment revealing the critical point (cf. Fig. 5.17) was way too small to accommodate a structure as large as the gas bubble appearing in Figs. 5.23(a) and 5.23(b), and the granular temperature could be assumed reasonably uniform in that setup.

We are thus faced with a new phenomenon which is not described properly by the theory we have developed in Section 5.3.1. As there is so far no standard theoretical framework for the treatment of collective phenomena far from thermal equilibrium, we fall back on computer simulations again. For a qualitative understanding of the basic mechanisms, it is sufficient to perform simulations in two dimensions, which are much more economical computationally. We thus represent the Petri dish by an area resembling a vertical cross-section, bounded from above and from below by a rigid wall. The distance of the walls shall be several grain diameters, resembling the height of the Petri dish. In the horizontal direction, we employ periodic boundary conditions, assuming the Petri dish to be essentially infinite in diameter. It should be noted that it is necessary to fully simulate the sinusoidal motion of the container walls, because simplified agitation schemes may give erroneous results.[6] In such setting, simulations with several thousand beads can be rapidly performed, such that a full phase diagram may be obtained in reasonable time.

A critical aspect is the choice of the interaction potential, U_{cb}, by which we model the capillary bridges. In Fig. 5.5, we have seen possibilities to choose from. For the free cooling, we used the square well (pseudo-) potential sketched in the bottom right of the figure, which enabled event driven simulations, and thereby a fast procedure. Although this form has both important ingredients, E_{cb} and s_c, we cannot yet be sure that such gross simplifications with respect to the true form, Eq. (4.27), are possible without loosing some essential aspects of the physics of the system. We will therefore consider here two different forms of $U_{cb}(s)$, a 'triangular' and a 'rectangular' one, and search for differences in the simulation results.

The result is shown in Fig. 5.24. Aside from some deformations we will

---

[6]If, for instance, one implements energy injection by imparting a certain average energy on each particle colliding with the wall, one looses the distinction between acceleration and energy, which we have seen to be of central importance. The full simulation of wall motion can be implemented by using appropriate lookup tables [175].

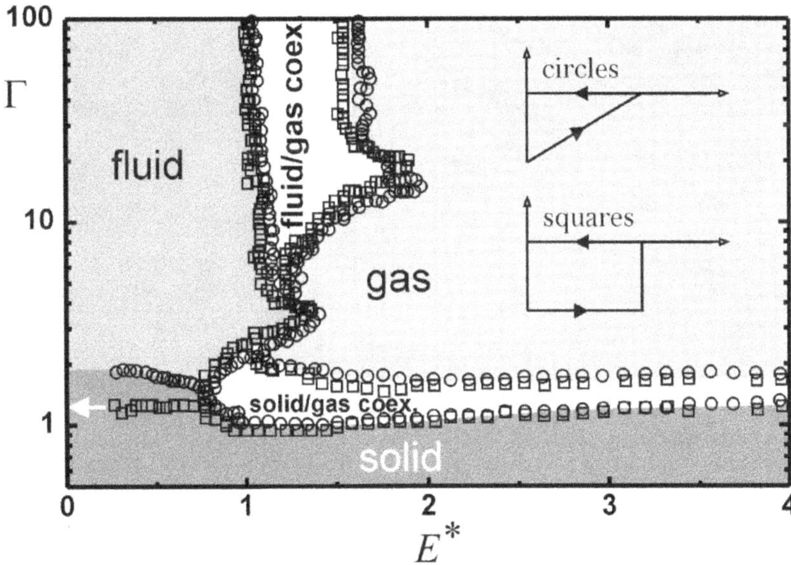

Fig. 5.24 Phase diagram of agitated wet granular matter, simulated with two different force characteristics. Remarkably, the phase boundaries obtained for a (hysteretic) square potential (squares) and for a triangular potential (circles) are very similar, even up to details. Obviously, the phase diagram does not depend significantly on the details of the interaction potential [148, 176]. The only exception is at small Γ and E^*, where the system is very dense (white arrow, see text). Adapted with modifications from [146].

shortly touch on below, we can clearly distinguish two systems of transition lines. First of all, there is a horizontal set of lines, which delimits the solid phase found at small acceleration. In the left part of the figure, this line corresponds to the transition indicated by the data in Fig. 5.22. While we connected the latter to the independence of the capillary force with respect to the liquid content, the horizontal orientation of the phase boundary in Fig. 5.24 indicates independence of kinetic energy. This corroborates the idea put forward above when discussing Fig. 5.22: fluidization takes place at a certain acceleration, corresponding to a certain force, which competes with the attractive capillary force.

Second, there is a set of vertical lines delimiting the region of fluid-gas coexistence (note the logarithmic scale). The vertical orientation indicates independence of acceleration, and hence of force, but instead singles out a certain energy at which the transition takes place. Indeed, this energy is of order unity if measured in units of the energy required to rupture the capillary bridge (definition of E^*). We thus have corroborated the

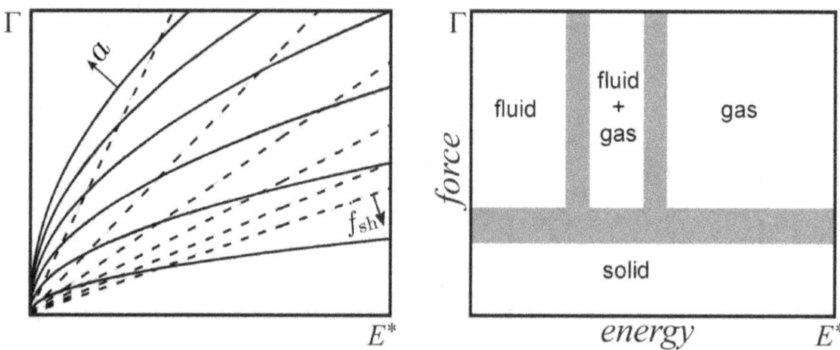

Fig. 5.25 The phase diagram of vertically agitated wet granular matter, in the plane spanned by the maximum acceleration (Γ) and the injected kinetic energy. (a) Sets of curves corresponding to constant shaking amplitude (solid curves) and constant shaking frequency (dashed curves), respectively. (b) Schematic representation of the phase diagram found in experiment as well as in simulations. There is a vertical transition structure, corresponding to an energy-driven fluid-gas transition, and a horizontal transition zone, corresponding to a force-driven solid-fluid transition.

evidence for a force-driven solid-fluid transition and an energy-driven fluid-gas transition again in a two-dimensional computer simulation. Figure 5.25 summarizes our findings again in a schematic fashion, along with a chart presenting the coordinate transform from (a, f_{sh}) to (Γ, E^*).

5.3.4 *Universal aspects of the phase diagram*

A particularly interesting aspect of Fig. 5.24 is that the two different phase diagrams, which have been obtained with different force characteristics, are lying almost perfectly on top of each other. The corresponding hysteretic 'interaction potentials' are sketched as insets. The top one, which corresponds to the open circles, is equivalent to a constant force (constant slope of the potential). Although the rectangular well potential, which corresponds to the open squares, is pronouncedly different from the triangular well, we see that the phase diagrams are in almost perfect quantitative agreement. All that seems to matter is the depth of the well, E_{cb}, and the rupture distance, s_c, which are the same for both potentials. This justifies, in retrospect, the simplifications made above with respect to the 'real' force characteristics shown in Fig. 4.9. For the rectangular potential, the force is a delta function located at $s = s_c$, but if the force is averaged over the interval $[0, s_c]$, corresponding to the triangular potential, the result is the same.

There is just one exception to this remarkable insensitivity of the system to the particular choice of the interaction potential. It concerns the position of the boundary between the solid and the fluid phase, which in the case of the rectangular potential is approximately at the same place where it would be in a dry system (white arrow in Fig. 5.24) [176]. This can be understood by appreciating that where the system is very dense, such that separations of order $s \approx s_c$ occur only very rarely, the system will not 'feel' any difference from a dry state. We will return to this transition later when discussing dense granular systems. Here we just note that we have to be careful when interpreting event-driven simulations in regimes with high packing density.

An independent experimental test of the connection of the energy driven transition to the capillary bridge rupture energy, E_{cb}, is again the variation of the liquid content. This is shown in Fig. 5.26. The kinetic energy at which the transition occurs is found to scale as the square root of W. If we assume that in our system the location of the phase boundary is determined by the cohesive properties of the dense phase, we should consider the dynamics of grains within that phase. First of all, we derive an expression for the mean

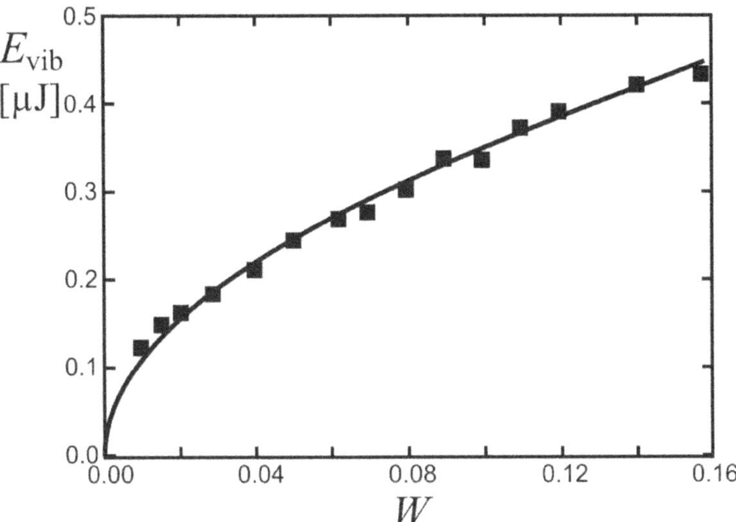

Fig. 5.26 The critical vibration energy, $E_{\mathrm{vib}} = \frac{m}{2}(a_0\omega)^2$ (where m is the mass of a grain), at which fluid–gas coexistence sets in, as a function of the liquid content of the sample. The solid curve indicates the variation of the energy required to rupture a liquid bridge (taken from [174]).

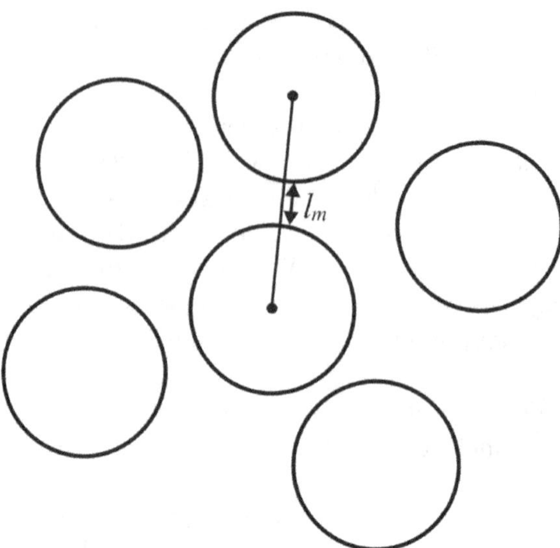

Fig. 5.27　Estimating the mean free path of grains in a strongly agitated dense granular phase.

free path at high volume fraction. From Fig. 5.27 we find

$$\left(\frac{2R + l_m}{2R}\right)^3 = \frac{\phi_j}{\phi}, \tag{5.47}$$

which immediately yields

$$l_m = 2R\left[\left(\frac{\phi_j}{\phi}\right)^{1/3} - 1\right]. \tag{5.48}$$

Now we turn to the rotations of the grains. It appears justified to assume local equipartition of translational and rotational energies. For spherical grains, we can then immediately write down the ratio of typical angular velocities, ω, divided by typical translational velocities, v. We find $\omega/v = \sqrt{5/2R^2} \approx 1.6/R$. In a dense granular gas, grains will collide with the same collision partner many times. The denser the gas, the smaller is the typical angle by which it can rotate between collisions. Combining this with Eq. (5.48), we can derive the typical angle by which a grain turns between two consecutive collisions. We obtain

$$\Delta\varphi = \frac{\omega l_m}{v} \approx 3.2\left[\left(\frac{\phi_j}{\phi}\right)^{1/3} - 1\right]. \tag{5.49}$$

The volume fraction of the dense phase is very close to its maximum possible value, ϕ_j. Hence $\Delta\varphi$ is small, and neighboring grains are likely to collide in the same contact region in consecutive collisions. As a consequence, the little heaps of liquid which remain after each collision, as sketched for $t > t_4$ in Fig. 5.2, will accumulate with time, such that in effect one has on the regions of collision a similar amount of liquid as V_f, instead of V_i. This is important to appreciate, since V_i and V_f scale differently with W. According to the above, we should expect the liquid volume at the areas of collision to scale linearly with W (as V_f does). Hence the critical vibration energy is approximately proportional to $\sqrt{V}$, just as the rupture energy E_{cb}. This is what we see in Fig. 5.26.

There is yet another important, and quite surprising, aspect to the experimental results shown in Figs. 5.22 and 5.26. A glance to the scale of the abscissa reveals that both plots extend over a very large range of liquid content. The complete saturation of the sample with liquid would correspond to $W \approx 0.4$ (or $w \approx 0.67$, respectively), and the abscissa reaches almost half that value. We have mentioned above that at a bridge angle $\beta = \pi/6$, adjacent capillary bridges merge forming larger liquid clusters. As we will see in Chapter 6, this corresponds to $W = 0.026$. We can thus be sure that for the majority of data displayed in both plots, we are dealing with liquid clusters larger than single bridges. The obvious commonality in the locations of the phase transitions, which are just as expected from systems with individual capillary bridges, calls for an explanation.

It is enlightening to study the forces occurring in a liquid cluster connecting three spherical grains, as opposed to the two grains considered so far in the discussion of liquid bridges. This is illustrated in Fig. 5.28. On the upper left, we see an optical micrograph of three glass spheres, connected by a liquid cluster. The forces which are present in this system can be studied by computing the total interfacial energy as a function of the liquid volume and the relative coordinates of the grains. Such computation can be carried out again by means of the Surface Evolver [95], which we have already used in Section 3.2.4. The results are shown in the three remaining panels of the figure. As it is shown, there does not seem to be any cross-dependence of the inter-granular forces. The force between grains 1 and 2 is independent of the distance between grains 2 and 3, or 1 and 3. Furthermore, there is no strong dependence of the rupture distance for a pairing upon the distances of one of the other pairings.

In other words, the result is consistent with the forces being additive to good approximation [177]. We will find in the next chapter that there are good reasons to believe that the energy landscape of larger clusters can be quite well inferred from the energy landscape of a trimer cluster,

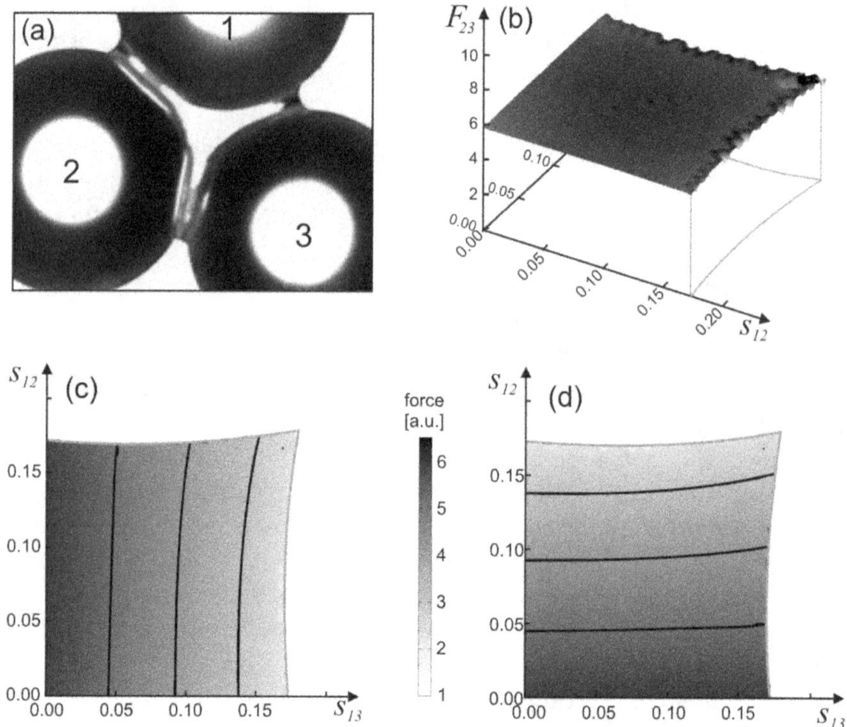

Fig. 5.28　Forces exerted by a trimer of merged capillary bridges. (a) Optical micrograph of a trimer of merged capillary bridges between glass spheres (image courtesy of Mario Scheel and Martin Brinkmann). (b) The force between spheres 2 and 3, F_{23}, calculated for an idealized system of three spherical grains with identical radius by means of the Surface Evolver as a function of the separations s_{12} and s_{13} (axis to the back). (c) and (d) F_{13} and F_{12}, respectively, as functions of s_{12} and s_{13}, encoded as grey levels. Their contour lines are almost parallel to the axes. A more complete investigation corroborates that the three forces are almost entirely independent, as if there were three individual capillary bridges (data and graphs courtesy of Martin Brinkmann and Ciro Semprebon, Göttingen) [177].

such as the one studied in Fig. 5.28. Hence we see that even for larger liquid content, a wet granulate seems to behave just as if all forces were two-body forces, following the characteristics derived in Chapter 4. This suggests that wet granular matter may in fact be simulated quite reliably with simple, two-body force models, even for rather high liquid content.

5.3.5 *The interplay of dissipation mechanisms*

So far we have concentrated on the effect of hysteretic capillary bridges as the cause of energy dissipation in the granular system. In a real granular gas, the inelasticity in the inter-grain contact, which is present as well in a dry granular gas, must as well be taken into account. Similar to the case of free cooling, it is very instructive to discuss the characteristic differences in the impact these two dissipation mechanisms have on the phase diagram.

Figure 5.29 presents again the phase diagram for a wet granular gas as obtained with two-dimensional simulations as in Fig. 5.24, using the 'triangular' pseudo-potential (circles in Fig. 5.24). The open symbols represent the same conditions as before, with perfectly elastic grains. For the closed symbols, inelasticity has been introduced ($\varepsilon = 0.8$). Clearly, the structure of the phase diagram as sketched in Fig. 5.25 is retained, with the two vertical phase boundaries between the liquid phase, liquid–gas coexistence, and

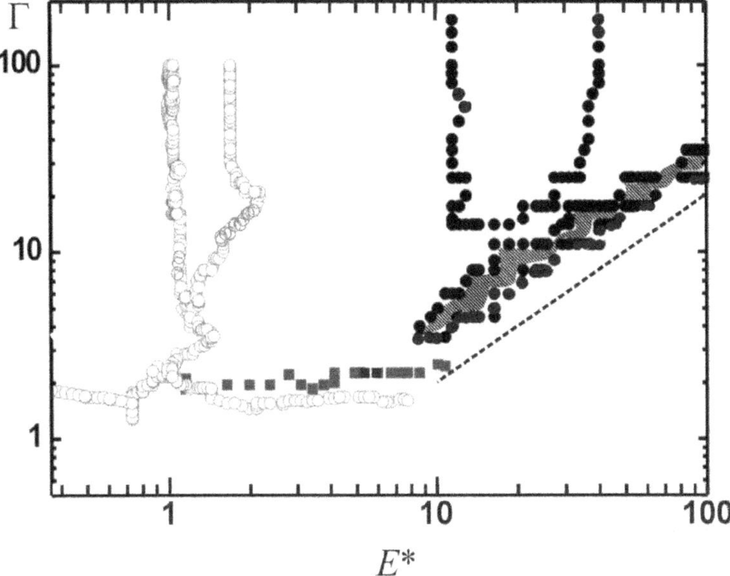

Fig. 5.29 The phase diagram of wet granular matter, determined by means of two-dimensional simulations. Open circles: perfectly elastic grains, $\varepsilon = 1$ (same as Fig. 5.24). Closed circles: inelastic grains, $\varepsilon = 0.8$. Clearly, the general structure as sketched in Fig. 5.25 is not changed by the inelasticity. The additional two-phase coexistence region (shaded in grey) roughly follows a line of unity slope (dashed line), thus representing a contour line of the excitation amplitude (data courtesy of Klaus Roeller).

the gas phase just shifted to higher excitation energies. While the boundary of the liquid phase is around $E^* = 1$ for the elastic case, it is shifted substantially above unity as inelasticity is introduced. A glance at Fig. 5.21 reveals that this is indeed what one observes in experiment. As the inset shows, the boundary between the liquid phase and liquid–gas coexistence lies well above unity here as well, between 2 and 3.

This behavior can be qualitatively understood as follows. The transition from the liquid phase to liquid–gas coexistence takes place when the energy injected into the pile of grains is sufficient to break the individual bonds and 'evaporate' the dense phase, thus forming the 'gas bubble'. At first glance, one might assume that in order for this to happen, it is sufficient to vibrate the bottom plate with a peak velocity equal to the velocity of a grain which has the kinetic energy E_{cb}. Since only few capillary bridges will break within the dense phase, the granular temperatures at the bottom of the pile and its top will rapidly equilibrate following the equipartition theorem, such that the grains at the top layer will as well acquire enough kinetic energy to leave the pile and form a dilute, gaseous phase. This reasoning is correct as long as the grains are perfectly elastic.

If, however, the grains are inelastic, it is clear that the energy injected at the bottom of the Petri dish will not propagate completely to the top of the pile, because a considerable amount of energy is dissipated in inelastic impacts between grains within the pile. This is sketched in Fig. 5.30. For the top layer to reach sufficiently high granular temperatures for the formation of a bubble, the excitation must be much stronger, which results

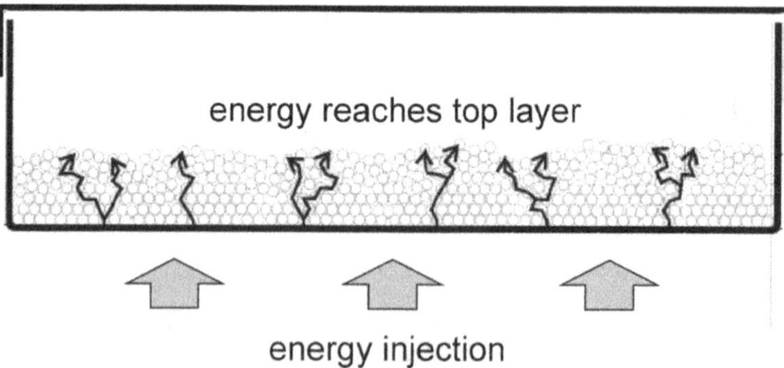

Fig. 5.30   Sketch of a granulate in a container, into which energy is injected by vertical vibration. Clearly, a significant amount of the energy is dissipated in inelastic collisions between grains before it reaches the top of the pile, where grains can 'evaporate' [148].

in the observed shift of the phase boundary. In fact, one observes that the shift increases when the depth of the pile increases, in accordance with the idea sketched in Fig. 5.30. When the injected energy falls below the value which assures sufficiently high granular temperature at the top of the liquid phase, the latter serves as an efficient sink to the particles in the gas phase, and the bubble collapses again.

This suggests that gas bubbles can nucleate even before the phase boundary is reached, since there are surface ripples being excited on the fluidized pile by the agitation, similar to thermally excited surface ripples on an ordinary liquid. Once the valley of one of these excitations reaches deeply enough into the pile, where the granular temperature is sufficient for rapid evaporation of grains into the gas phase, a bubble can form spontaneously. In fact, this is observed in experiments. Bubbles sometimes nucleate before the agitation energy corresponding to the phase boundary is reached, with a typical nucleation time which strongly increases away from the boundary, towards lower agitation energy.

As we have mentioned, the shift of the phase boundary, and thereby its exact position, depends upon the depth of the dense granular pile. Since the latter is an extensive variable concerning the geometry of the experiment, the position of the phase boundary is not as intrinsic to the system itself as one is used to in equilibrium phase transitions. However, the overall structure of the phase diagram is very robust, and does not depend on the details of the setup.

Interestingly, there is an additional feature emerging in the phase diagram as inelasticity is introduced. Close to the dashed line in Fig. 5.29, and roughly following that line, a second region of liquid–gas coexistence opens up, which is indicated in dark grey. This is entirely absent in the elastic case, where we just find an anomaly in the right boundary of liquid–gas coexistence. That this is not just an artifact from the simulation can be shown by performing experiments in an appropriate range of parameters and comparing them with simulations. Figure 5.31 shows simulations where the 'new' coexistence region has merged with the one discussed before, resembling the foot of a snail below its shell. As the inset shows, the same 'snail's foot' structure, as we will henceforth call it, is observed in the experiment as well. Note that Fig. 5.31 is semi-logarithmic, which gives rise to a curvature of the 'snail's foot', as opposed to Fig. 5.29.

It turns out that the snail's foot is present as well in the dry system. It had been overlooked for some time, because this region of coexistence is very narrow if the grains are not extremely inelastic. We will discuss this

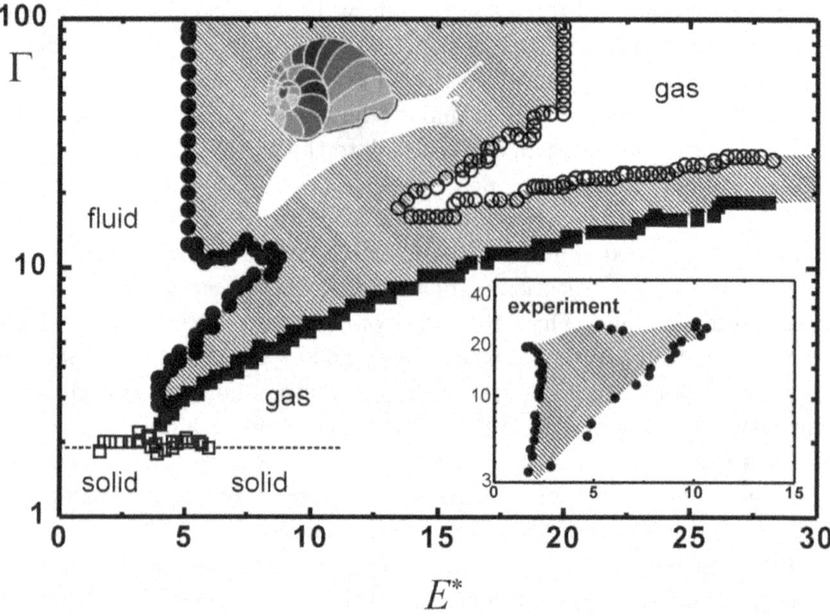

Fig. 5.31   The phase diagram of vertically agitated, wet granular matter with inelasticity. Main panel: event driven simulations with finite inelasticity. Inset: experimental results. Hatched: fluid–gas coexistence. In both cases, the 'snail's foot' is clearly visible (simulation by Klaus Roeller, experimental data by Kai Huang).

feature of the phase diagram in the next section in more detail. Here we restrict ourselves to rationalize why it follows so closely the iso-lines of the vibration amplitude. The answer is in fact quite straightforward. In the dry system, there is no inherent energy scale, nor a force scale. The only characteristic quantities of a dry granular gas are the restitution coefficient and the size of the grains. The former is dimensionless, so the only scaling parameter which can couple to experimental conditions is the grain size, which has the dimension of a length. Hence in the plane spanned by the frequency and the amplitude, such as the phase diagrams we have discussed (e.g., Fig. 5.21), only the amplitude can be a relevant parameter. That the amplitude at which the transitions occur happens to be of the same order as the grain diameter is not so readily explained.[7]

---

[7]Strictly speaking, the argument is incomplete, since the presence of the acceleration of gravity introduces both an anisotropy into the system and a time scale, $\sqrt{R/g}$. Hence a dependence on frequency would in principle be conceivable. As long as Γ is large as compared to unity, however, this is not expected to be physically significant.

5.3.6 *Interfaces and interfacial tensions*

So far, we have mainly discussed simulations in two dimensions, for reasons of computational economy. In order to compare in some more detail, and more quantitatively, with the experimental findings, it is necessary to perform simulations in three dimensions. These are possible with the event driven scheme, which we have seen to work well as long as we are careful when interpreting details of the emerging structures at high density. Figures 5.23(c) and 5.23(d) show the corresponding result. The circular Petri dish has been replaced with a square shaped box, with periodic boundary conditions in both horizontal directions. We see that the boundary between the dense and the dilute phase seems to exhibit a surface tension, as the gaseous bubble in the center acquires a circular shape. While in the experiment (top panels of the figure) one may still suspect the circular bubble to simply reflect the circular shape of the Petri dish, this cannot

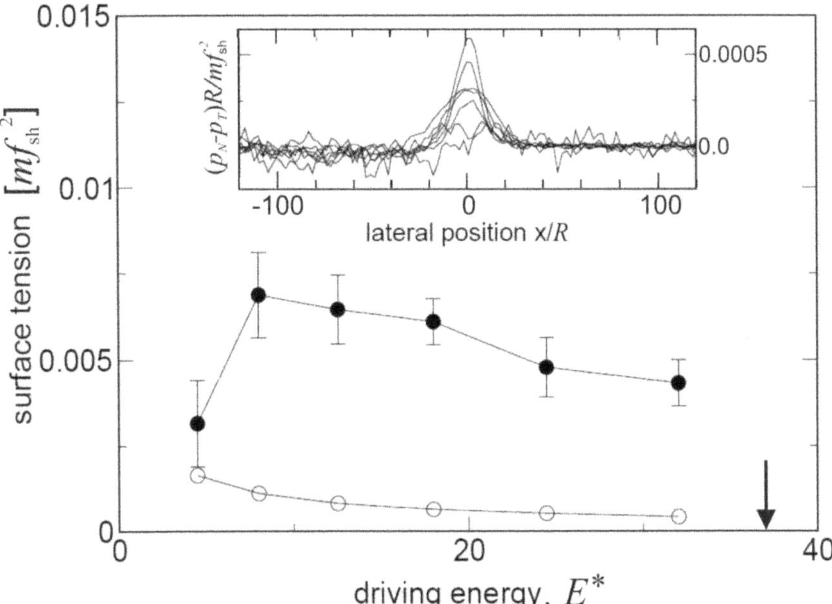

Fig. 5.32 Surface between the dense and dilute phase, determined by simulation (filled circles). The open circles provide an upper bound for the contribution from the capillary forces. The main contribution clearly originates from another source. The arrow indicates the driving energy at which both phases become indistinguishable (critical point). Inset: difference of the normal and transverse components of the pressure tensor (integral over the normal coordinate yields the surface tension [178]).

be the cause in the simulation, since the latter is performed in a square shaped domain with periodic boundary conditions. Figure 5.23(d) shows what happens when the dilute bubble is large enough to reach the boundary of the simulation box. Due to the periodic boundary conditions, we have now individual dense 'droplets' sitting on the corners of the simulation box, with one quarter of a droplet being visible in each of the corners. Obviously, these droplets as well attain a circular shape, thus minimizing their interface with the dilute phase.

The surface tension of this interface can be calculated from the components of the pressure tensor, P_{ij}, which can be obtained directly from the simulation data [178, 179]. The inset of Fig. 5.32 shows the difference between the normal and transverse components of the pressure tensor (p_N and p_T), plotted as a function of the distance from the interface between both phases. This must be integrated to yield the corresponding surface tension [180],

$$\text{surface tension} = \int (p_N - p_T)\, dz, \tag{5.50}$$

where $p_T = \frac{1}{2}(P_{xx} + P_{yy})$ and $p_N = P_{zz}$. The z-direction is normal to the surface. It is plotted as the full symbols in Fig. 5.32, as a function of the driving energy, in units of $m f_{sh}^2$.

It is instructive to compare this with what we would expect on the basis of the capillary bridges. The excess energy per unit area can be estimated form the number of bridges which must be broken when a dense pile representing the condensed phase it being dissected into two parts. This is easy to calculate, since we have the density of the bridges from the simulation data. The result is plotted as the open circles. As we clearly see, the actual surface tension is, for almost all agitation energies investigated, much larger than the contribution from the attractive capillary force alone. It obviously contains non-trivial contributions the origin of which is hidden more deeply in the microscopic dynamics of the system.

This is a very interesting point, since in equilibrium systems the surface tension is defined as the excess free energy per unit area. However, in a system far from thermal equilibrium, there is no straightforward definition of a free energy, and it is of great interest in what sense an analogous quantity can be defined here. This also touches the question if there is a law of corresponding states, or even some form of universality in phase transitions far from thermal equilibrium, similar to the well established universality classes known from equilibrium phase transitions. We will discuss these aspects in more detail in the next section.

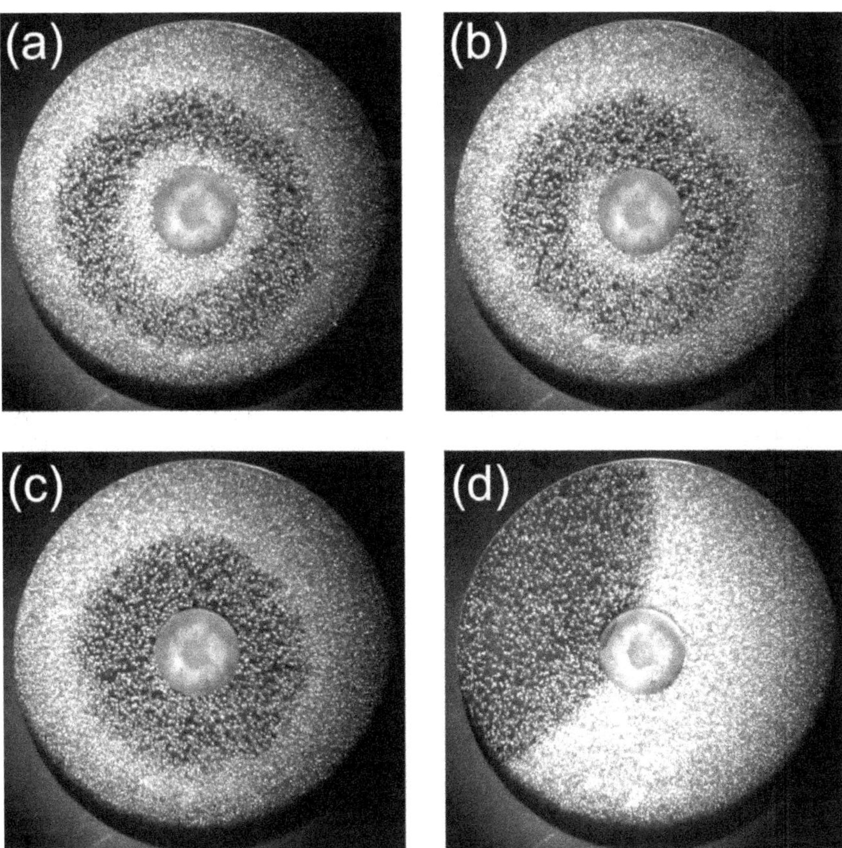

Fig. 5.33   Meta-wetting: configurations of the interface between the dense and dilute phases in a Petri dish with a cylindrical post as extra geometric feature. (a) A bubble nucleates which tries to stay away from the walls, forming a thin layer at the perimeter of the post. (b) The layer at the central post evaporates, by virtue of its convex shape, and thus higher vapor pressure. (c) The thin layer at the post is gone. (d) Here an asymmetric bubble has formed. Note that the contact angle the interface makes with the walls is nowhere exactly equal to 90 degrees (images courtesy of Kai Huang, Göttingen/Bayreuth).

That we are really dealing with physically meaningful interfacial tensions can be corroborated experimentally again in a rather simple way. In Fig. 5.33, an experiment is shown similar to the one of Fig. 5.20, but with a cylindrical intruder placed in the center of the Petri dish. When the agitation amplitude is raised such that the system is driven into the fluid-gas coexistence region, the bubble nucleates preferentially away from the walls.

As time proceeds, however, the 'wetting layer' on the intruder shrinks, while condensed phase keeps accumulating on the outer walls. This is understood easily with the help of a surface tension. According to the Kelvin equation, the saturated vapor pressure of a convex surface of a condensate is larger than that of a concave surface. Consequently, material will evaporate from the inner wetting layer and condense on the outer one, until the inner wetting layer vanishes (Fig. 5.33(c)).

Finally, we note that the interfacial tensions we are encountering in the system are far from trivial quantities. This is appreciated by a glance at Fig. 5.33(d), which shows a late stage of an experiment similar to the one in the three previous panels. Here the interface is pinned to the central intruder. As there should be no tangential force exerted by the capillary bridges on the walls, one might naively expect that the liquid–gas interface should hit every wall at a right angle. However, the fact that the fluid–gas boundaries do not extrapolate precisely to the center of the intruder, as well as the slight curvature of these interfaces, suggest that the contact angle the interface makes with the container and intruder walls is not equal to ninety degrees, but slightly smaller.

5.3.7 *Binodals and spinodals, wet and dry*

We have seen above that in an agitated wet granular gas we can under certain conditions observe phase separation into a dense and a dilute phase, and that there seems to be a surface tension associated with the interface between these phases. In order to investigate this finding in some more depth, it is instructive to vary the total (average) density of particles in the Petri dish. What we expect is to find a well-defined region of two-phase coexistence, and some analogy with the binodal and spinodal lines which are well known from equilibrium systems. It is then of particular interest how the absence of thermal equilibrium affects the overall behavior.

The result, obtained again with three-dimensional computer simulations, is presented in Fig. 5.34. It shows the phase diagram in the plane spanned by the spatially averaged density (i.e., the volume filling fraction) and the agitation energy. The data indicated by the squares correspond to the binodal line in a regular binary fluid, i.e., these represent the two densities of the dense and dilute phases, respectively, which are found in the steady state. If we want to reconcile this with the phase diagrams in the form of, say, Fig. 5.24, we have to consider a fixed average volume fraction of the sample cell. If we choose, for instance, $\phi \approx 0.1$, we see from Fig. 5.34

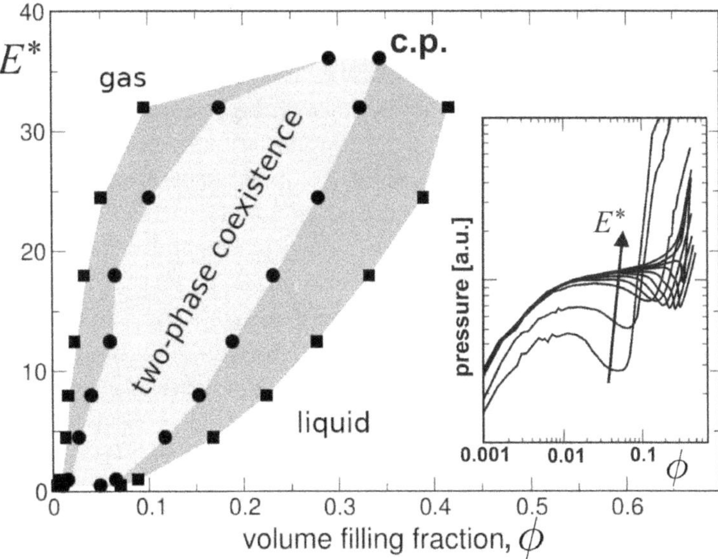

Fig. 5.34   Phase diagram of a wet granular gas, obtained from simulations in three dimensions. The abscissa is here the average filling fraction of the sample. When the
sample is prepared with a filling fraction corresponding to the light grey area, spontaneous phase separation via unstable fluctuations is observed (spinodal decomposition).
From the dark grey areas, phase separation rather resembles a nucleation scenario. c.p.,
critical point. Inset: the pressure as a function of filling fraction, for different strength
of driving (redrawn with modifications from [178]).

that we enter the coexistence region when $E^* \approx 1$. This corresponds to
the vertical phase boundary in Fig. 5.24 at $E^* \approx 1$. As E^* is increased
further, we leave the two-phase coexistence again, entering the gas phase
at a somewhat higher agitation energy. In Fig. 5.34, this corresponds to
the upper boundary of the coexistence region, which for $\phi = 0.1$ lies at
about $E^* \approx 35$. Depending on the amount of granulate in the sample
cell, the width of the coexistence region can vary dramatically. This can
be appreciated as well from the strongly different widths of the liquid–gas
coexistence regions in Figs. 5.24, 5.29, and 5.31. We see that although the
phase diagrams of the type sketched in Figs. 5.25 and, on the other hand,
Fig. 5.34 look very different, they are easily reconciled. As Fig. 5.34 shows,
we obtain a structure similar to the equilibrium case, with a critical point
(c.p.) at a certain excitation energy, E^*_{cp}. If the initial density is prepared at
a value lying between the filled circles (light grey area), fluctuations in the
density are found to be unstable, leading to immediate phase separation.

If however, the initial density lies within one of the stripes between the squares and the circles (dark grey areas), a minority phase must nucleate for phase separation to proceed. Hence the two boundaries indicated by the full circles represent the spinodal lines of the phase diagram.

All this seems to indicate that our system is very reminiscent of a regular molecular fluid close to its liquid–gas phase separation. Specifically, we expect the phase diagram shown in Fig. 5.34 to correspond directly to the equation of state we have derived and experimentally corroborated in Section 5.3.1. In order to test this expectation, we can perform simulations in just a small volume, which does not leave enough room for phase separation to occur. This resembles the small (tubular) container we have used in the corresponding experiment (cf. Fig. 5.17). The inset of Fig. 5.34 shows simulation results obtained for the pressure, plotted in log-log fashion as a function of the density. Qualitatively, we obtain indeed the same structure as displayed in Figs. 5.14 and 5.15: a monotonous increase for high excitation energy (or granular temperature, respectively), and a distinctly non-monotonous behavior for smaller values of E^*, corresponding to temperatures below the critical point. In some region of densities, the pressure decreases with increasing density, which leads to phase separation.

However, a closer look reveals an important difference with respect to Fig. 5.14. The densities at which the slope of the curve becomes negative are extremely small. Negative slopes occur first at densities well below 0.1, which must be compared to the density of random close packing, which is at about 0.64. In Fig. 5.14, which shows the equation of state in the two-dimensional system, the dip in the pressure occurs very close to the maximum density, which lies at about 0.9. The physical reason in that case was easy to see: only when the grains are very close to each other, such that the typical separation of adjacent grain surfaces are smaller than the rupture distance, $s_c \ll R$, there can be many capillary bridges, and thus a noticeable influence of their attractive force on the pressure. In the experiments we performed for corroborating the critical point, we did not only use a small container to prevent large-scale phase separation; we also prepared the system at a high volume fraction, thus enabling the detection of the clustering effects as the temperature dropped below its critical value, $T_c \approx E_{cb}/4$.

But what then is causing the pressure to decrease with volume fraction, as seen in the inset of Fig. 5.34? In order to try to answer this question, we shall remember that our system is dissipative. The granular temperature will thus not be determined by equilibration and equipartition, as for an energy conserving system. In the latter, we can assume that the particles attain the characteristic kinetic energy per degree of freedom (*vulgo*: temperature) from equilibration with the walls, i.e., attain their temperature. In a dissipative system, however, we have to replace energy equipartition with a balance between energy injection and energy dissipation, and the granular temperature will adjust such as to meet this balance condition. We have chosen the dimensions of the simulation box used to obtain the inset in Fig. 5.34 such that its lateral dimensions were small, in contrast to the 'Petri' dish simulations as for Fig. 5.24. This allowed to prevent lateral phase separation. The height of the simulation box was chosen large enough to resemble the natural height of the Petri dish. Thus the bulk phase is sufficiently decoupled from the walls, such that the vibration does not directly determine the temperature, but rather the rate of energy injection.

We shall try to derive what we should expect in this case. If the granular gas is very dilute, the grains collide virtually only with the container walls. If they receive an extra energy of order E_{vib} at each such collision, they must attain a granular temperature of order $E_{\mathrm{vib}}/(1 - \varepsilon^2)$, which can be much larger than E_{vib}. Now let us increase the density such that the mean free path of the grains, $l_m \approx R/6\phi$, becomes smaller than the vertical extension of the container. Collisions between grains then start to dominate over impacts with walls, and the granulate is cooled much more effectively due to the higher collision rate. The total frequency of collisions with walls, f_{wall}, is related to the total number of collisions in the bulk, f_{coll}, via

$$f_{\mathrm{wall}} = \frac{2l_m}{H} f_{\mathrm{coll}} = 2 f_{\mathrm{coll}} \mathsf{Kn}, \tag{5.51}$$

where $\mathsf{Kn} = l_m/H$ is the Knudsen number of the granular gas in the simulation box of height H. For the energy injected per unit time, we make the ansatz

$$E_{\mathrm{inj}} = f_{\mathrm{wall}} E_{\mathrm{vib}}, \tag{5.52}$$

while for the dissipated energy, we can write

$$E_{\mathrm{diss}} = f_{\mathrm{coll}}[(1 - \varepsilon^2)T_g + E_{cb}]. \tag{5.53}$$

Since we consider now a regime of rather low density, we can assume that every collision will be accompanied by the rupture of a capillary bridge, resulting in a unity coefficient of E_{cb} in this expression. Demanding that $E_{\mathrm{diss}} = E_{\mathrm{inj}}$, we immediately obtain

$$T_g = \frac{2E^* \mathsf{Kn} - 1}{1 - \varepsilon^2} E_{cb}, \tag{5.54}$$

for the granular temperature which will be established in the bulk in steady state. Since $\mathsf{Kn} = R/6H\phi$ strongly decreases with increasing ϕ, it follows from Eq. (5.54) that as the density increases, the system shifts from a high-temperature branch of the equation of state to a lower temperature branch. As a consequence, the pressure is reduced. In this picture, the negative slope in $p(\phi)$ should occur when the mean free path of the granular gas becomes significantly smaller than the vertical extension of the container or, in other words, as the Kundsen number falls well below unity.

It has probably not escaped the reader's attention that the above argument is not restricted to wet granular gases. It works equally well for the dry case! We then have

$$E_{\text{diss}} = f_{\text{coll}}(1 - \varepsilon^2)T_g, \qquad (5.55)$$

which leads to

$$T_g = \frac{2\mathsf{Kn}E_{\text{vib}}}{1 - \varepsilon^2}. \qquad (5.56)$$

Obviously, we have again the possibility of a strongly decreasing temperature (and thereby decreasing pressure) with increasing density, and therefore phase separation. All that has changed with respect to Eq. (5.54) is that the scaling with respect to E_{cb} is now removed, and the second term in the numerator has vanished. This does not change the qualitative behavior, which is determined by the decrease of Kn.

We should take Eqs. (5.54) and (5.56) with a grain of salt. Their derivation is very simplistic indeed, and serves only to rationalize the idea of a strong decay of T_g with increasing density. A more accurate derivation [173], which leads to an even stronger decrease of T_g with density, is tedious and beyond the scope of this book. It can, however, be qualitatively appreciated from the difference between the expressions for the mean free path, l_m, we derived for small ϕ (Eq. (2.5)) and for large ϕ (Eq. (5.48)) [11].

The behavior of the temperature as just discussed takes us back to the 'snail's foot' we have discussed above (Fig. 5.31), which represents a liquid–gas phase separation phenomenon in its own right, without capillary bridge forces being at work. In fact, Fig. 5.35 shows that by varying the overall density, as we did for obtaining Fig. 5.34, we obtain a very similar structure. The most prominent difference is that we now plot the vibration amplitude on the ordinate instead of the vibration energy, for reasons we have explained already above (end of Section 5.3.5).

The inset of Fig. 5.35 shows again the pressure as a function of volume fraction. For amplitudes below ($a = 0.5R$) and above ($a = 4.26R$) the

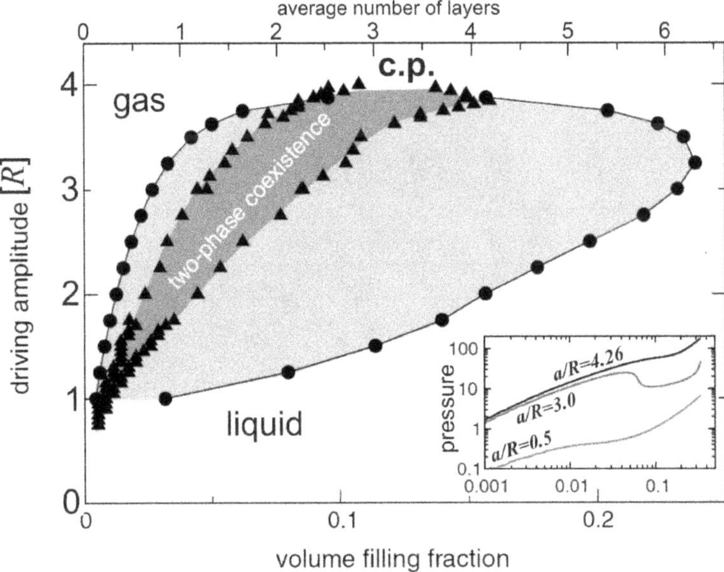

Fig. 5.35 Phase diagram of a dry granular gas, obtained from simulations in three dimensions. The abscissa is again the average filling fraction of the sample. The ordinate is the amplitude of agitation in units of the grain radius. The meaning of the light and dark grey areas is as before. Inset: the pressure as a function of filling fraction, for different strength of driving [181] (redrawn with modifications from [178]).

coexistence region, the pressure curve is monotonous. In the intermediate range (e.g., at $a = 3R$), it is non-monotonous as before. A pronounced difference with respect to the wet case, however, is the absence of the very sharp dip occurring at high densities. This was attributed to the pressure reduction from the capillary bridges, which is of course absent in the dry granular gas.

Let us now try to put everything we have into one diagram. The three important variables we can control in experiments and simulations are the acceleration of the vibration, Γ, the rescaled vibration energy, E^*, and the filling fraction of the sample cell, ϕ. Figure 5.36 shows a schematic representation of the full phase diagram of a wet, inelastic granular gas under external agitation in the space spanned by these three parameters. We find two distinct regions where liquid–gas coexistence occurs. They show up as two prisms[8] with cross-sections of distorted circles. One lies parallel to

---

[8]We use here the mathematical notion of a prism, as a body resulting from an arbitrary shape in the plane, extruded into the third dimension.

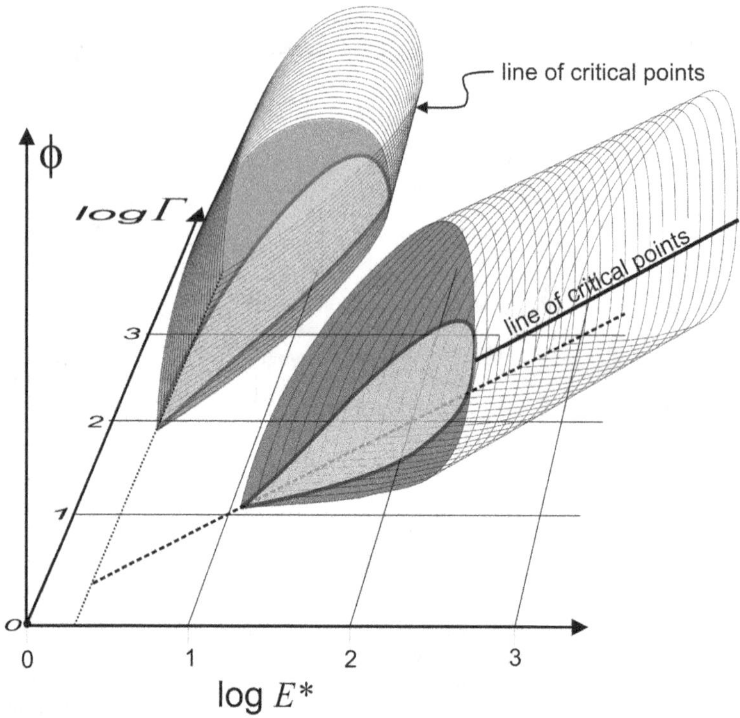

Fig. 5.36 Schematic representation of the full phase diagram of a wet, inelastic granular gas under external agitation (no thermostat). Note the log–log scale of the (E^*, Γ) plane. There are two regions of liquid–gas coexistence, which show up as the two prisms with cross-sections of distorted circles. One lies parallel to the Γ axis, the other lies parallel to the dashed line, which represents a line of constant vibration amplitude. The prisms have been cut at the fore end, showing the phase diagrams from Figs. 5.34 and 5.35, with the spinodals delimiting the brighter grey regions. The exact position of the two prisms depend on ε and s_c. In the left foreground, they merge forming a complex structure, which has so far not been investigated.

the Γ axis, and represents phase separation due to the attractive capillary forces. The other lies parallel to the dashed line, which represents a line of constant vibration amplitude. This is governed by the balance between injected and dissipated energy. It represents the phase separation which is observed as well in dry granular gases provided ε is smaller than unity. The prisms have been cut at the fore end, showing the phase diagrams from Figs. 5.34 and 5.35, respectively. The spinodals are shown on the sections, as lines delimiting the brighter grey regions.

The exact position of the two prisms depend on ε and s_c. The former shifts the prism parallel to the Γ axis, the latter shifts the prism following lines of constant amplitude. In the left foreground, both prisms merge, forming a complex structure. It is clear that for vanishing s_c as well as for $\varepsilon = 1$, one of the two prisms does not show up. Each of them has a line of critical points on one of its vertical slopes. The phase diagrams shown in Figs. 5.24, 5.25, 5.29, or 5.31 can be easily imagined as horizontal sections through Fig. 5.36.

We should mention explicitly that the phase transition and two-phase coexistence described by the equation of state, Eq. (5.41), cannot appear here. This is for a system with thermostat, where the granular temperature can be assumed to be the same all over the system. The setup used for the experiment described above to corroborate the critical point (Section 5.3.2) was designed such as to come as close as possible to a thermostated system: not only was the separation of the walls as small as feasible, but there was also the inner electrode wire, which was vibrating in phase with the walls and thus imparting the same oscillatory motion on the grains in its neighborhood. Here, in contrast, we are dealing with a completely different geometry. The container is large in the lateral directions, and the top and bottom walls are farther away. This ensures what we have assumed above when deriving expressions for the bulk temperature which will establish itself: that there is an energy injection from the walls, but the system is free to react on that injection, adjusting the granular temperature in the bulk itself by balancing energy injection with energy dissipation.

5.3.8 *Surface tension of granular gases*

Let us finally have a look at the surface tension in the dry system, which might be present in analogy to what we have observed in the wet granular gas. In fact, a glance at the inset of Fig. 5.37, which shows again a three-dimensional simulation, demonstrates the same tendency to form a circular gas bubble, thereby minimizing the total area of the liquid–gas interface. As before, we can integrate the difference of the normal and tangential parts of the pressure tensor to determine the surface tension quantitatively. For the simulation shown in the inset, we find a value of 0.94 ± 0.05 mN/m. In order to test again whether this can be considered a surface tension in the true physical sense, we can measure (i.e., calculate from simulation data) the pressures inside and outside the bubble. As discussed in Section 3.1.6. The difference of these pressures should scale linearly with the curvature of the phase boundary, with the surface tension as the prefactor (Laplace

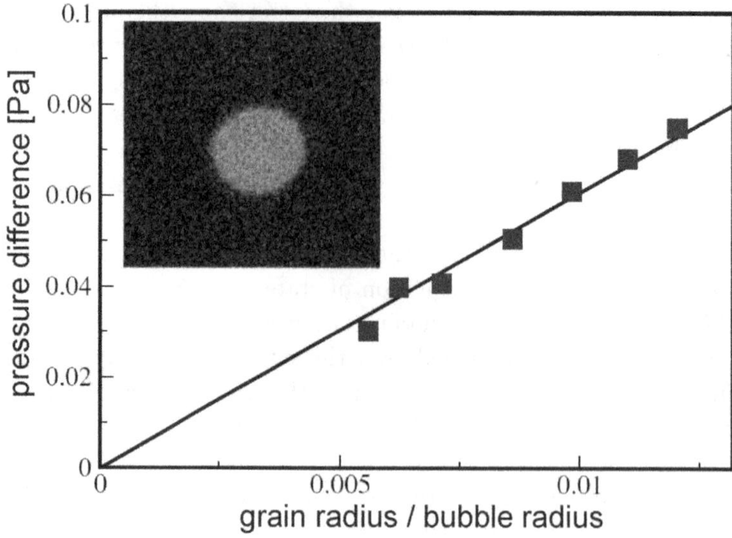

Fig. 5.37   Pressure difference between inside and outside of a gas bubble in a computer simulation, plotted as a function of inverse bubble radius. According to Laplace's equation, the slope yields the interfacial tension. Redrawn with modifications after [181]. Inset reproduced with permission. Copyright (2012) American Physical Society.

pressure). In the main panel, the pressure difference is plotted as a function of the inverse bubble radius, and thus the (mean) curvature of the phase boundary. The data are well fitted by a straight line through the origin, as expected for the Laplace pressure. The slope of the line corresponds to an interfacial tension of 1.0 ± 0.1 mN/m, in very good agreement with the value above.

Even though the physical origin of the surface tension is even more elusive here than it was in the wet case, since there is no cohesive force in the dry system, the good agreement reported above gives us enough confidence to push the concept of surface tension even farther. From equilibrium systems it is well known that when one quenches the system into the coexistence region, represented by the grey shades in Figs. 5.34–5.36, characteristic spatial density patterns emerge. Their morphology strongly depends on whether one has quenched the system into the region between the spinodals, or into the region between a binodal and a spinodal. In the former case, one obtains what is called spinodal decomposition, which is characterized by a well defined dominant spatial frequency of the evolving density pattern. This pattern will of course decay as time proceeds, partitioning the sample into just two singly connected regions (one with the dilute and one with the dense phase). This process is governed mainly by the effect of interfacial tension, its late stage dynamics is known as Ostwald ripening.

In fact we can observe spinodal decomposition as well as Ostwald ripening in the dry granular gas, in much the same way as it is known for molecular systems. Figure 5.38 shows evolving density patterns in a vertically agitated dry granular gas, in both experiment (top) and simulation (bottom). Not only is their overall appearance, with the strongly dominating spatial frequency, very much reminiscent of analogous results in molecular fluids. The coarse graining of the structure, the first steps of which are shown in the figure for equivalent time delays, show the same scaling with time as known from standard thermodynamic systems [181]. This gives us strong confidence that the concept of surface tension applies very well to agitated granular matter (both wet and dry), although its precise physical origin is not clear.

As surface tension is commonly defined as an *excess free energy* per unit surface area, this raises again the question what we shall call a free energy in the system under study, since free energy is so far defined only for equilibrium systems, and we are far away from that. Hence we shall

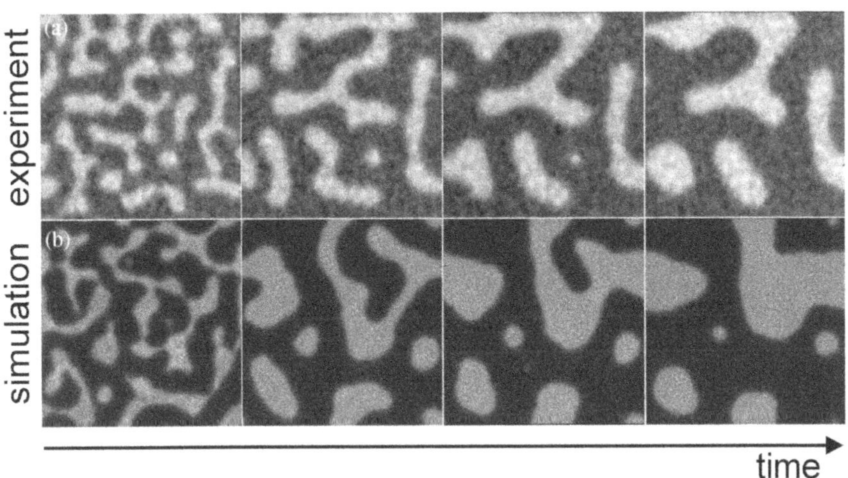

Fig. 5.38 Phase separation of a dry granular gas under vertical oscillation; quenching into the coexistence region. Top: experiment, imaged with a standard CCD camera. Bottom: event-driven simulation of the same system [181]. Reprinted Fig. 1 of [181]. Copyright (2012) American Physical Society.

first have a general discussion of collective phenomena far from thermal equilibrium, and then revisit the question of how to interpret interfacial tensions again.

5.4   Collective phenomena far from thermal equilibrium

Collective phenomena, like the swarming of birds, bacteria, or plankton, the transition from flowing traffic to stop-and-go waves on a freeway, the spontaneous emergence of a 'Mexican wave' in a soccer stadium, or just the transition from frozen to a molten state, have been fascinating mankind for centuries. It is indeed completely counter-intuitive that the molecules in an ice cube all seem to conspire to collectively melt and move around when a precisely defined temperature of $0°C$ is reached, although their thermal vibration energies follow a broad distribution. The same is true for a ferromagnet, which loses its spontaneous magnetization at a precisely defined temperature (the Curie temperature, T_c), although the thermal energy is broadly distributed among the degrees of freedom of the system. The spontaneous magnetization plays the role of the so-called *order parameter* [182], which is a concept generally used in phase transition physics. It is a quantity defined such that it vanishes on one side (usually the high temperature side) and is finite on the other side (usually the low temperature side) of the phase transition. It can be a real scalar (in liquid–gas phase separation is the difference of densities of the two phases), a tensor (as in liquid crystals and in liquid ^{3}He), a complex number (as in liquid ^{4}He and in superconductors), or other. Its mathematical structure generally reflects the structure of the microscopic sub-systems (mostly the particles) and their mutual interaction. In our system of phase separating granular gases, the grains are considered spherical, and in time average this is true even for real granulates. Furthermore, their interaction has no intrinsic angular dependence, with the forces depending only on the particle distance. Hence the order parameter will be isotropic and scalar, i.e., it will be just a real number, as in simple liquids.

It was one of the major achievements in physics during the late 20th century to develop a comprehensive theory of phase transitions and critical phenomena for systems at thermal equilibrium. This was completed by the development of renormalization group theory. It allowed to fully understand the influence of thermal fluctuations, which become dominant in the vicinity of second order transitions. One of its most striking and fascinating results is to correctly account for the *universality* of critical exponents. If $X(T)$ is an arbitrary intensive variable of the system which continuously vanishes (or diverges) at the transition,

$$X(T) \propto \left(\frac{T - T_c}{T_c}\right)^{\alpha}, \tag{5.57}$$

where T_c is the critical temperature, then the *critical exponent* α depends *only* upon the number of dimensions of the system and the number of degrees of freedom of the order parameter [182]. These latter two numbers thus define universality classes, such that all systems belonging to one such class exhibit the same critical behavior, i.e., their critical exponents are *quantitatively* the same.

In essence, the theory of phase transitions follows from statistical mechanics under the assumption of homogeneity and thermal equilibrium, much in the spirit of Chapter 2. Under the constraints imposed by the pertinent conservation laws, like energy conservation, it derives the most probable value of any observable under consideration, and its distribution around that value. A particularly important quantity in this context is the free energy, $\mathcal{F}(X_1, X_2, X_3, \ldots)$, where the X_i are the observables of the system. Particular values of the observables are found with probability

$$\mathrm{p}(\mathbf{X}) = \mathcal{N} e^{-\mathcal{F}(\mathbf{X})/k_B T}, \tag{5.58}$$

where $\mathbf{X} = (X_1, X_2, X_3, \ldots)$, and $\mathcal{N}$ is a normalization factor which assures

$$\int \mathrm{p}(\mathbf{X}) d\mathbf{X} = 1. \tag{5.59}$$

It is clear from the form of Eq. (5.58) that $\mathrm{p}(\mathbf{X})$ is strongly peaked at the minimum of $\mathcal{F}(\mathbf{X})$. Hence the structure of that minimum tells us a lot about the most probable state in which we should expect to find the system.

Consider, for instance, a system which consists of a closed cell containing a molecular liquid at an average density corresponding to its critical density, and the temperature shall be variable. If then $X(\mathbf{x})$ is the local density at $\mathbf{x}$, we should write $\mathcal{F}$ as a functional, $\mathcal{F}\{X(\mathbf{x})\}$. When the system is cooled to a temperature below its critical point, phase separation occurs, just in the spirit of the discussion in Section 5.3.1 above. In the language of the free energy, there are two minima of equal depth in $\mathcal{F}(X)$ if X is assumed spatially homogeneous. One minimum corresponds to the condensed (X_+, liquid), the other to the dilute phase (X_-, gas). Since the average density is fixed, which represents one of the constraints (or conservation laws), the system has to accommodate both phases, as observed in the experiment. Hence there must be regions in space where X crosses over from X_- to X_+. These are the liquid gas interfaces between, say, the droplets which have nucleated as T came below T_c and the surrounding gas phase. Since $\mathcal{F}$ turns out to always contain terms of the form $|\nabla X|^2$ with positive coefficient, the system will, seeking the minimum of the free energy, try to minimize the extent of these regions. Hence the minimum of the total free energy of the system corresponds to a configuration containing only two singly connected regions, one filled with liquid, the other with gas, and an interface between them which is as small as geometrically possible in the given shape of the cell.

Can we find a similar theory for systems far from thermal equilibrium? Is there, e.g., some functional, replacing the free energy of the equilibrium system above, the minimization of which naturally yields, e.g., the hexagonal pattern developing in a Bénard convection cell? Can we find, inspired from equilibrium statistical physics, a *general prescription* for the construction of such functionals, such that the structure of bacterial colonies, clouds of locusts, a school of fish, or just the Mexican wave in the stadium come out as naturally as the geometry of partitioning our cell into two regions filled with liquid or gas, respectively?

It is clear that such theory would have to be quite complex, as it has to yield the full theory of equilibrium phase transitions as a 'trivial' limiting case. Nevertheless, there have been numerous attempts to come up with such concepts. Notably the approach by Glansdorff and Prigogine [183] has gained considerable visibility, who claimed that entropy production would provide such functional. In the language of dynamical systems, entropy production was claimed to be a Lyapunov function [184][9] and would approach a minimum at late times. However, although a certain class of systems indeed minimize entropy production in their non-equilibrium steady states, it was later shown that this is by no means a general property [185–187].

In order to appreciate the difficulties involved in such endeavors, it is quite instructive to turn back to the equilibrium theory, and consider the minimization of the free energy from the perspective of its possible generalization to systems far from equilibrium. Can we call this minimization a deeper *principle* of equilibrium statistical physics, which we then may try to generalize, or adapt in some sense? Let us have a closer look at the conceptional basis of free energy minimization. What it means is to determine

---

[9]A Lyapunov function $\mathsf{L}[\mathbf{X}(t)]$ is a function on a bounded open set $\mathcal{B}$ which obeys $d\mathsf{L}/dt \leq 0$ under the dynamics of the system, and $\mathsf{L} > 0 \; \forall \mathbf{X} \in \mathcal{B}$ except one.

the subset of microstates of the system, corresponding to a certain value of the variable (observable) under consideration, in which the system has the largest probability to be found. This does of course not mean that it will never be found in states with other values of the variable. Only the distribution is sharply peaked, and the other states can be safely neglected for many practical purposes.

The statement 'system S will most probably be found in a state corresponding to the value X_0 of the variable' thus does not contain any more information than the sentence: 'most probably, you will find system S in its most probable state (which corresponds to X_0)'. This blunt tautology reflects the fact that in computing the free energy, which provided us with the probability distribution, Eq. (5.58), we have not done anything more deeply than grouping the microstates of the system into macrostates corresponding to certain values of the variables, or observables, which we have chosen ourselves,[10] and to count the number of microstates in each such macrostate. The deeper reason why this procedure yields such powerful tools as those of equilibrium statistical physics is not that there is a general *principle* hidden inside. It is rather the *conservation laws* of the microscopic evolution of the system, like energy conservation, which provide us with mathematically convenient tools to dissect the space of possible system configurations, and thus to sort out efficiently what values of the observables correspond to the largest set of microstates. The minimization of free energy is thus not to be seen as some kind of an arcane principle, but rather as a mathematical trick for sorting and counting microstates, which works well only due to the extra structure the conservation laws bestow to the configuration space of the system.

In systems far from thermal equilibrium, some of these conservation laws are necessarily breached. If energy is dissipated, it must be replenished form the outside. There is thus a continuous flux of (at least) energy through such systems, which are therefore also called 'open systems'. As a consequence, the mathematical procedure applied in the case of equilibrium systems is not generally applicable in systems far from equilibrium. There may be no general *principle* for the selection of non-equilibrium steady states just because there is no such principle even in equilibrium systems. It is just that the mathematical tricks developed for equilibrium do not work similarly well for systems far off equilibrium.

We can escalate this point even further. It is indeed possible to show

---

[10]That is, for the sake of convenient description.

that a Lyapunov functional must exist for *any* dynamical system, no matter how many degrees of freedom it may have. This can be appreciated by considering the space of all possible microstates of the system. From each of these states to any other, there is a certain transition probability per unit time, and the totality of these transition probabilities can be cast into a master equation with as many terms as there are microstates in the system. If detailed balance would be fulfilled within this master equation, one could immediately write down (if all transition probabilities were known) a potential, defined over the entire space of microstates, the minimum of which would correspond to the state the system would be inclined to approach [188]. The central problem of searching for such potentials in systems far from thermal equilibrium is that detailed balance is typically breached in these systems if considered in the space of their microstates. It turns out, however, that detailed balance can be regained by appropriate coordinate transformation. If the system is characterized not by considering its microstates, but instead by considering the set of all possible closed orbits within the space of its microstates, it can be shown that detailed balances holds in this space of orbits [189, 188]. Hence in the space of closed orbits in state space, there is detailed balance and hence a potential the minimum of which assigns the non-equilibrium steady state which is most probable to be visited by the system [188].

Although at first glance this seems to pave the way to the formulation of a 'free energy' for systems far from thermal equilibrium in general, a closer look reveals that there is a deeper problem behind. It is easy to see that the number of closed orbits in the space of microstates scales roughly as (in fact, even more rapidly than) the factorial of the number of microstates. Hence the transform to the space of orbits does not help in any way to arrive at a useful function reminiscent of what is called a free energy in equilibrium physics, as its evaluation in any case of application would be way more costly (computationally) than finding the full solution of the dynamical problem (with all microscopic degrees of freedom). One would then rather do the latter. A 'principle of minimization' which does not provide a 'trick' (see above) to predict the properties of a system with less effort (such as the free energy in classical statistical physics) than necessary to find its full microscopic solution is in fact no 'principle' at all, and does not tell us anything about the system under study. In the light of these findings, the search for a 'general principle' governing collective dynamics far from equilibrium appears as fundamentally ill-posed.

5.4.1 *Surface tension revisited*

The striking similarity of phase separation in agitated granular gases and phase separation in molecular liquids, however, strongly suggests to search for a description which is at least in some aspects analogous. We have seen above that the dynamics of granular gases leads to a quantity with properties very close to an equilibrium surface tension. The setup of the system directly suggests to replace the (abandoned) energy conservation with the balance condition

$$E_{\text{diss}} = E_{\text{inj}}, \tag{5.60}$$

which is to be understood per unit time. As we could see, this condition allows to predict the qualitative structure of the non-equilibrium steady state, which is represented by phase separation with minimized surface area. We will now try to derive some more detailed predictions, exploiting Eq. (5.60) and the properties of the surface tension outlined above.

In Fig. 5.39 the qualitative behavior of the pressure as a function of density is sketched. It largely resembles what is shown in the insets of Figs. 5.34 and 5.35. Since the space in our sample cell is translation invariant (except

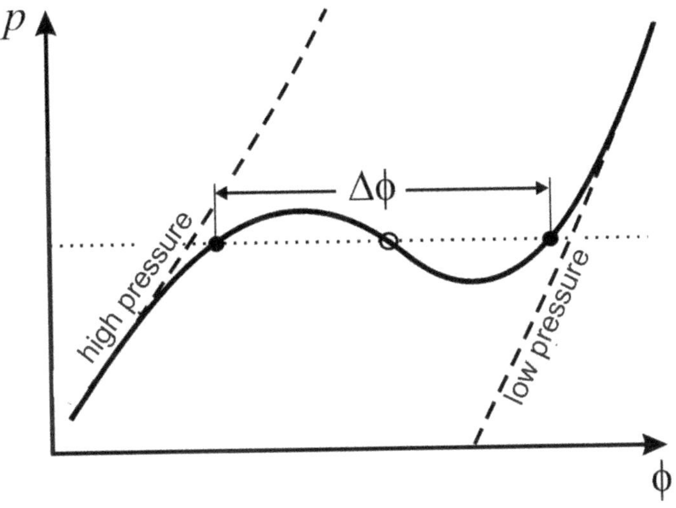

Fig. 5.39 Schematic representation of the pressure as a function of density for a granular gas as an open system. The non-monotone shape of the solid curve resembles the behavior shown in the insets of Figs. 5.34 and 5.35. The dashed curves indicate the low pressure and high pressure branches of the characteristic. Dotted line: balanced pressure between the dilute and the dense phase. Filled circles: stable solutions for pressure. Open circle: unstable solution. The density difference, $\Delta\phi$, serves as the order parameter.

for the boundaries), momentum is conserved, and the pressure must be the same everywhere. Hence the system will select the two densities of the separating phases such that they lie on a horizontal line in this sketch (dotted line). The negative slope entails there being three possible solutions. The solution with the negative slope (open circle) is unstable, so the solutions corresponding to the full circles will be selected. The horizontal distance between these points, $\Delta\phi$, is the order parameter of our system. At the critical agitation strength, E_c^*, where phase separation is just about to set in, the three intersection points merge into one, and $p(\phi)$ attains a saddle point. Hence in the vicinity of the critical agitation strength the pressure curve has the form

$$p(\phi) - p_c = A(\phi - \phi_c)^3 + B(E^* - E_c^*), \tag{5.61}$$

where A and B are constants. As it is easy to see, this immediately entails the form

$$\Delta\phi \propto (E^* - E_c^*)^\beta, \tag{5.62}$$

where $\beta = 1/2$ is the critical exponent of the order parameter. This is the same value as for equilibrium systems in mean field approximation [182], which is clear from the similarity of the mathematical constructions.

Now let us turn to the surface tension again. This has been calculated as an integral over the anisotropy of the pressure tensor close to the surface. It has furthermore become apparent above that it is completely dominated by kinetic energy contributions in our system. Even for the wet granular gas, the cohesive forces were found to be of minor relevance. In other words, the surface tension is due to the anisotropy of granular motion in the vicinity of the interface. It is therefore expected to scale in the same way as the difference in the temperatures of the two phases. This can be exploited when we appreciate that the pressure is a non-monotone function of density only because the temperature depends strongly on density. When the temperature is fixed, the pressure is a monotone function of density in the range of interest (cf. Fig. 5.14), and for fixed density it is a monotone function of temperature. Close to the critical point, p does not change to first order in ϕ, such that the points represented by the full circles in Fig. 5.39 move along the contour line defined by $p(\phi, T_g) = p_c$. Hence the temperature difference is proportional to $\Delta\phi$, with the prefactor given by the slope of that contour line. We can thus conclude that the surface tension should vanish at the critical point in the same way as the order parameter,

$$\text{surface tension} \propto (E^* - E_c^*)^{1/2}, \tag{5.63}$$

or, in other words, the surface tension should have the same critical exponent as the order parameter. This is distinctly different from equilibrium systems, where the critical exponent for the surface tension is universally ≈ 1.28 [182], leading to a qualitatively different critical behavior. In equilibrium systems, it vanishes at the critical point with zero slope, while we predict here that it vanishes with an infinite slope (since $1/2 < 1$) at $E^* = E_c^*$. It is interesting to note that the simulation results presented in Fig. 5.32 tentatively corroborate our prediction. The arrow indicates the location of the critical point. Although there are no data closer to this point, it is already apparent that the critical exponent will be substantially smaller than unity, in agreement with our crude model, but in clear contrast to the universal finding in equilibrium systems.

Given the similarities not only of the phenomena, but also of the concepts which obviously lead to a successful description, it suggests itself to study agitated dissipative gases in a larger framework. Specifically, we can plot classical systems like the van der Waals gas together with dry and wet granular systems in a single diagram, spanned by the 'dissipativity' and the cohesiveness of the mutual interaction of the individual particles. This is

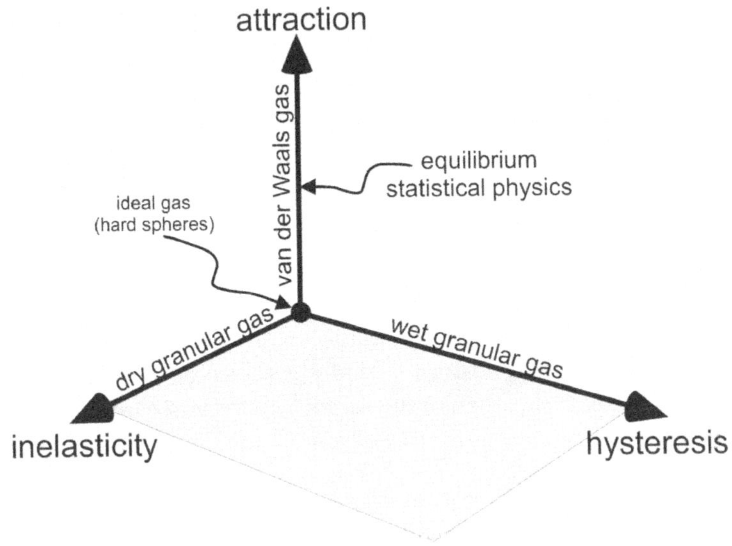

Fig. 5.40 The systems under discussion, depicted in the plane spanned by the amount of cohesion and dissipation, respectively, in the inter-particle interactions. Along the vertical axis, standard statistical mechanics is valid. It is of great interest to study how its principles and universalities break down as one moves away from that axis. Wet and dry granular gases appear as promising systems for such endeavor.

sketched in Fig. 5.40, where the classical equilibrium systems known from statistical mechanics text books are lying on the vertical axis. The dry granular gases, which are not cohesive, but dissipative, are located along the axis annotated 'inelasticity'. The wet granular systems lie along the axis 'hysteresis', referring to the fact that the interaction pseudo-potential differs between approach and withdrawal. Exploring the space spanned by these three axes, one might find out what concepts can be carried over from equilibrium statistical mechanics to systems far from thermal equilibrium, which aspects of universality break down, and in which way they do so. This is to be taken as a suggestion for future research and goes well beyond the scope of the present book.

5.4.2 *Chaoticity of the wet granular gas*

Let us close this chapter with a few aspects which are suitable to set wet granular gases into a larger perspective of complex systems in general. With the separation s of two impacting grains as a spatial coordinate, we can easily construct a qualitative phase portrait of the impact process, as shown in Fig. 5.41. As in dynamical systems, we plot momentum (p) versus position (s). For a dry granulate (top left panel), the motion is completely free of force, except for a single point on the time axis, when the impact takes place. The trajectory jumps from $(0, p_i)$ to $(0, -p_f)$ here, where p_i and p_f are the relative momenta before ('initial') and after ('final') the collision. For reasons which will become clear in a minute, we consider here only the magnitudes of the momenta for, $|p|$, for the phase portraits. So far we note that the trajectories after the impact are closer together than those before the impact.

When we replace the inelastic collision with a capillary bridge as the dissipative element, we obtain the phase portrait shown in the top right panel. The most striking feature is that the trajectories after the impact are now farther apart than before the impact. If d_i and d_f are the transverse distances of the neighboring trajectories before and after the impact, it is elementary to show that

$$\frac{d_f}{d_i} = \frac{1}{\sqrt{1 - \frac{\Delta E}{E_i}}}, \tag{5.64}$$

which is independent of the form of $F_{cb}(s)$. Obviously, the phase flow is transversely expanding, in strong contrast to what one usually encounters in dissipative systems.

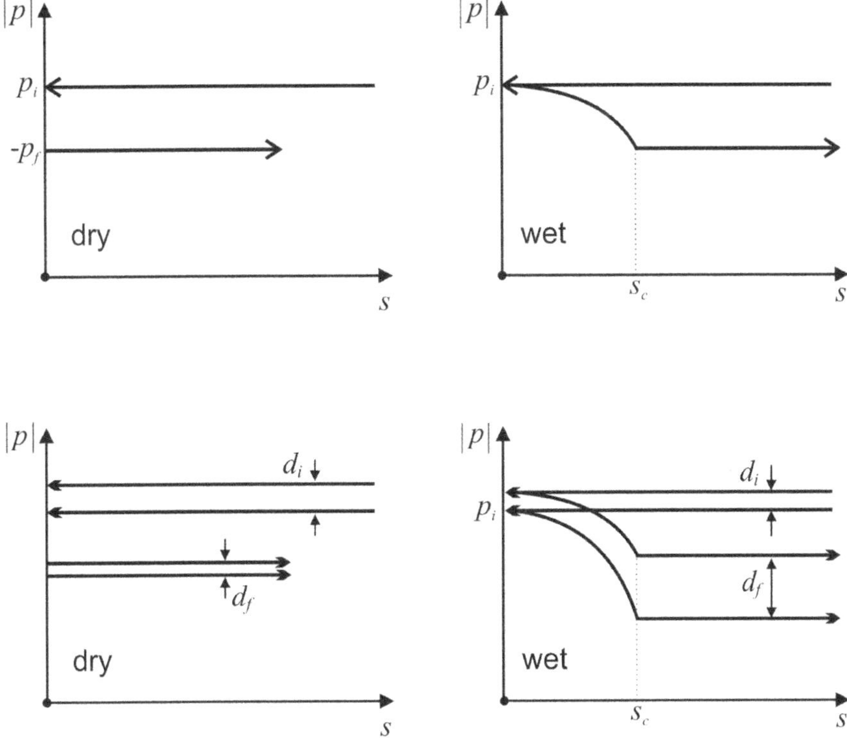

Fig. 5.41 Schematic phase portrait of an impact of two grains in a dry (left) and a wet (right) granular gas. Note that on the vertical axis the absolute of the momentum is plotted, which is positive before and after impact. While the top row shows single trajectories, the bottom row explores the fate of two neighboring trajectories and their mutual transverse distance.

A more formal presentation of the transverse spread of the phase flow uses the Lyapunov spectrum of the system. The Lyapunov exponents, λ_i, describe the temporal evolution of the distance $d(t)$ of two closely neighboring points in phase space. Since for sufficiently small distances the dynamics can be regarded as linear, the distance is exponential in time, $d(t_1) = d(t_0)exp[\lambda(t_1 - t_0)]$. We thus can write

$$\lambda \sim \frac{\ln d(t_1) - \ln d(t_0)}{t_1 - t_0}, \tag{5.65}$$

and bear in mind that λ is to be defined as a time average. It is clear that in $6N$-dimensional phase space, there are $6N$ such exponents, and these are the eigenvalues of the linear map of the distance vector, $\mathbf{d}(t)$, provided by the phase flow along the trajectory.

For both the dry and the wet granular gas, we will have spatial contributions to the λ_i as known from the hard sphere gas [190, 191] (which are positive), plus in addition the effect of the dissipative impact kinetics represented pictorially in Fig. 5.41. From Eq. (5.64), we can derive the average transverse contribution to the Lyapunov exponents as

$$\langle \lambda_T \rangle = -\frac{f_{\text{coll}}}{6N} \ln \left(1 - \frac{\Delta E}{E_i} \right) \tag{5.66}$$

which are obviously positive [192].[11] The divergence, which also appears in Eq. (5.64), will be smeared out due to the statistical distribution of impact energies. The central quantity to study here is the so-called Kolmogorov-Sinai entropy, h_{KS}, which is defined as the sum of all *positive* Lyapunov exponents of the system. It determines the rate of information entropy production within the system, i.e., the chaoticity of its dynamics. It is tedious but possible to evaluate h_{KS} for the wet granular gas [192]. As one may already anticipate qualitatively from Fig. 5.41, the general result is that h_{KS} increases monotonously when a dry granular system is being wetted.

5.4.3 *Capillary bridges as active networks*

For a somewhat more speculative look at the collective dynamics in an agitated wet granular system, let us consider the capillary bridges as individual entities coupled to each other by the kinetics of the grains. Similar as in Fig. 1.6, Fig. 5.42(a) shows an individual gap between two adjacent grains as a system which can be in three different states. The separation of the grains surfaces can be either zero or finite, and there can be either a capillary bridge or not. This yields four different states, of which one is irrelevant: if the grains are in contact, there is immediately a capillary bridge, such that one state (contact but no bridge) is extremely short lived and needs not to be considered. As the bold arrows in the figure indicate, not every transition is allowed between these three states. In the language of the center figure (b), the two transition probabilities a_{21} and a_{13} vanish. As a consequence, a single gap between adjacent grains, under the dynamics induced by some external agitation, would lead, on average, to a clockwise circular motion in the system of three states as depicted in (a) and (b). Hence there would be finite probability currents in the system, which is the

---

[11] When $\Delta E/E_i \leq 1$, there are additional positive contributions due to the rotation of two collision partners about a common axis.

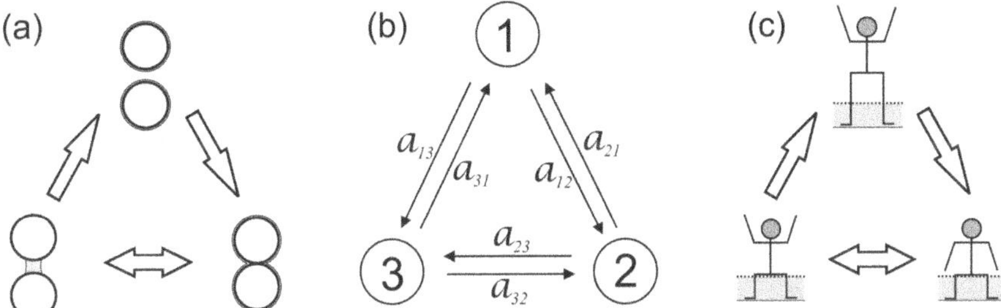

Fig. 5.42   Capillary bridges as systems with broken detailed balance. (a) The three distinct states of a contact region between adjacent grains. There can be just an empty gap (top), the grains can be in solid state contact (bottom right). It is clear from the dynamics of the liquid film that not every transition is allowed between these states. Bold arrows indicate allowed transitions. (b) Formal representation of a three-stats system as the one shown in (a), with transition probabilities between them. In our system, the transition probabilities a_{21} and a_{13} vanish. (c) Spectators in a (e.g.) soccer stadium are formally equivalent to the system in (a). They can be in a state of excitement (bottom left), from which they may rise (top), or in a refractory state (bottom right), from which they can become more excited, but not go directly to the top state. The significance of this (very imaginative) model is its equivalence to, e.g., starving *Dictyostelium* cells in a colony, or heart muscle cells.

defining property of a system with *broken detailed balance*. In general, this is a prerequisite of non-equilibrium systems, as equilibrium systems always fulfill detailed balance, i.e., there are no probability currents in the steady state in equilibrium.

The collective behavior of a system consisting of many coupled sub-systems of the kind sketched in Fig. 5.42(b) can be very rich, as one may appreciate considering another, completely analogous example which is depicted in Fig. 5.42(c). This shall indicate a spectator in, e.g., a soccer stadium, mostly sitting on the bench (bottom right). Only from a state of baseline excitement (bottom left) will the person consider to stand up (top). After some time standing, the most probable way is to sit back down again, with some reluctance to stand up soon again. In the stadium, the spectators are coupled, in the sense that one stands up very likely from a state of baseline excitement when the neighbors stand up. As it is well known, this can lead to what is known as the Mexican wave, a coherent front of standing-up action which may go around the stadium several times.[12] The same is true for starving colonies of cells of *Dictyostelium discoideum*,

---

[12]It is not essential whether the spectators are doing this unconsciously or play it as a game. In the latter case, they just consciously 'simulate' the system depicted in Fig. 5.42(c), with the same result.

which start sending out chemical wave fronts of a substance called cAMP. These signals call the cells to gather and form a stalk, which can survive food-deprivation for extended time. These cells can as well acquire basically three states, an excitable one (corresponding to the bottom left in Fig. 5.42(c)), an active one, sending out a burst of cAMP (top in Fig. 5.42), and a refractory one, in which they are not easily excitable (bottom right Fig. 5.42(c)). The result is a periodic sequence of cAMP bursts across the cell colony, much like a Mexican wave. Heart muscle cells do essentially the same when they undergo fibrillation instead of following the (usual) low-pace stimulus of regular heart beat. We see that the model depicted in Fig. 5.42 is a rather general and paradigmatic one. It is interesting to realize that the capillary bridges in agitated wet granular matter belong to the same class.

Would we thus expect 'Mexican waves' in the wet granular system? We would not, as we will see in a moment. Let us have a closer look at the coupling of neighboring sub-systems. Coupling comes about by the dependence of the transition probabilities, a_{ij} on the state of the neighbors. In the Mexican wave (or cAMP wave, or heart fibrillation), a neighbor in state 1 (cf. Fig. 5.42(b)) greatly increases the transition probability towards state 1, which is a_{31}. This is the main ingredient of the next-neighbor interaction. Since a neighbor in state 1 increases the tendency towards state 1, we may call this interaction 'ferromagnetic'. In the system of

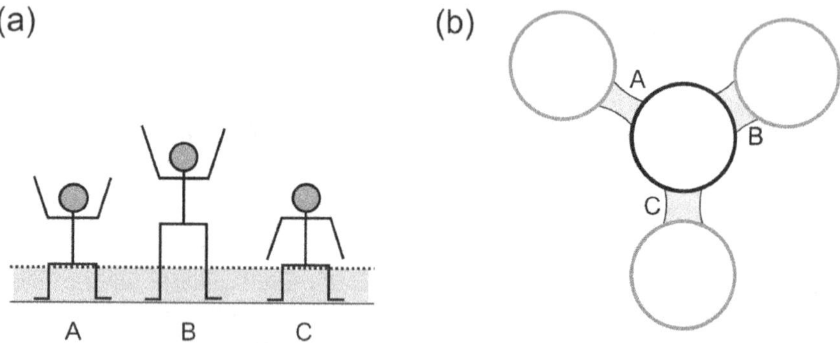

Fig. 5.43   Next neighbor coupling in systems like those of Fig. 5.42. (a) Spectators in a soccer stadium. The proximity of spectator B (excited) to spectator A will increase the probability that A (sensitive) becomes excited, too. Spectator C is presently in the refractory state and will not react, nor will it reduce A's excitement. (b) The force exerted by capillary bridges A and B tends to increase the probability that A ruptures, and vice versa.

capillary bridges, we readily appreciate that most of the time the opposite is the case. As the neighboring bridges are likely to be on the opposite side of a grain, their presence adds a force which pulls that grain away, tending to extend the surface separation of the gap under consideration. Hence a neighboring capillary bridge in state 3 will be likely (depending on its relative angular position on the grain) to suppress the transition probability a_{12}, which would lead to the formation of a new bridge. We may call this interaction 'anti-ferromagnetic'.

Hence we do not expect a feature as prominent as a Mexican wave, or a fibrillation front, in the dynamics of formation and rupture of capillary bridges, as we know that a phase transition in an anti-ferromagnet needs lower temperatures and is much harder to observe than the spontaneous magnetization in a ferromagnet. Furthermore, the frequent occurrence of triangular neighborhoods in a granular pile reminds us that the phase transition of an anti-ferromagnet tends to be frustrated in such lattices. Nevertheless, it appears worthwhile to investigate the details of collective capillary bridge dynamics in the light of the many interesting critical phenomena which do take place in anti-ferromagnets, and about which a vast body of literature is available.

5.5   Conclusions

The free cooling of wet granular gases in three dimensions has been found to yield aggregates of grains with a fractal dimension of $D_f \approx 2$. The fundamental difference in dry and wet granular aggregation could be cast in a simple concept: wet gases condense in real space, dry gases in momentum space. For many purposes, the collision dynamics of wet grains can be conveniently expressed in terms of an effective restitution coefficient. We found that the kinetics of energy dissipation resembles a system with hysteretic capillary bridges even when the viscosity of the liquid is large enough to prevent their hysteretic formation and rupture.

We have found that despite the inherent complexity and polydispersity of a wet granular gas, one can capture a large fraction of its behavior by means of the rather simple concept of hysteretic capillary bridges. One of the most striking is the phase separation into a dense and a dilute phase if the system is physically agitated. Only two parameters seem to be relevant for the description of this phase transition: the distance at which a capillary bridge ruptures, and the energy which is to be afforded to accomplish this rupture. Interaction characteristics which are entirely different but agree only in these two parameters lead to almost identical phase diagrams. An even more striking commonality concerns the liquid content of the sample. Although capillary bridges will merge forming larger aggregates as more liquid is added to the sample, the system still behaves as if only two-body forces were involved. This can be tentatively traced down to the peculiar energy-landscape of liquid aggregates between more than two grains, which appears as if the mutual forces were almost exactly additive. This eases computer simulations of wet granular matter considerably.

Our study of phase transitions in agitated wet granular gases revealed that agitation energy and agitation acceleration provide two entirely independent experimental parameters. Only if they are controlled independently, can the full phase diagram of wet granular matter be explored. It comprises two regions of coexistence between a cold dense phase and a hot dilute phase. Each of these coexistence regions contains two spinodals and a critical point (or, more precisely, a line of critical points) at which the density difference vanishes, much like in equilibrium phase transitions. One of the phase transitions is due to the presence of the capillary bridges (i.e., the wetness of the system), the other is due to the finite dissipation of the collisions ($\varepsilon < 1$) and is therefore present as well in the dry system.

It is in particular due to the excellent agreement between experimental data and simulation that these transitions recommend granular gases as well suited model systems for the study of collective behavior far from thermal equilibrium. We have found that by replacing energy conservation with a balance between energy injection and energy dissipation, we are led to a qualitative understanding of phase separation in granular gases, both wet and dry. This concept allows to compute critical exponents in a way similar to mean field theories of phase transitions in equilibrium systems.

Further reading

We had to be particularly short on the equation of state in a particulate system, when the temperature is well defined. To get into this matter, a very early [193] and a more recent [194] paper may help. Since we touched the field of collective behavior far from thermal equilibrium, it should be worthwhile to read through some classic literature treating phase transitions and pattern formation in nonlinear systems [184, 195], and some more recent account of steady states far from thermal equilibrium in general [189, 188].

Chapter 6

Wet Granular Piles

Fig. 6.1 Mold casting ceramic vases. Both the mold and the vase are made from wet granular material, exploiting its shape stability (image courtesy of Luc van Hoeckel[1]).

The present chapter takes us back to the pasty, moldable material we started from in the introduction, alluding to childhood experiences in the sand box, and to sophisticated sculptures at the beach. While the consistency of dense, wet granular matter may be a nuisance in land slides or mud flows, it is exploited in millennium-old technologies, such as ceramics or expendable-mold casting techniques. In the latter, one uses a cohesive granular material to imbed a wax model of the sculpture to be cast. The model is then removed by melting the wax away, thus leaving a void within the granulate, the shape of which matches that of the wax model. The void is stable by virtue of the mechanical stiffness of the cohesive granulate. Since the grain material can be chosen such as to withstand high temperatures, this void can be filled, e.g., with molten bronze. When the metal has cooled and solidified, the granulate can be easily removed by virtue of its brittleness, unveiling the freshly cast bronze sculpture.

It has become clear by now that the wet granular material owes its mechanical stiffness to a network of liquid bridges connecting adjacent grains.

---

[1]The image provides a view into the artist's atelier, showing a step in the production of his *Turtle Vase*.

219

The connectivity of this network of liquid bridges depends upon the packing geometry of the grains, which is a complex topic in itself [196–198]. Even for a random close packing of spheres of equal size, which is a comparably 'simple' case, the precise coordination and nearest-neighbor relations are still under investigation [199–201].

There are two distinct ways by which the attractive forces exerted by the capillary bridges can affect the mechanical stiffness of a dense granular medium, i.e., be counteracting a shear force applied to the whole granular pile. On the one hand, the applied shear may have to do work directly against the capillary forces, because of site exchange processes among the grains which cause the extension of capillary bridges. On the other hand, the capillary forces give rise to tangential friction forces by virtue of the static friction coefficient, μ, which depends upon the roughness and surface composition of the grains. Since granular piles are always mechanically frustrated (cf. Section 4.4.1), tangential friction will generally impede deformation. It will be one of the goals of the present chapter to provide some tools for unravelling these two contributions wherever possible.

6.1 Geometrical aspects of granular piles

Before we discuss the effects of capillary bridges, and more complex liquid structures, onto the mechanical properties of a dense granular material, we will first say a few words about the geometry of dense packings of grains with variable size and shape. There are two quantities of major interest which will be discussed. The first one is the contact number, k_c, which is defined as the average number of mechanical contacts a grain has with neighboring grains. The second is the packing density, ϕ, which we have encountered already before. It is defined as the volume fraction of solid grain material found in the sample.

6.1.1 *Random piles of equal spheres*

The first quantitative experimental study of the coordination, i.e., the nearest-neighbor relations and contact numbers, of randomly packed spheres of equal size has been performed by Bernal and Mason in 1960 [197]. A random pile of steel balls of equal size was immersed in black Japan paint. The paint was then drained out of the pile, such that capillary bridges formed close to the contact points. After the paint had solidified, the pile was taken apart. The points of contact were revealed as annular stains of paint on the balls, and could thus be counted. For the random packing which was achieved, with a density of $\phi \approx 0.62$, an average number of about six contacts was obtained for each sphere.[2]

We can easily verify that we should expect a contact number of that order. First we appreciate that it is possible to find for each contact point a second one which is approximately on the opposite position on the same sphere. These two contacts can be imagined to fix the sphere with respect to one translational coordinate, which coincides with the line of connection of the two contacts. Since the position of the sphere is determined by the three coordinates of its center, we need three such contact pairs to fix the position of the sphere in three-dimensional space. Hence we need approximately six real contacts on each sphere in order to obtain a mechanically stable pile.

Refined estimates have been tried by means of computer simulation, following the above idea, that a stable pile needs to have just enough contacts to fix all of its internal degrees of freedom. Such piles are called *isostatic*, and isostaticity has been a major paradigm in research on random piles for some time. However, it has become increasingly clear over time that isostatic packings are not as relevant for real piles as it may seem. This is mainly because friction plays an important role in the piling process as well as for the pile stability. Most packings of relevance in fact turn out to be hyperstatic (i.e., they have more contacts than required for an isostatic pile) [202]. Hence it is advisable to treat current opinions on random piles with a grain of salt: this is still a field of ongoing research and lively debate. What can be taken as a rather robust result, however, is that the contact number is *around six* for piles of equal spheres in three dimensions, and generally about $k_c \approx 2D$ in D dimensions. In the literature, k_c is frequently referred to as the coordination number.

One of the major sources of uncertainty in the field of research on random piles is that their characteristics may depend significantly on the protocol of their creation. This is most easily appreciated in the case of equally sized spheres: with just a little diligence, one can pile them into a hexagonal close-packed (hcp) or a face-centered cubic (fcc) arrangement. This has a packing density of $\phi_{hcp} = \phi_{fcc} \approx 0.74$. If instead one throws the spheres into a container at random, one is likely to achieve a packing density around 0.60. The densest achievable random packing of equal spheres, called random close packing (rcp), has a volume fraction of $\phi_{rcp} \approx 0.64$. Several

---

[2]The authors first reported an average contact number of $k_c = 6.4$, but later found that many of these 'contacts' still had a finite distance.

different protocols have been invented in order to find packing densities and contact numbers in computer simulations, with subtle but significant differences in their results. The value of 0.64 given above for ϕ_{rcp}, however, can be taken as quite well established.

Since the generation of a random pile of grains is always a dynamical process, it is not surprising that its result may depend on grain properties such as friction and inelasticity, or incomplete restitution. Both affect the piling process, which ever may be the protocol. Consequently one observes that both the contact number and the packing density depend significantly on the static friction coefficient, μ, between the grains, and on their restitution coefficient, ε. Some numerical results supporting this view are shown in Fig. 6.2. As typical friction coefficients are around 0.5, we can expect that in a real pile of spheres, it will take some effort to reach a packing fraction better than about 0.62.

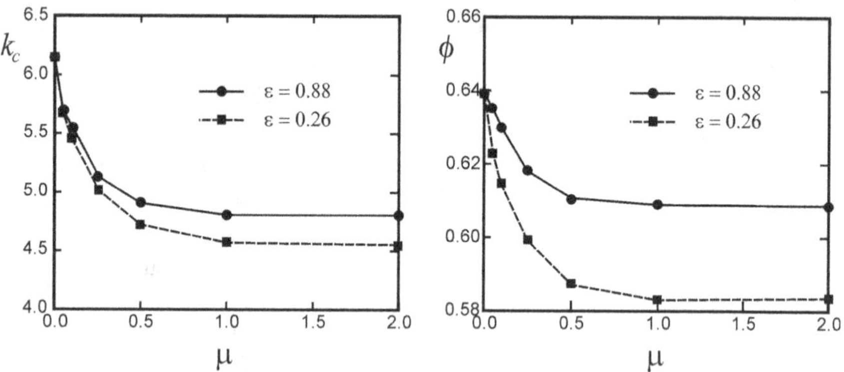

Fig. 6.2   The average number of real contacts (left panel) and the packing density (right panel) of random piles of spheres for variable friction and inelasticity, as obtained from computer simulations [203] (figure redrawn).

There is a simple argument which provides a qualitative understanding of the overall behaviour shown in Fig. 6.2. Both the inelasticity of the grains and the friction between them leads to dissipation of energy. It is clear (and we have seen this impressively above in the discussion of clustering) that dissipation on the grain scale leads to effective cooling of the pile, i.e., the random motions of the grains are slowed down. As a consequence, there will be less time for the grains to rearrange into a packing of highest density, and hence both the packing density and the contact number must decrease as dissipation increases. In place of the maximum packing density of frictionless, elastic spheres, $\phi_{\mathsf{rcp}} \approx 0.64$, we will hence use the somewhat less strictly defined term *random dense packing* (rdp) for the densest packing achievable under the conditions of the respective experiment. For the spherical glass beads frequently used for experiments (ballottini) on granular matter, one usually obtains $\phi_{\mathsf{rdp}} \approx 0.62$, which is close to dilatancy onset, ϕ (Section 2.2.5). Consequently, this takes already some effort to achieve.

6.1.2 *Effects of grain size: Polydispersity*

While the discussion so far considered mono-disperse samples, granular systems always exhibit a certain polydispersity, i.e., not all grains have the same size. An experimentally very favorable effect of polydispersity is that it prevents crystallization in piles of even perfectly spherical particles, which would otherwise lead to poorly controllable side effects. As introduced in Section 2.1.4, we quantify polydispersity as the width of the size distribution of the grains, divided by the average grain size. In what follows, we will still assume that the grains are spherical, but see what effects we have to expect when their sizes are not equal.

Obviously, the overall density of a pile of spheres can be easily increased if small spheres are added which fit into the nooks between the larger spheres. Let R_1 and R_2 be the radii of the large and small spheres, respectively. Then if the second species of spheres are very small, $R_2 \ll R_1$, we can assume that all of the interstitial space will finally be filled by a random close packing of the smaller spheres. The overall maximum binary packing density is then

$$\phi_{\mathsf{bin}} = \phi_{\mathsf{rdp}} + \phi_{\mathsf{rdp}}(1 - \phi_{\mathsf{rdp}}) = \phi_{\mathsf{rdp}}(2 - \phi_{\mathsf{rdp}}) \approx 0.86, \qquad (6.1)$$

which is well above ϕ_{rcp}. The most straightforward way to produce such packing is to first pack the large spheres as densely as possible and then let the small ones trickle into the interstices of the large spheres packing. For this to work out, the small spheres must fit in between the triangular opening left between three adjacent large spheres in contact. This requires $R_2 < R_1(2/\sqrt{3} - 1) \approx R_1/6.46$. In principle, we could go on and fill even smaller spheres, with radii $R_3 < R_2/6.46$, into the interstices of the small spheres, and then continue with spheres with radii $R_4 < R_3/6.46$, and so on. The result will then be a packing of spheres with a density arbitrarily close to unity, depending how many different sizes we use. Spherical grain

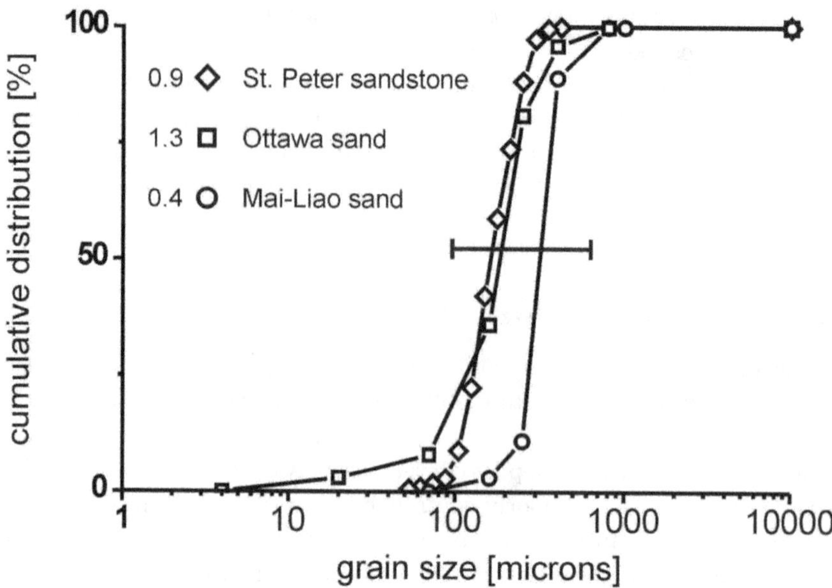

Fig. 6.3 Cumulative size distributions for the three sand samples as in Fig. 2.6, but on logarithmic scale. The scale bar indicates the size ratio at which the smallest grains would fit in the nooks between three of the largest grains in mutual contact (see text further below). The numbers next to the symbols in the inset indicate the respective polydispersity.

packings of this kind are commonly called Apollonian packings[3] [204]. With a quaternary system consisting of four different grain sizes, a packing fraction as high as 0.951 has been obtained. The grain diameters varied from 40 μm to 13 mm.

This is a very wide distribution. In granular systems, the distribution of grain sizes is usually much narrower than that, even for most natural granulates, like sand. Figure 6.3 shows a few cumulative size distributions of natural sands for reference. The scale bar indicates the size ratio of 6.46, when the smallest grains fit in the nooks left between large grains in mutual contact. Clearly, the widths of the distributions are smaller than the bar, such that Apollonian packings are not expected to play any significant role here. Furthermore, it is essential that the size distribution be discrete for reaching the reported high densities.

---

[3]This term is not due to some arcane relation to greek mythology, but merely to the fact that the first systematic study of this type of systems was done in Apollo, Pennsylvania (USA) in 1961 [204].

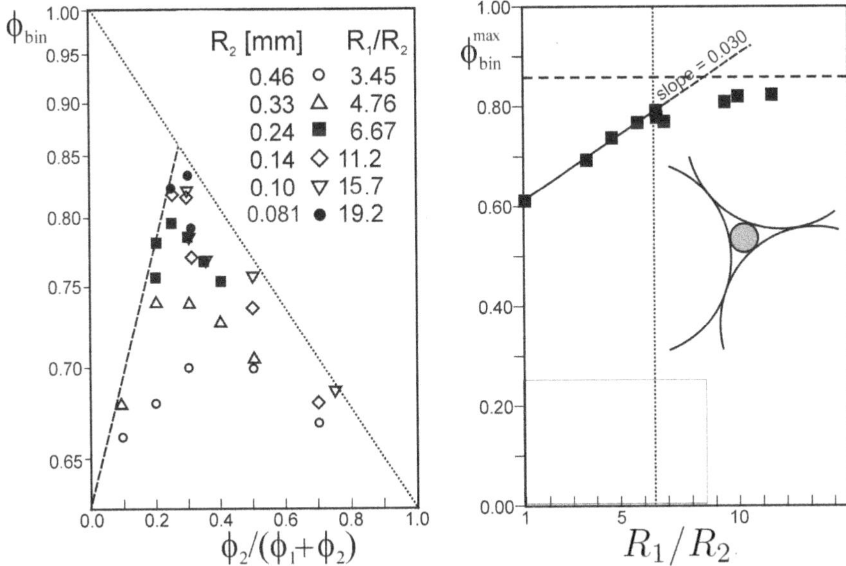

Fig. 6.4   Packing density for binary mixtures of spherical grains, with radii R_1 and R_2, respectively. $R_1 = 1.57$ mm. Left panel: The packing density, ϕ_{bin}, has a distinct maximum at some finite mixing ratio. It falls off towards the 'pure' piles, where it reaches $\phi = \phi_{\text{rdp}} \approx 0.62$. The maximum becomes sharper as the ratio R_1/R_2 increases. the dashed and dotted lines indicate the limit of infinite size ratio. Right panel: The maximum achievable packing density, $\phi^{\text{max}}\text{bin}$, as a function of the size ratio. The vertical dotted line indicates the size ratio above which the smaller spheres fit into the nooks in between three larger spheres in mutual contact (inset sketch). At smaller size ratio, $\phi^{\text{max}}\text{bin}$ seems to vary approximately in a linear fashion. Note that the poly-dispersities encountered with balottini or other typical samples for granular matter experiments are less than the symbol size. Data taken from [204].

For our purposes, the results obtained for smaller differences of grain radii are more significant. Figure 6.4 shows the data obtained by McGeary for various size ratios and packing fractions [204]. For very large ratios R_1/R_2, the system would follow the straight lines on the left panel of the figure. The dotted line corresponds to single large spheres embedded in a dense pile of small spheres. The dashed line corresponds to small spheres gradually filling the interstitial space of a dense packing of large spheres. The intersection of the two lines is at $\phi_{\text{bin}} = 0.86$, corresponding to Eq. (6.1). As the sizes of the grains come closer to each other, a smoother variation is obtained (open circles in Fig. 6.4, left panel). However, it is invariably observed that the packing density of a bi-disperse system is superior to the maximum packing density attainable with a mono-disperse system of either grain size.

The right panel of Fig. 6.4 shows the maximum packing fraction of binary mixtures, $\phi_{\text{bin}}^{\max}$, which is reached at some intermediate value of the mixing ratio (cf. left panel). We see that this maximum varies approximately linearly with the ratio of radii, R_1/R_2, until it bends off at about $R_1 = 6.46R_2$ (dotted line), when the smaller grains fit in between the larger ones, and approaches the horizontal dashed line at $\phi_{\text{bin}}^{\max} = 0.86$ (Eq. (6.1)) for $R_1 \gg R_2$. In the present discussion we focus of rather narrow size distributions (small polydispersity), such that only the range of small ratios of radii is relevant. The widths of the distributions shown in Fig. 6.3 are smaller than the symbol size in the right panel of Fig. 6.4, such that we can safely assume the maximum packing density to vary about linearly with the size ratio in the relevant range. The straight solid line in the figure suggests an approximate form,

$$\phi_{\text{bin}}^{\max} \approx \phi_{\text{rdp}} \left[1 + 0.05 \left(\frac{R_1}{R_2} - 1 \right) \right], \tag{6.2}$$

which is to be considered as an upper bound if the sample is not bi-disperse, but consists of spherical grains with a continuous distribution of radii between R_2 and R_1. If we then define polydispersity as

$$\text{Pd} = \frac{R_1 - R_2}{\langle R \rangle}, \tag{6.3}$$

and consider $\text{Pd} \ll 1$, we can identify the expression in parentheses in Eq. (6.2) with Pd in good approximation.

We can see that packing density is not strongly affected by polydispersity, as long as the width of the size distribution remains small as compared to the average grain size. What is dramatically affected, however, is the *structure* of the pile: polydispersity is very effective in preventing crystallization (which is of course easy to achieve in piles of spheres of equal size). A polydispersity around 0.1 is sufficient to suppress crystallization reliably [36, 37]. This is why piles of natural granulates are generally free of any crystalline regions, even in the case of well sorted sediments. In experiments, one deliberately uses slightly poly-disperse samples in order to prevent crystallization, which would give rise to substantial side effects.

6.1.3 *Effects of grain shape*

We have seen above that friction, which mainly stems from a certain degree of roughness of the grain surfaces, leads to a reduced packing density. Since roughness may be seen as a form of deviation from the ideal spherical

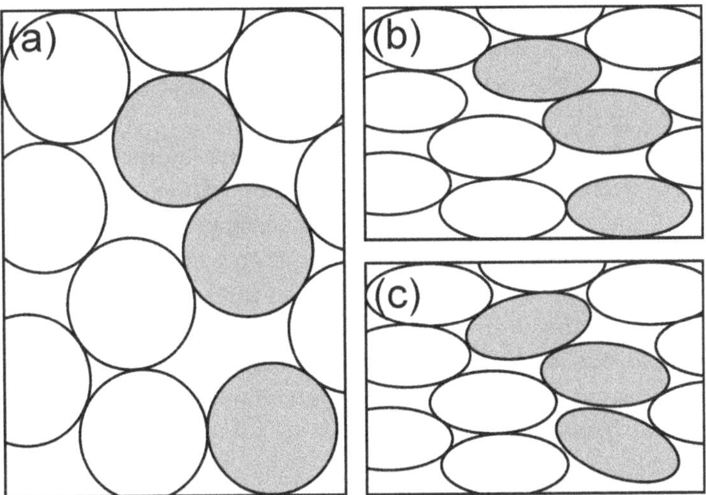

Fig. 6.5 Sketch of a two-dimensional random pile of disks, to illustrate the impact of grain asymmetry on the maximum packing fraction. (a) Pile of un-deformed (circular) disks. (b) Same pile as in (a), but vertically compressed by a factor of two. The volume fraction of the pile is strictly conserved under this transformation. One can anticipate that the grains shaded in grey are now not completely locked in their position. (c) Same pile as in (b), with the grey disks moved within the freedom they gained when going from (a) to (b). The white disks have been kept in place. Note that the disk touching the top disk in (b) becomes free only when the top disk is moved. Since some grains can be moved which were arrested in (a), a higher packing fraction can be achieved.

shape, one might be tempted to conclude that deviations from the sphere in general reduce the packing fraction and the contact number. However, in Chapter 2, we have prudently distinguished between roughness and shape, and we will see now that (roughness-induced) friction and non-spherical shape even tend to have opposite effects. As an illustration, Fig. 6.5(a) shows a sketch of a random arrangement of circular disks. All disks shown are arrested in their position by the contacts to their neighboring disks.

Figure 6.5(b) shows the same as Fig. 6.5(a), but after vertical compression by a factor of two. All disks have now been converted to ellipses with an axis ratio of 0.5. By virtue of the properties of the compression process, the packing density has remained identical. A closer look at the three 'grains' which are shaded in grey, however, reveals that they are now not anymore arrested. The topmost of the grey disks can be rotated counter-clockwise and moved to the left, yielding some space to move as well to the grey disk below. The bottom grey disk can be rotated clockwise and moved up substantially. The result is shown in Fig. 6.5(c). The fact that the elliptical disks can be moved means that they can better optimize the piling, thus increasing the packing density with respect to its original value for circular disks. It is obvious that the same will hold in three dimensions for elliptical particles, compared to spherical ones.

One may be tempted to falsify this argument by reversing the whole procedure. Starting with a vertical expansion of Fig. 6.5(c) by a factor of two, one may try to convert the arrangement of Fig. 6.5 back into a denser arrangement of circular disks, with then elevated packing fraction. However, expansion of Fig. 6.5(c) will not result in an arrangement of circular disks, since the grey disks have been turned by some finite angle, such that an expansion in the vertical direction will not reconvert them into circles. We thus can summarize that the deviation of the grains away from the spherical (here: circular) shape leads to a larger, not smaller, packing density.

Not only the packing fraction is affected by shape in the opposite way as by friction: the same is true for the contact number, as one readily appreciates by alluding to the argument we had put forward above in connection to Fig. 6.5(c). As long as one considers symmetric grain shapes such as circles or spheres, arresting the positions of their centers is sufficient for obtaining a stable pile. However, any irregularly shaped grain must as well be arrested with respect to its angular position. Hence for non-circular disks, we need not only four contacts (arresting two translational degrees of freedom), but six, arresting the rotational degree of freedom as well. In three dimensions, we then expect to find contact numbers k_c around 12 rather than 6, for three translational and three rotational degrees of freedom. That this is indeed the case has been shown in a beautiful experiment using (non-spherical) candies as the grains [200]. Again, this is expected to hold only for frictionless bodies, as a contact point with friction is capable of arresting tangential motion as well, and thus rotational degrees of freedom. In practice, the effect on k_c is not that strong, but it is noticeable. Figure 6.6 shows experimental data obtained from two samples with grains of different shape. The grains in one sample were almost spherical and could be approximated as ellipses with an axis ratio of 0.97. In the other sample, which consisted of placebo pills, the axis ratio was 0.40. Obviously, the number of mechanical contacts, k_c, is systematically larger for the non-spherical grains at fixed ϕ, while the overall dependence of k_c upon ϕ is virtually the same.

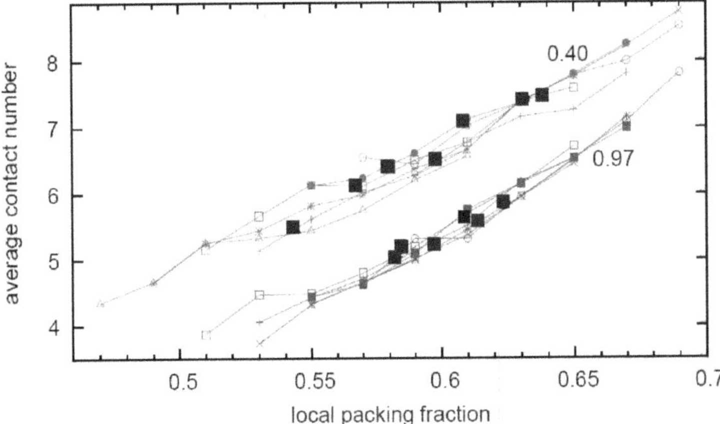

Fig. 6.6   The relation between contact number and packing density for non-spherical and (almost) spherical particles, as determined experimentally through X-ray microtomography. The non-spherical particles were placebo pills of approximately ellipsoidal shape, with an axis ratio of 0.40 (upper data set). The almost spherical particles had an axis ratio of 0.97 (lower data set). Each solid square represent an average over an entire sample, while the grey symbols indicate sub-sets of data obtained via Voronoi-tesselation and appropriate binning. Data kindly provided by Fabian Schaller, Göttingen/Erlangen [205].

6.2   Regimes of wetness

It is a well known experience that the humidity of the surrounding air has a sensible impact on the behavior of fine powders. While they can be easily poured from one container into another when the air is dry, they tend to form clumps, and do not run as easily down a slope, on a damp summer day. This is due to minute amounts of liquid adsorbing onto the grain surfaces from the vapor phase. As we have seen in Chapter 3, typical liquid film thicknesses well off coexistence, i.e., if the humidity is noticeably below 100%, are just a few molecular layers. Although there are then no fully developed capillary bridges between the grains, some liquid collects at the contact points and affects the mechanical behavior of the material. In order to understand the impact of liquids on the properties of granular piles, we therefore have to include considering even minute amounts of moisture. In what follows, we will classify certain regimes of wetness, starting from the dry state, and gradually increasing the liquid content.

6.2.1 *The humidity regime*

Exposing a dry granulate to humid air corresponds physically to the adsorption isotherms discussed in Section 3.1.2. On the grain surfaces, a molecularly thin liquid film forms which is bound to the surfaces mainly by van der Waals forces. Its thickness adjusts such that the film is in equilibrium with the vapor in the surrounding gas phase. As the humidity is further increased, the thickness of the adsorbed film increases, and the troughs of the roughness on the grains surfaces are filled with liquid, according to the discussion in Section 3.2. It is not trivial to predict what precisely happens in this regime at the points of contact between two rough grain surfaces. In particular, we expect that the liquid will start to form macroscopic capillary bridges at some partial pressure corresponding to the filling transition on the grain surface topography. In principle one can predict, on the basis of roughness topography data, *a priori* when this will happen. Alternatively, one may want to observe this transition experimentally. It is therefore desirable to find a way of looking inside a granular pile, in order to determine the morphology of the spatial distribution of liquid, and to detect and image the emerging liquid bridges directly. A rather simple way to achieve this goal is to replace the interstitial air with a liquid whose refractive index equals that of the grain material. This enables a deep look inside the pile by optical means, such as an optical microscope. For reference, we shall describe this method in some detail.

As the wetting liquid, one may use water with a fluorescent dye for staining, such as fluorescein. A mixture of toluene (CH_5OH) and di-iodomethane (CH_2I_2) may be used to replace the interstitial air. At a volume fraction of 11.9% of di-iodomethane, the refractive index of this mixture is 1.51, matching that of the glass spheres. The aqueous phase wets the glass beads completely ($\theta \approx 0$) in the presence of the index matching liquid, and forms capillary bridges which can be directly observed and counted. Perfect index matching cannot be easily achieved, since the refractive index of commercially available glass spheres varies slightly from bead to bead. However, sufficient transparency can be readily achieved to demonstrate how the morphology of the liquid structures within the granular pile changes as more liquid is added.

Samples can be prepared by adding ballotini (cf. Fig. 2.7), and subsequently the stained water, into a glass cuvette containing the immersion liquid. The samples are then shaken for a few minutes until the color (due to the added dye) appears evenly distributed. Immediately after sample

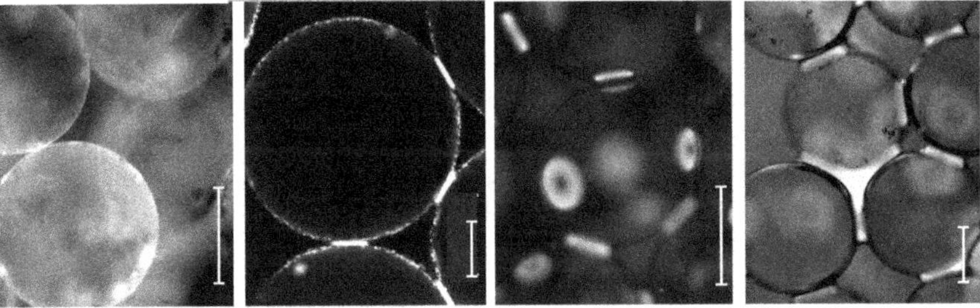

Fig. 6.7   Fluorescence microscope images for different liquid contents W. (a) $W < W_{cb}$, no capillary bridges have formed yet; all liquid resides within the roughness of the glass surface, which therefore appears more or less homogeneously illuminated. (b) $W \approx W_{cb}$, confocal image. Capillary bridges begin to form. Their content of liquid (and hence of dye) is still small enough for not completely overwhelming the liquid occupying the surface roughness. The latter is still clearly discernable. (c) $W_{cb} < W < W^*$, capillary bridges are fully developed. The sensitivity of the camera had to be reduced with respect to (a) to avoid overexposure, such that the liquid on the rough surface is not visible anymore.   (d) $W > W^*$, some of the capillary bridges have already merged forming larger clusters. Images taken by Mario Scheel, ESRF Grenoble. Scale bars: 200 μm.

preparation, the volume fraction of the glass beads is typically well below 0.6, but by gently tapping against the sample, packing densities of about 0.62 can be quite readily achieved.

Figure 6.7 shows optical micrographs of the interiors of samples prepared as described, with variable liquid content. Images (a), (c), and (d) have been taken by conventional fluorescence microscopy, while image (b) was recorded with a confocal fluorescence microscope. The images are compiled in ascending order of the liquid content. In the left image, (a), we have $W < W_{cb}$, and no capillary bridges have formed yet. However, as we have discussed in Section 3.2, in the context of capillary forces between rough surfaces, liquid resides in the troughs of the surface roughness (roughness regime, cf. Fig. 4.14), or is trapped at small protrusions wherever these come close to the opposite grain surface (asperity regime). This (inhomogeneous) liquid coverage is visible as the faint glow covering the grains. Image (b) is a confocal image taken at a liquid content corresponding roughly to W_{cb}, which depends of course on the roughness. Capillary bridges are now clearly visible, but their content of liquid (and hence of dye) is still small enough for not completely overwhelming the liquid filling the surface roughness.

As W is increased further, we enter what is called the *pendular regime*, where capillary bridges are the dominant morphology in which liquid is found within the sample. This will be discussed in detail in the next section.

For the moment, we stay with the *humidity regime*, which is represented by Figs. 6.7(a) and 6.7(b). By means of optical microscopy, we cannot reliably pin down the transition from the asperity regime to the roughness regime. This is, however, not to be considered a major setback, since for most real samples there will anyway be no clear separation between both regimes. What we can do is characterize the effect of humidity on the properties of the granular pile.

This has been done in a number of simple but elegant experiments. A quantity already introduced in Chapter 2 which is particularly sensitive to the addition of moisture is the angle of repose, Θ_R, of the pile. It can be measured in the 'rotating drum' already discussed (cf. Fig. 2.9), but also by simpler methods. A frequently used setup consists of just a cylindrical container with a central hole in the bottom, placed at some height above a horizontal plate (see right panel of Fig. 6.8). The granulate, which we may imagine to consist of glass beads again, is premixed with a well-defined amount of liquid, poured into the container and allowed to drain through the hole out of the container and onto the bottom plate. As the wetting liquid, a low-vapor-pressure liquid such as silicone oil should be used in order to prevent evaporation. On the bottom plate, a conical heap forms with a rather well defined cone angle. As it continues to accumulate grains on its tip, the cone angle exhibits a steady increase until it reaches a maximum angle of stability, Θ_M, and then decreases abruptly in an avalanche which takes it back to the repose angle, Θ_R. This continues as long as sand is added, with the actual angle oscillating between Θ_M and Θ_R, similar to the behavior observed with granulates in the rotating drum. A similar phenomenon appears in the upper container, with a funnel-shaped void forming in the granular heap. The flow eventually comes to rest when the funnel angle equals the angle of repose, and the funnel surface matches the rim of the hole in the bottom of the container. It is easy to imagine that both Θ_R and Θ_M are very sensitive to cohesion between the grains, i.e., the presence of liquid. They can be measured from the geometries of both the heap on the bottom plate and the funnel-shaped void formed in the container.

Typical results are shown in the main panel of Fig. 6.8. Starting from its value for the dry granulate, the angle of repose increases approximately in a linear fashion as liquid is being added. The overall behavior is very similar although it has been obtained in different laboratories with different samples. On the other hand, we see that absolute values may differ noticeably even if the sample parameters are very similar. This is not surprising since

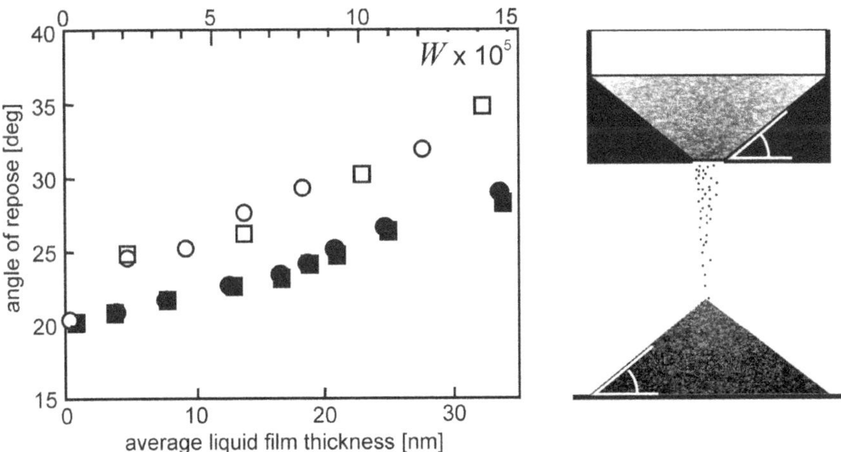

Fig. 6.8 The linear variation of the repose angle with the liquid content, when the latter is very small. Data are taken from [124] ($R = 400$ μm, closed symbols) and [42] ($R = 450$ μm, $\delta = 1$ μm, open symbols) where the impact of wetting sand piles with small amounts of non-volatile oils was investigated.

glass spheres with very similar radii may nevertheless differ substantially in their roughness topographies, which we have found to play a significant role in the wetness regime investigated here. Experiments of this type have been reported by several authors [124, 42, 108, 109], using glass spheres throughout. The results were generally along the lines of Fig. 6.8, and not qualitatively different from findings obtained with non-spherical granulates, like regular sand.

It should be noted, however, that experimental results obtained for the repose angle are difficult to link to a quantitative theoretical treatment. Halsey and Levine have shown that the repose angle should depend upon the height of the pile, and the exact form of that dependence is affected by the (unknown) constitutive relation of the granulate [123]. We will come back to this point in Chapter 7, where we will discus, among others, the specific physics applying to sand sculptures and similar structures.

6.2.2 *The pendular regime*

Let us return for a moment to the paradigm of gradually increasing the vapor pressure of the wetting liquid in the surrounding gas phase. When we have reached a vapor partial pressure leading to a moisture equivalent of W_{cb} in the pile, the system is already very close to liquid–vapor coexistence.

The curvature of the outer meniscus of the capillary bridges which just have formed is of order δ, and thus at least on the order of a micron (equivalent to a freshly polished surface) or more. It is illuminating to calculate what pressure difference to the saturated vapor pressure, p_s, this corresponds to. According to the Kelvin equation, Eq. (3.28), and considering water as the liquid, we obtain that the vapor pressure of a liquid structure with a mean curvature of a micron is reached about seven meters above the water table due to gravity (barometric formula). This corresponds, obviously, to a minute pressure difference which will be difficult to control in experiments. We therefore leave the paradigm of adsorption isotherms at this point, and demand henceforth that the liquid content of the sample, W, be controlled by just adding the desired amount of liquid, followed by some appropriate stirring process.

If the amount of liquid we add to the grains is sufficient to finally produce a situation as the one shown in Fig. 6.7(c), the sample is said to be in the pendular regime. This term refers to the fact that since capillary bridges always have a certain rupture distance, s_c, they can exist at places where the surfaces of neighboring grains come very close, but have no mechanical contact. The capillary bridge may then be called to 'hang' between the grains.[4] Let us consider the wet pile of grains to be composed at time $t = 0$ by bringing together a large number of individual grains, each with wetness w. The wetness of the resulting pile is then $W = w\phi$, and we assume now that $W \geq W_{cb}$. According to the discussion in Section 5.1, capillary bridges will form immediately with a volume of $\tilde{V}_i$. Once these initial bridges have been established, liquid starts to accumulate near the points of grain contact by virtue of the gradient in Laplace pressure. This can proceed either through the adsorbed liquid film [206, 207], or through the gas phase [208, 209] by evaporation and recondensation. It is difficult to assess the relative contributions of these two mechanisms in the real system. In particular, the roughness may influence the transport within the film noticeably [210–212]. It is therefore indispensable to study this process empirically. This has been done by Kohonen *et al.* [118], who investigated the growth of capillary bridges between glass beads by optical microscopy with a number of different liquids. It was found that the temporal evolution of the bridge volume followed, within experimental scattering, an exponential:

$$V(t) = V_f - (V_f - V_i)e^{-t/t_0}, \qquad (6.4)$$

---

[4]The latin word *pendere* means 'to hang'.

where t_0 is a constant (cf. Fig. 4.4). For V_i and V_f we have given expressions in Chapter 4. The time scale t_0 scales with the viscosity of the liquid, η. In piles of glass ballottini of a few hundred microns diameter, it was found to be on the order of five minutes for water at room temperature as the wetting liquid.

For several simple liquids, it has been shown that the equilibration dynamics through liquid wetting layers on surfaces with some roughness is remarkably weakly dependent on the adsorbed film thickness [211, 212]. It was found that effects of roughness and of dispersion forces partly cancel each other, such that the decay time of lateral variations in film thickness is quite constant over a wide range of adsorbed liquid coverage. For simple liquids like ethanol and propane, the lateral transport coefficient was found to be of order 10^{-10} to 10^{-11} m^2/s. For grains with a radius of few hundreds of microns, the corresponding time scale for liquid transport is then of order 10^2 to 10^3 s. This is remarkably close to t_0 as reported above. It is then natural to propose that t_0 should scale as ηR^2, but be only weakly dependent on the amount of adsorbed liquid.

It is furthermore important to check whether transport through the gas phase may be a viable route for the equilibration of capillary bridges. It was shown before that the transport within a wetting film and through the vapor phase which is in equilibrium with that film may be comparably effective [212]. In a wet granulate, however, the gas phase in the interstice is usually not the vapor of the liquid, but mainly consists of a different gas at much higher pressure. In the case of a granulate wetted with water at room temperature, the water vapor accounts only for a few percent of the gas molecules. Hence the diffusivity of gas molecules, which is proportional to the mean free path and thus inversely proportional to the pressure, is about two orders of magnitude smaller. We therefore do not expect transport through the gas phase to be noticeable in most cases of relevance. This is even more evident when the primary fluid (cf. Fig. 1.9) is not a gas but a liquid. This situation is encountered in oil reservoirs, but also in experiments which employ an immersion liquid as the primary fluid for index matching. Another effect which applies to many systems and impedes equilibration through the gas phase is osmotic stabilization, which is active when the wetting liquid contains a non-volatile solute in appreciable concentration.

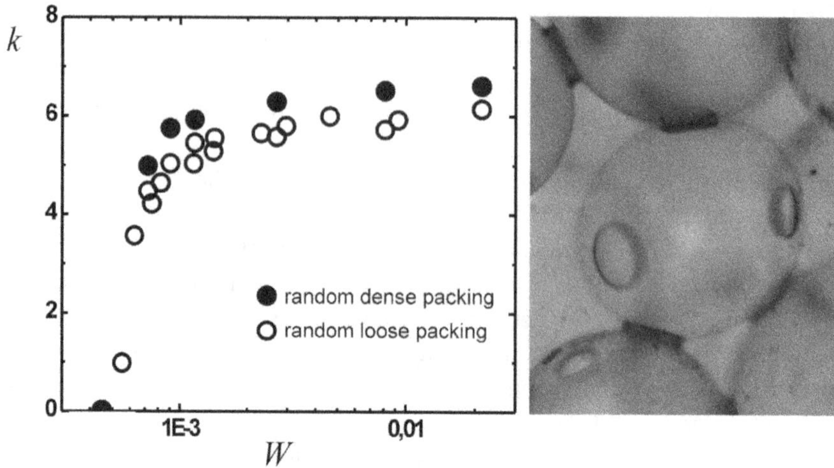

Fig. 6.9 The average number of capillary bridges on a spherical grain, as obtained by means of optical microscopy in a sample as those used for Fig. 6.7. On the right-hand side, an optical microscopy is shown under visible liquid illumination (corresponding to Fig. 6.7(c)) for illustration. The main panel shows a well-developed plateau at around six bridges on a grain, with some variation according to the packing density. Open circles: random loose packing, $\phi_{\mathsf{rlp}} \approx 0.57$. Closed circles: random dense packing, $\phi_{\mathsf{rdp}} \approx 0.62$.

As soon as the bridges are fully developed, their average number per sphere, k, can be determined by zooming with the microscope through the sample.[5] For samples like those shown in Fig. 6.7, this can be carried out by optical microscopy. Figure 6.9 shows a typical result for random loose packing ($\phi_{\mathsf{rlp}} = 0.57$, open circles) as well as for random dense packing ($\phi_{\mathsf{rdp}} = 0.62$, closed circles). We observe a well developed plateau at about six capillary bridges being (on average) in contact with each sphere. From this result, we can estimate the average capillary bridge volume as a function of the liquid content, W. We have

$$\tilde{V} = 8\pi W/3\phi k, \tag{6.5}$$

if we assume all liquid to go into the capillary bridges. Using $W = w\phi$, we check that this is identical to the expression for V_f we obtained in Section 4.2.1 with reference to w, cf. Eq. (4.21).

The robustness of the above result, that there are about six capillary bridges on each sphere, can be tentatively understood with the help of Fig. 6.10, where the number of capillary bridges is plotted as a function

---

[5]k is frequently referred to as the 'wet coordination number'.

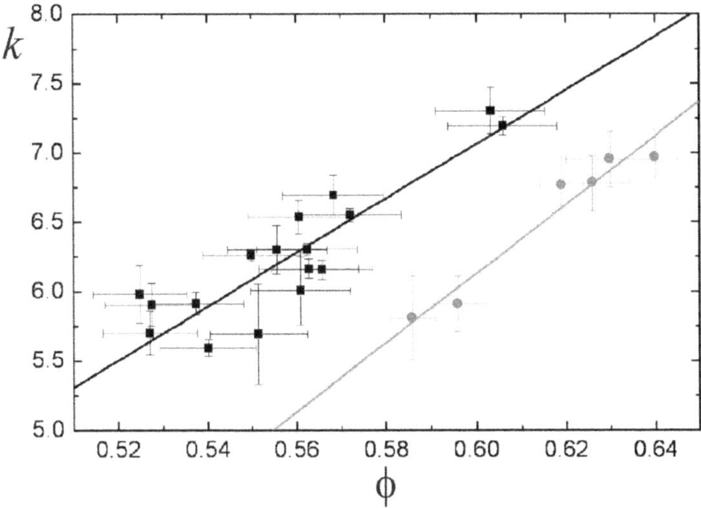

Fig. 6.10 The average number of capillary bridges in contact with each sphere in a wet pile (black), and the average number of contacts in a dry pile (grey) for reference. Although both data sets follow different overall characteristics, both numbers are in the same range. Data taken from [213].

of packing density (black squares), along with the number of contacts, k_c, for a dry pile of the same particles (grey disks). We see that two effects come into play which have opposing consequences. On the one hand, the addition of liquid reduces the overall packing density. This is well in line with the argument we raised in discussing Fig. 6.2, since the addition of liquid adds new channels of dissipation, and hence impedes the search of the pile for the densest packing geometry. On the other hand capillary bridges can span a certain distance ('pendular bridges'), such that they can persist at places where grain surfaces come very close, but do not form real mechanical contact. Hence there are many more candidate places for capillary bridges than are there real mechanical grain-to-grain contacts. As we see from Fig. 6.10, the result is that the range of contact numbers in the dry case agrees amazingly well with the range of capillary bridge numbers in the wet case, although the packing densities are significantly different.

So far, we can summarize that for not too large deviations from spherical shapes of the grains and for moderate polydispersity, packing densities are quite generally around 0.6, and there will be on average about six capillary bridges in contact with each grain in the pendular regime of wetness.

6.2.3   *The funicular regime*

As more liquid is added to the pile, liquid bridges occasionally coalesce to liquid aggregates. An example is shown in Fig. 6.7(d) for $W = 0.03$. This regime of wetness, where capillary bridges coalesce to larger aggregates, is commonly called the *funicular regime*.[6]

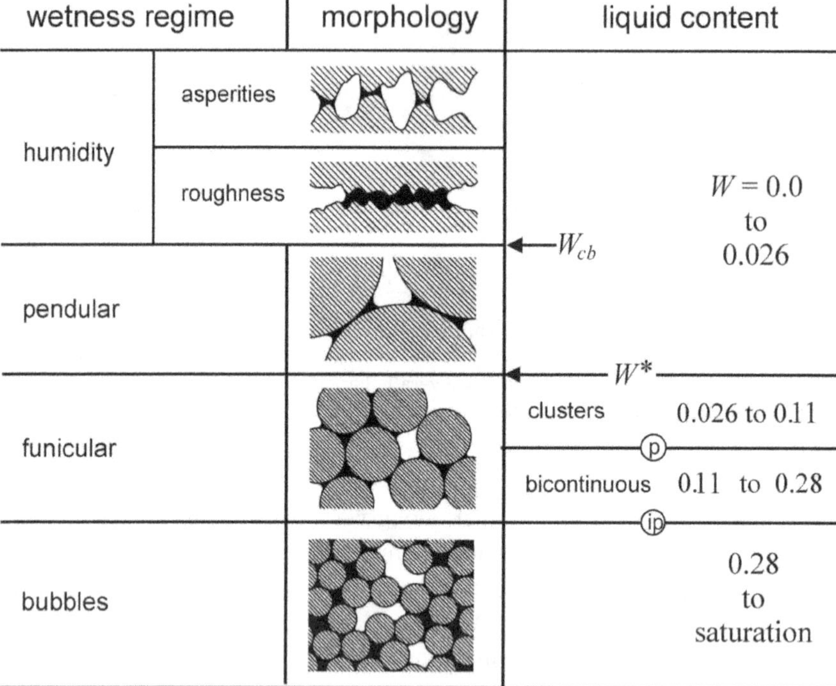

Fig. 6.11   The different regimes of wetness of a granular pile. The nomenclature has been chosen such as to optimize overlap with standard nomenclature, but to account as well for important new aspects, like the clearer distinction between the pendular, funicular, and bicontinuous regimes. In the right-hand column, p and ip indicate percolation and inverse percolation, as discussed in the main text.

---

[6]The term 'funicular' appears somewhat elusive. It refers originally to the latin word *funiculus*, which means 'a little bundle of ropes'. The allusion to ropes has, on the one hand, lead to the term 'funicular' being used in material sciences as synonymous for 'catenary', which is conceptually connected to minimal surfaces (cf. first paragraphs of Chapter 4), and thus to equilibrium morphologies of liquid interfaces. On the other hand, it alludes to something 'hanging together', providing another possible route to the use of 'funicular' for connected liquid clusters within a granular pile. In spite of its delphic pedigree, we use this term here, simply because it has become customary in the field of wet granular matter.

The different regimes of wetness are compiled in Fig. 6.11. We know already the regimes governed by asperities and by roughness, which together form the regime dominated by humidity. This term alludes to the fact that here we can still control the wetness via the vapor pressure. The next regime is then the pendular regime discussed in the previous section, where capillary bridges (or pendular bridges) are the dominant morphology in which liquid is found in the pile. The present section will be devoted to the regime where most of the liquid is found in clusters larger than a single capillary bridge.

In order to investigate the morphology of liquid clusters in the funicular regime with reasonable statistics, very many clusters must be characterized and imaged with high resolution. Optical methods rapidly come to their limits here, but modern X-ray micro-tomography provides enough resolution to collect data of sufficient quality. A sketch of a standard X-ray tomography setup is shown in Fig. 6.12. As the X-ray source, one uses a tightly focused electron beam impinging at an energy of order 100 keV on a cooled metal target which forms the window of the X-ray tube. This target consists of a thin metal film coating a high thermal conductivity substrate, such that the X-rays emerge from only a very tiny spot, directly into the space outside the tube. Spot sizes of the order of one cubic micron are available in state-of-the-art, stand-alone machines. The radiation emerging from this volume is then passed through the sample and collected on a camera chip, as sketched in the figure.

The diverging beam provides a natural means of magnification. If a small sample is placed close to the X-ray source, a large geometric magnification can be achieved. The spatial resolution is then limited by the size of the X-ray source. Since this can be as small as about one micrometer, the spatial resolution can be comparable to that of an optical microscope. This is comparable to the resolution of an optical microscope. Three-dimensional information is gained by recording a large number of X-ray shadows at different angular orientations of the sample. The latter is mounted on a rotation stage, and imaged in, e.g., 300 equidistant angular positions. It is then the task of a large computer to calculate the three-dimensional profile of X-ray absorbance within the sample from the large amount of collected data. Taking an X-ray tomogram of useful resolution with commercial stand-alone, in-house equipment usually takes two hours, during which the sample must be kept still with respect to the rotation stage.

With the vast amount of photons available at a synchrotron source,

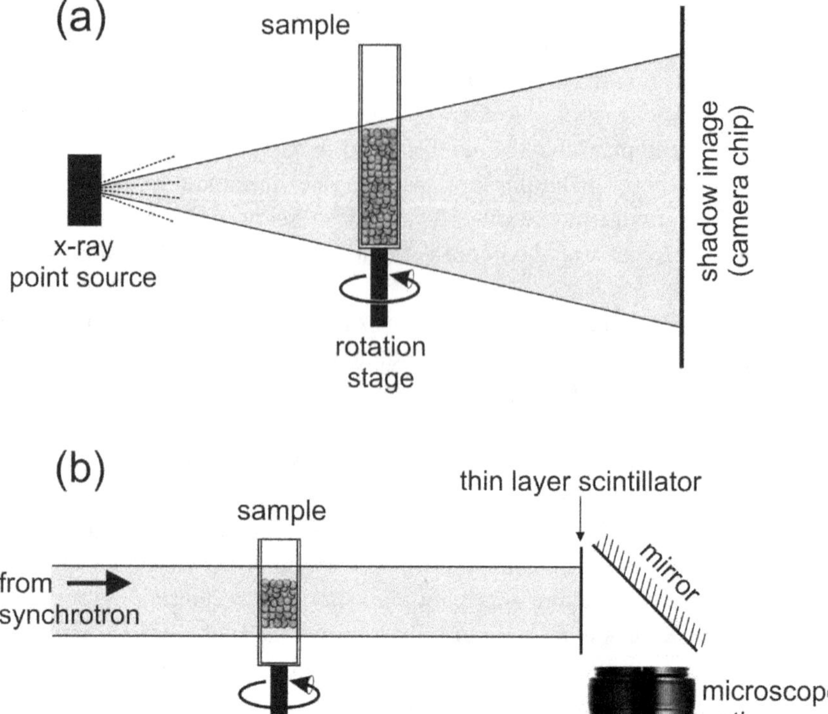

Fig. 6.12 Schematic representations of experimental setups for X-ray micro-tomography. (a) For smaller equipment which can be used in standard laboratories, bremsstrahlung from an electron point source is used. The diverging beam provides adjustable magnification over a wide range. Many (typically a few hundred) shadow images are takes at different rotation angles of the sample. The three-dimensional profile of X-ray absorption constant within the sample can then be reconstructed numerically. (b) Rapid image acquisition is possible when the X-ray beam emerging from a synchrotron is used. Owing to the high photon flux, sample damage must be reckoned with, and the camera and sensitive optics must be kept out of the beam. Acquisition of a full three-dimensional image may take less than a second.

three-dimensional images can be recorded much faster. The synchrotron beam is a parallel beam, however, such that there is no geometrical magnification, and the resolution is limited by the pixel size of the detector screen (scintillator). At the same time, the sample size is limited to the beam size, which is usually of order one centimeter. The spatial resolution is furthermore limited by blurring effects in the scintillator layer, which serves as a target and converts X-ray photons into visible light. Time resolution, however, is impressive: at ID15 in Grenoble, a three-dimensional image can meanwhile we recorded in about 100 ms. The observation of dynamics is thus mainly limited by the physical sample properties, such as by the impact of centrifugal forces on the characteristics of the rapidly rotating sample.

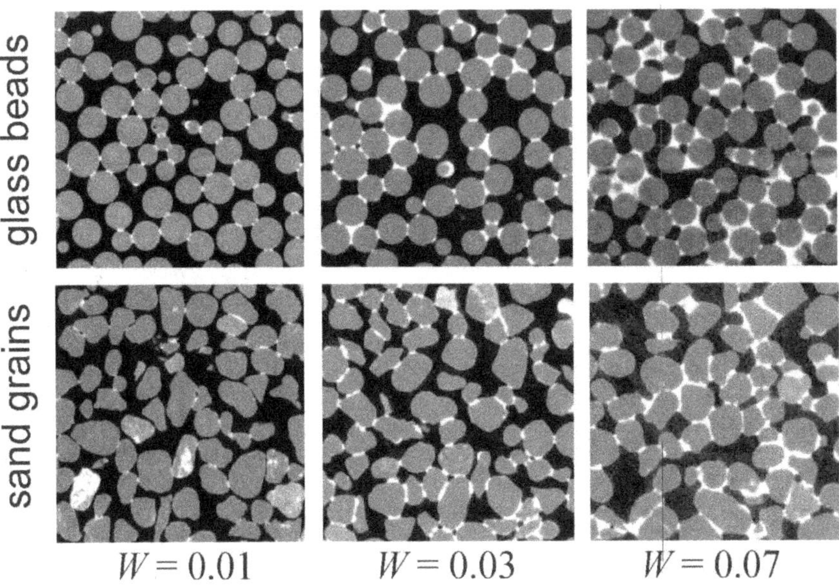

Fig. 6.13   Two-dimensional sections through X-ray tomographic renderings of wet granular piles at variable liquid content, W. Top row: glass beads. Bottom row: sand grains. Contrary to what can be found in some text books, the 'funicular state' is already found at $W = 0.03$.

In Fig. 6.13, X-ray tomography sections for three different values of the liquid content are shown, for glass beads as well as for sand grains. Obviously, many capillary bridges have merged, forming larger liquid clusters (funicular state) [214], already at $W = 0.03$. A closer examination shows that for spherical grains, the transition from the sole presence of individual capillary bridges to an abundance of larger liquid clusters (funicular state) takes place at $W = W^* \approx 0.025$ [214]. In what follows, we will chiefly discuss findings for spherical grains, and then return to the irregularly shaped sand grains for comparison.

Figure 6.14 shows a number of typical liquid structures encountered within a random pile of spherical beads, as obtained experimentally by means of X-ray micro-tomography. We immediately recognize the capillary bridge to the upper left. The 'roughness' of the surface comes about from the finite resolution of the method. It corresponds to the size of the image elements, called *voxels* (as the three-dimensional analog of the pixel, as the element of a two-dimensional picture). Clearly, it will be difficult to derive details such as the mean curvature from such renderings. The overall size of a capillary bridge, $2r_2$ (cf. Fig. 4.3), however, should be easily accessible. For the larger clusters, we see that most of them appear as being composed of individual capillary bridges, 'glued together' at their perimeters. The

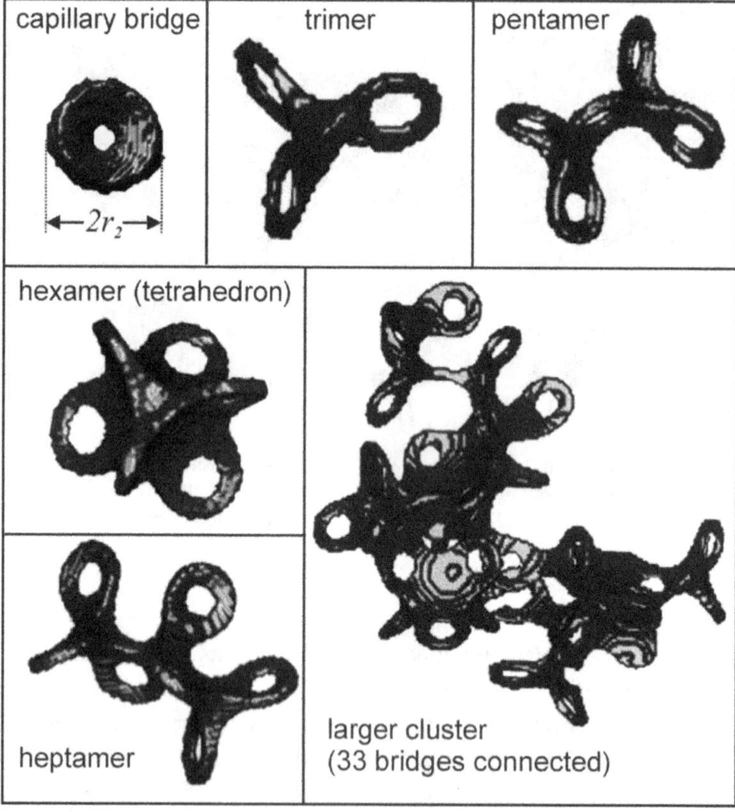

Fig. 6.14 A collection of liquid cluster morphologies found by X-ray micro-tomography within a wet pile of spherical glass beads. The names are given according to the number of capillary bridges out of which they can be thought of as being composed.

names of the clusters are formed according to the number of the bridges out of which they appear to be composed. Hence the pentamer, e.g., seems as composed of five individual capillary bridges. The hexamer corresponds to the bulk filling of the interstice within a tetrahedral packing of six spheres, connecting six bridges.

Further information can be gained if the liquid clusters detected in the X-ray micro-tome are sorted according to their volume and interfacial area [215]. In Fig. 6.15, each liquid cluster found in a sample of glass beads at a liquid content of $W = 0.03$ are represented as a dot in the plane spanned by its total surface (liquid–gas plus liquid–grain) and its liquid volume. Note the logarithmic scale. Clearly, all clusters fall approximately on a single 'scaling curve'. It lies well above the solid line, which represents the interstitial space. While there are still many individual capillary bridges (cb), many larger liquid clusters have already formed. They correspond to the distinct clouds of points at larger volumes. The width of the clouds is most probably due to the polydispersity of the glass beads (about 10% spread in diameter, corresponding to a polydispersity of 0.1).

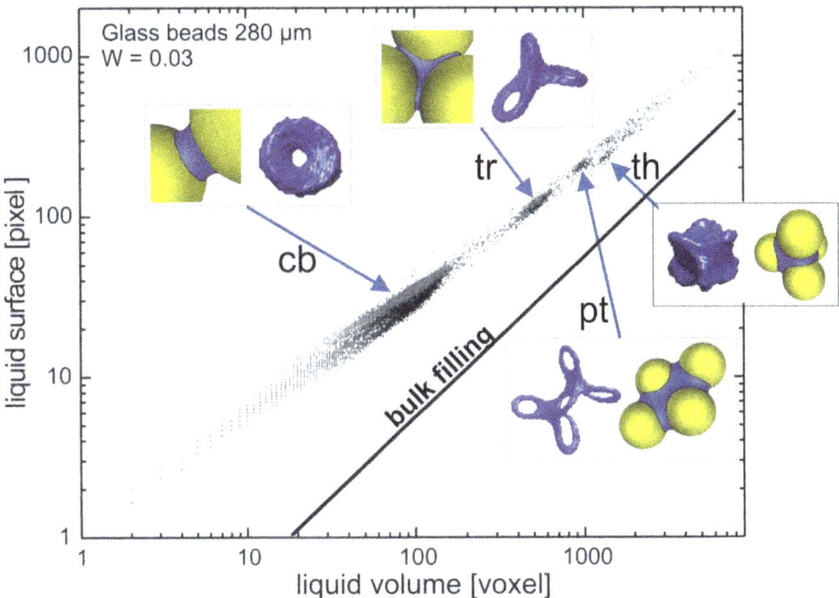

Fig. 6.15 Cumulative plot of liquid clusters found in a wetted pile of glass spheres by X-ray micro-tomography. Each liquid cluster is represented by a single dot (cb: capillary bridges; tr: trimers; pt: pentamers; th: tetrahedral filling). Data taken from [213].

Let us try to understand why the different clusters within the granular pile, which in principle have completely different morphologies, tend to align so nicely in the graph displayed in Fig. 6.15. It is illuminating to consider the Laplace pressure of these liquid structures. We start with the individual capillary bridge. Its Laplace pressure is given by

$$p_L \approx \gamma \left(\frac{1}{r_2} - \frac{1}{r_1} \right), \tag{6.6}$$

where we use the nomenclature of Chapter 4 (cf. Fig. 4.3). With a little trigonometry, we find

$$p_L \approx \frac{\gamma}{R} \left[\frac{1}{\sin\beta} - \frac{\cos(\theta + \beta)}{1 - \cos\beta} \right]. \tag{6.7}$$

It is useful to compare the various expressions for capillary forces and pressures we have meanwhile encountered. In Fig. 6.16, the capillary bridge force between spherical grains of radius R is plotted as a function of the bridge angle β, and normalized with respect to $2\pi R\gamma$. The solid horizontal line represents Eq. (4.26), which is just a constant and accounts quite well for experimental data [119]. The solid curve represents the Laplace pressure according to Eq. (6.7), times the projected area covered with liquid, $\pi(R\sin\beta)^2$. The dash-dotted curve emerges from the solid by adding the

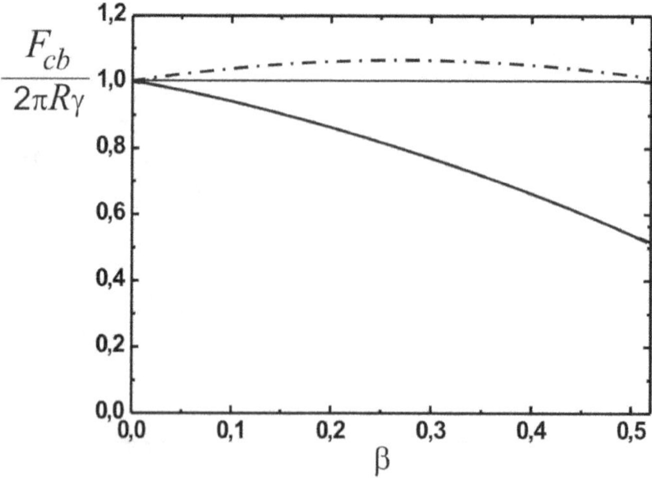

Fig. 6.16   Comparison of several expressions concerning the capillary force. Horizontal straight line: $F_{cb} = 2\pi R\gamma$ (Eq. (4.26)). Solid curve: wetted area times Laplace pressure, as derived form Eq. (6.7). Dash-dotted: as solid curve, but with addition of the perimeter force term, $2\pi\gamma R\sin\beta$.

'perimeter force term', $2\pi\gamma R\sin\beta$. The agreement with Eq. (4.26) (horizontal straight line) is good. We should keep in mind that the perimeter term becomes significant as β comes close to $\pi/6$, such that quantitative estimates of capillary forces should not rest on considerations of Laplace pressure alone.

We see from Eq. (6.7) that the Laplace pressure depends, aside from material constants like the surface tension, the contact angle, and the grains radius, only on the 'bridge' angle, β. This angle, however, is geometrically constrained by the arrangement of the spheres in the packing structure. The corresponding geometry is shown in Fig. 6.17 in a simplified two-dimensional representation. Consider first just a single contact between neighboring grains, and the capillary bridge formed between them. We may assume that in a three-dimensional arrangement there is a high probability that one of the grains in the immediate vicinity touch both grains, such that there are three grains in mutual contact (such a geometry is shown, e.g., in Fig. 3.5(b)). We will call such arrangement a *contact trimer* below. In the sketch in Fig. 6.17, this is the case for the three leftmost grains. It is obvious that none of the capillary bridge in such contact trimer can ever reach a volume corresponding to $\beta > \pi/6$, for geometric reasons. As this angle is reached, neighboring capillary bridges will merge their three-phase contact lines and coalesce, forming a larger liquid cluster. If we insert $\beta = \pi/6$ in

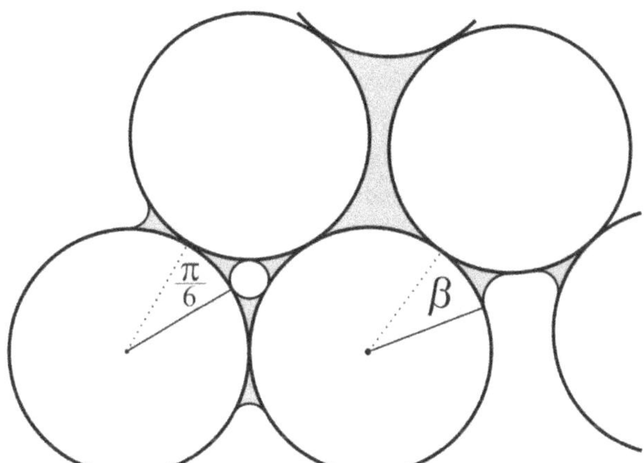

Fig. 6.17 Two-dimensional sketch of a liquid cluster and capillary bridges (grey shading) between piled spheres (white circles). The angle β is used to characterize the size of capillary bridges or outer loops os clusters. β cannot exceed $\pi/6$ in an environment of three spheres in mutual contact ('contact trimer', lower left).

Eq. (6.7), we obtain

$$p_L\left(\frac{\pi}{6},\theta\right) = \frac{2\gamma}{R}\left[1 - \frac{\cos(\theta + \pi/6)}{2 - \sqrt{3}}\right] = \frac{2\gamma}{R}\left[1 + 1.87\sin\theta - 3.23\cos\theta\right],$$

(6.8)

which for complete wetting ($\theta = 0$) yields

$$p_L\left(\frac{\pi}{6},0\right) = -4.46\frac{\gamma}{R},$$

(6.9)

which should be a good approximation for aqueous liquids wetting, e.g., a pile of glass beads. It then denotes the Laplace pressure within the capillary bridges when these start to coalesce.

The corresponding value of W will be called W^*. This marks the transition from the pendular to the funicular state (cf. Fig. 6.11). From Eq. (6.5) we have

$$W^* = W\left(\beta = \frac{\pi}{6}\right) = \frac{3k\phi}{8\pi}\tilde{V}_f\left(\beta = \frac{\pi}{6}\right).$$

(6.10)

For $\phi = 0.6$ and $k = 6$, this yields $W^* = 0.026$. This agrees with the finding reported above that at $W = 0.03$ there is already a large number of clusters, thus identifying this regime of wetness as already funicular.

Returning to the discussion of Laplace pressure, we should remark hat in the engineering and geo-scientific community, one customarily uses [216, 217] the term[7]

$$\text{matric suction} = -p_L.$$

(6.11)

In fact, as p_L is negative in most cases, the use of the matric suction to describe the state of the interstitial liquid is sometimes less confusing, since it is represented by a positive number. We continue the discussion here using the Laplace pressure instead, for the sake of consistency.

The significance of $p_L(\frac{\pi}{6},\theta)$ extends in fact well beyond the capillary bridges. We see this when we look more closely at the morphology of the larger clusters. As evident from Fig. 6.15, they are terminated at their periphery invariably by liquid loop structures, each resembling a capillary bridge. We can characterize such loops by the opening angle β of their contact line, just as we did with the individual capillary bridges. This is evident again from Fig. 6.17. If the two spheres between which the loop or bridge is located belong to a contact trimer, it is clear that this angle cannot exceed $\pi/6$, just as for single capillary bridges. Since the Laplace pressure is governed by the radius of curvature at the perimeter

---

[7]This refers to the negative pressure (suction) between the scaffold of grains (matrix).

of the loop, r_1 (cf. Fig. 4.9), its maximum value (i.e., the minimum of the matric suction) corresponds to the geometry at $\beta = \pi/6$. As the clusters grow, the probability that at least one of its terminal loops belongs to a contact trimer approaches unity. Thus the Laplace pressure corresponding to $\beta = \pi/6$ (Eqs. (6.8) and (6.9)) becomes an upper bound for the pressure (or a lower bound for the matric suction) in large clusters. At the same time, the addition of wetting liquid drives the systems towards this limit as it tends to increase the pressure. The inner geometry of the pile thus pins the pressure in the clusters to $p_L(\pi/6)$, which is given solely by the size of the grains, the surface tension of the liquid, and the contact angle.

In order to test this far reaching prediction, it is be desirable to measure the Laplace pressure of liquid clusters independently. This can be done again by means of X-ray micro-tomography. In principle, the Laplace pressure can be derived if the interfacial tension is known and if one can measure the mean curvature of the liquid interface. Since the tomography allows to locate the interface, this should in principle be possible. However, as the calculation of the curvatures involves differential operators which are sensitive to any kind of noise in the data, this turns out not to be feasible with sufficient resolution in a real system.

We thus need to follow a different strategy, which conceptually proceeds in two steps. First of all, we measure the time evolution of the volume of a number of capillary bridges after sample preparation. Typical results are shown in Fig. 6.18 for a pile of glass beads (left panel) and a pile of ruby beads (right panel). Clearly, the volumes of different capillary bridges within the same sample immediately after sample preparation ($t = 0$) is dramatic in both cases. Volumes easily vary over an order of magnitude. In the glass pile, these differences are strongly reduced over time, such that after about ten minutes, the volumes of all capillary bridges lie within a factor of two. We recognize the (roughly) exponential decay discussed above (Eq. (6.4)). In strong contrast, there does not seem to be any such equilibration within the ruby pile, even though the time span of observation is more than twice as long as for the glass beads.

The explanation for this marked difference is straightforward. The aqueous liquid wetted the glass beads well, while it made a contact angle of about 55 degrees with the surface of the ruby beads. According to the discussion in Section 3.2, we should under these circumstances not expect any percolated wetting layer in the troughs of the surface roughness on the ruby beads, which could sustain any noticeable lateral transport of liquid. The absence of equilibration in the case of large contact angle corroborates the

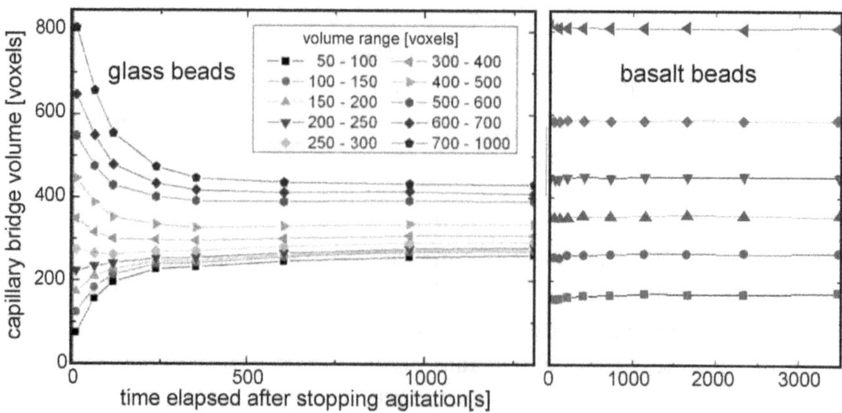

Fig. 6.18   Time dependence of the volume of several capillary bridges after sample preparation, as revealed by X-ray micro-tomography. Left: Glass beads, which are well wetted by the aqueous liquid phase, such that the liquid film on the grains is percolated. Right: Ruby beads, same vertical scale. The contact angle of the liquid phase is about 55 degrees here, and the liquid film in the troughs of the roughness (if there is any) is certainly not percolated. Data taken from [213].

idea put forward above that transport through the gas phase is negligible. In the pile of glass beads, however, the time constant of equilibration is very much in line with what has been found before for single capillary bridges by optical microscopy [118].

The driving force for the equilibration can be only the Laplace pressure, which depends strongly on the volume of the individual bridge. The explanation for the finite spread in capillary bridge volumes which remains at late times is straightforward as well. The Laplace pressure depends not only on the bridge volume, but also on the separation of the grains surfaces and on the exact size of the grains between which a capillary bridge is located. Hence a variation in these latter parameters necessarily leads to a variation in volume if one demands the Laplace pressure to be the same for each bridge. If the Laplace pressures of different liquid capillary bridges equilibrate with each other, so must the Laplace pressures of all clusters within the granular pile. It is then sufficient to find one cluster morphology for which the Laplace pressure can be reasonable well measured, and the measured value can be assumed to be valid as well for all other cluster morphologies.

In search for such easily accessible morphology, we arrive again at the individual capillary bridge. It turns out that even after most capillary bridges have merged into larger clusters, there are still a large number of them remaining, which can be used for pressure measurements. For capillary bridges, it is easy to measure the size (i.e., the radius of the contact line) with sufficiently high precision (cf. Fig. 6.14). From this we can infer the Laplace pressure, because the overall geometry of the capillary bridge is known. It follows from the above discussion that the matric suction is well defined for well wetted grains, but may be completely ill-defined if the contact angle is too large. According to Fig. 6.18, a contact angle of 55 degrees may already be 'too large'.

The results of a series of measurements of the Laplace pressure of liquid clusters in a well wetted pile are compiled for variable liquid content in Fig. 6.19. The horizontal grey bar indicates the prediction according to Eq. (6.9), its width corresponds to the (small but finite) contact angle hysteresis. As predicted, the Laplace pressure indeed clings to the grey bar within experimental error, for all $W > W^*$. It is enlightening to notice the

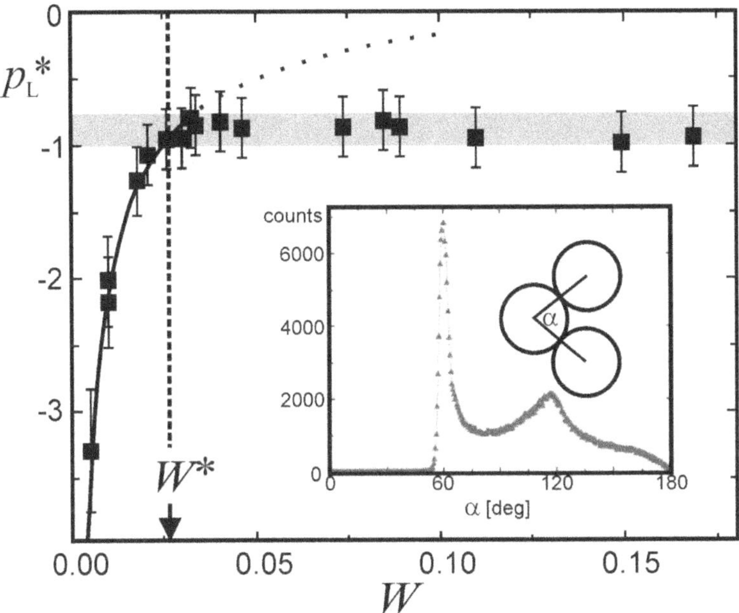

Fig. 6.19 The Laplace pressure of the liquid clusters as a function of the liquid content. The solid curve indicates what is expected for single capillary bridges. At W^*, the capillary bridge angle reaches $\beta = \pi/6$. The grey bar indicates what is expected for p_L on the basis of Eq. (6.8), taking contact angle hysteresis into account as a source of uncertainty. Inset: Probability distribution of angles found in trimers of grains. Clearly, the angle of 30 degrees ($\pi/6$) dominates. Data taken from [213].

inset, which shows the distribution of angles between triples of contacting spheres. The sharp peak at $\alpha = 30$ degrees corresponds to the contact trimers, which are the decisive geometry for determining the pressure. The high intensity of this peak demonstrates the dominance of this geometry in the pile, and thereby helps to appreciate the robustness of the result expressed in Eq. (6.9).

We thus arrive at the important result that in the wide range of liquid content where capillary bridges can coalesce, the pressure in each liquid structure, including capillary bridges, is given by the universal value $p_L(\pi/6)$, which only depends on the surface tension of the liquid, the grain size, and the contact angle, but not on the details of the packing geometry, or the size of the liquid clusters. This result pertains to all piles which are sufficiently well wetted by the liquid. No such result can be derived for poorly wetted piles, in which therefore the matric suction must be concerned as ill-defined.

Let us follow the development of the number of different species of clusters as we gradually increase the liquid content. By means of X-ray micro-tomography, we can identify and count each species. The result for a pile of about 10^4 glass beads wetted by an aqueous liquid is shown in Fig. 6.20. At $W = 0.01$, which lies between W_{cb} and W^*, there are about 3×10^4 individual capillary bridges (cb) being detected, corresponding to $k \approx 6$ (full circles). When $W \approx W^*$, larger clusters are massively generated. That this is not strongly visible in the number of capillary bridges owes to the logarithmic scale.

As W is increased beyond W^*, more bridges coalesce forming small clusters (tr, pt, th, ht). At the same time, these small clusters coalesce into larger ones. As a result, only the number of capillary bridges decreases, and only the number of clusters larger than heptamers (ht), which is indicated by the solid diamonds (larger), shows a significant increase. Beyond about $W = 0.05$, this exchange between different cluster species has reached some kind of a steady state, and all numbers gradually decrease as W increases further.

Between $W = 0.075$ and $W = 0.11$, however, there seems to be some kind of a transition in this coalescence process. While for smaller W the number of large clusters roughly follows the number of trimers (tr), it follows the number of pentamers (pt) for larger W. Note that the difference between the number of trimers and the number of pentamers is almost an order of magnitude.

A tentative explanation can be gained if one does not discuss the

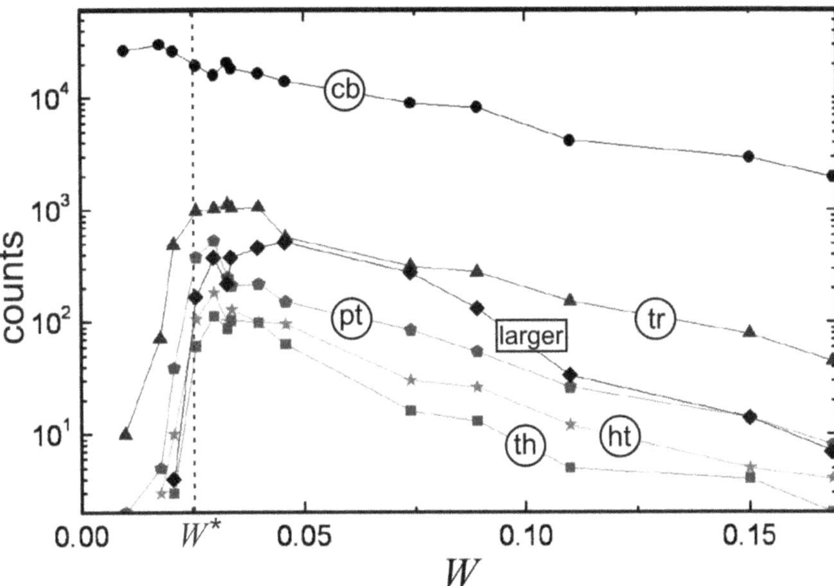

Fig. 6.20 Statistics of the individual morphologies of clusters found in a pile of about 10.000 spheres, at variable morphology. The most frequent species is the capillary bridge (cb), followed by trimers (tr), pentamers (pt), heptamers (ht), and tetrahedral fillings (th). Furthermore, larger clusters are found, which undergo a percolation transition between $W = 0.075$ and $W = 0.11$ (modified after [213]).

number of cluster species, but their volume. In Fig. 6.21, the volume of the largest cluster is plotted, normalized with respect to the total volume of all clusters. The solid symbols correspond to clusters of liquid in an interstitial space which is filled with air elsewhere. The open symbols correspond to clusters of air in an interstitial space which is filled with liquid elsewhere. The squares and circles represent random dense and random loose packing, respectively.

As we can see, there is a marked increase in the volume of the largest liquid cluster between $W \approx 0.075$ and $W \approx 0.11$, where this volume increases by almost two orders of magnitude. We assign this transition to the percolation transition, where most of the liquid within the pile becomes connected to each other.[8] A large cluster of liquid encountered in this

---

[8]Strictly speaking, percolation is defined as the emergence of one cluster spanning the whole sample [218]. Here we consider the point where all (or most) of the liquid belongs to the same large cluster. In the experiments presented here, it was difficult to reliably resolve the difference between these two transitions. With the help of electrical conductance measurements, one could probably do so if desired.

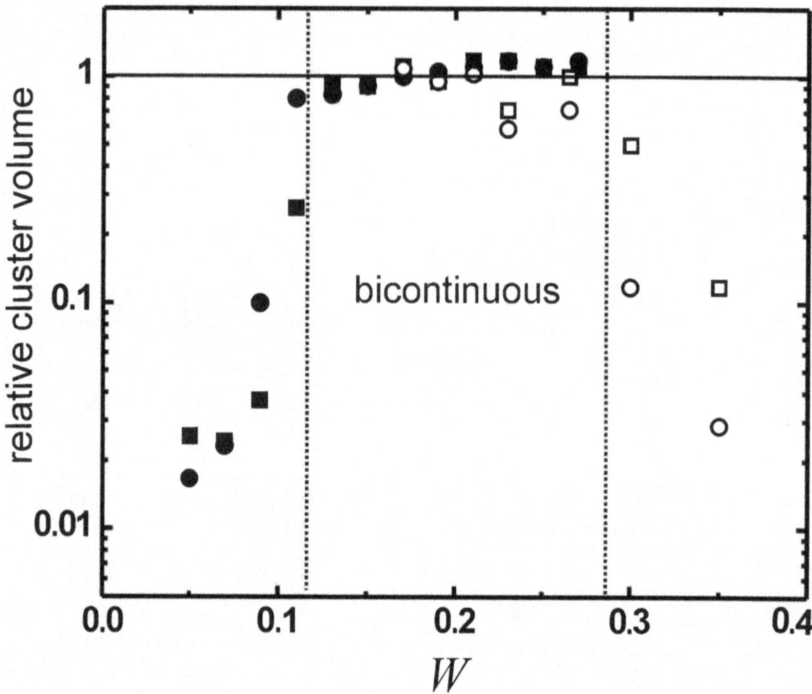

Fig. 6.21   The volume of the respectively largest cluster, divided by the total volume of the corresponding phase. Closed symbols: wetting liquid (secondary phase). Open symbols: interstitial phase (primary phase). Squares: rdp; circles: rlp. In the range of $W \in [0.11, 0.28]$, both phases are almost exclusively stored in their largest cluster, which is then percolated. Hence the system is bicontinuous in this regime.

regime is shown in Fig. 6.22. When this has happened, a *Paramaecium* cell would be able to swim from one end of the sample to the other without having to traverse a dry area.

In addition, the connectedness of the liquid entails a particularly rapid transport across the sample when variations in the matric suction occur. This may be the case, e.g., when shear bands appear in a sheared granulate [219]. In the shear band region, the volume fraction tends to be reduced, and hence the matric suction. As a result, one may expect the liquid will drain out of the shear band rapidly, with immediate consequences on the local mechanical properties. That this is indeed the case has been shown by Mani *et al.* both experimentally and by computer simulation [220]. For the sake of conciseness, we focus in the present book on processes which leave

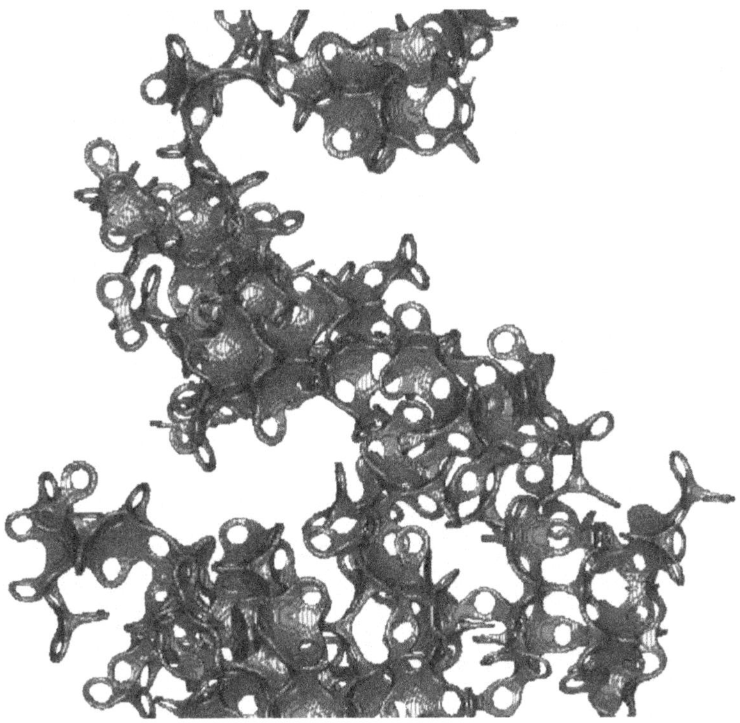

Fig. 6.22   A particularly large cluster found at intermediate liquid content, above the percolation transition ($W \approx 0.12$). Image from [213].

the granulate homogeneous, and do not dwell any further on superstructures such as shear bands.

When percolation of the liquid has been reached at $W = 0.11$, the largest air cluster seems to span almost all of the space filled with air. Hence the latter represents a percolated structure as well: a mite or bug could crawl from one end of the sample to the other without the danger of drowning. We therefore call this regime the *bicontinuous regime* (cf. Fig. 6.21). Only when the liquid content exceeds $W \approx 0.28$ is a marked deviation from this state observable. At $W \geq 0.3$, the air forms individual bubbles in interstitial space filled with liquid everywhere else. Remarkably, neither of the two transitions, the percolation of the liquid clusters or the 'inverse percolation' of the air, seems to depend strongly on the packing density.

Note that in the bicontinuous regime, air (or the primary fluid) can still flow freely through the sample. It is directly connected to the outside, such that the pressure is kept strictly constant, at least under quasi-static conditions. As soon as W grows beyond about 0.028, this changes qualitatively. Here we have the primary fluid contained in occluded bubbles, and its pressure is set by the local conditions within the medium. Flow of primary phase through the pile is not possible anymore. One may expect significant changes in the rheology of the granulate at these transition points. Furthermore, one may expect that soils behave completely differently concerning their role as a habitat for many different species, depending on whether they are in the funicular, the bicontinuous, or the bubble regime.

6.2.4   *The bubble regime*

It is instructive to discuss the behavior of the interfacial free energy of the system as a function of the liquid content. As long as the liquid content is small such that there are only isolated capillary bridges in the pile, the free energy of one such bridge is γ times the area of the free liquid surface at its perimeter, minus $\gamma \cos\theta$ times the wetted area. The second term directly follows from the Young-Dupré equation, Eq. (3.4). We readily find

$$\mathcal{F}_{cb} = 2\pi R^2 \gamma (1 - \cos\beta) \left[\frac{(\pi - 2\theta - 2\beta)\sin\beta}{\cos(\theta + \beta)} - 2\cos\theta \right]. \tag{6.12}$$

The total free energy of the capillary bridges is then given by $\mathcal{F}_{\text{tot}} = \mathcal{F}_{cb}nk/2$, where n is the number density of grains in the pile. If we plot this as a function of the total liquid volume stored in the bridges, we obtain a curve like the one sketched in Fig. 6.23 in bold solid. As W is increased, the free energy is reduced, reflecting the fact that the Laplace pressure is negative and tends to suck more liquid in. For fixed W, the Laplace pressure is the slope of the solid curve. When $W = W^*$, we arrive at the end point of the curve (open circle, $W = W^*$).

Beyond this point (i.e., for larger liquid content) the capillary bridges cannot anymore be the only morphology present in the sample, since their volume is limited by the maximum bridge angle, $\beta = \pi/6$. As we have seen, larger clusters will form as we add more liquid. We found that their Laplace pressure is largely constant due to their particular morphology. Hence they must be represented in Fig. 6.23 by curve segments which all touch the dotted line. This has the same slope as the solid curve at its end point, and must emerge from that point. T he dotted line, which (at least approximately) represents the free energy of the larger clusters, $\mathcal{F}_{cls}$, is therefore given by

$$\mathcal{F}_{cls} = \mathcal{F}_{\text{tot}}(\beta = \pi/6) + (W - W^*)p_L, \tag{6.13}$$

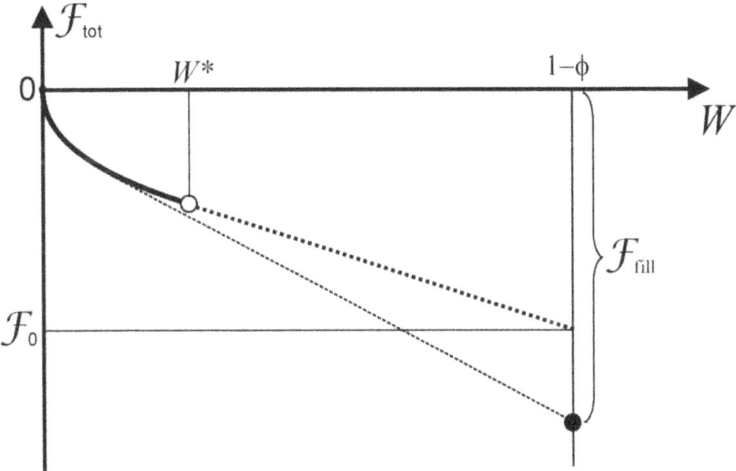

Fig. 6.23 Variation of the total free energy of the system with the liquid content (see text).

where according to Eq. (6.12) the first term is given by

$$\mathcal{F}_{\text{tot}}(\beta = \pi/6) = \pi n k R^2 \gamma (2 - \sqrt{3}) \left[\frac{(\frac{\pi}{3} - \theta)}{\cos(\theta + \frac{\pi}{6})} - \cos\theta \right], \qquad (6.14)$$

and W^* is given by Eq. (6.10).

When the liquid content approaches saturation, $W = 1 - \phi$, we would naively expect that the free energy, labelled $\mathcal{F}_0$ in the figure, must then be equal to the total free energy gained upon filling the entire interstitial volume. This is obviously given by

$$\mathcal{F}_{\text{fill}} = -n 4\pi R^2 \gamma \cos\theta = -3\phi \frac{\gamma}{R} \cos\theta. \qquad (6.15)$$

This does not depend on k while the above expression for $\mathcal{F}_{tot}$ does. Hence the two quantities cannot be identical. More specifically, if $\mathcal{F}_{\text{fill}} > \mathcal{F}_0$, we expect that the open morphologies we have compiled in Fig. 6.15 are the stable morphologies throughout, and we will gradually fill their pockets until the whole interstitial space is filled. If $\mathcal{F}_{\text{fill}} = \mathcal{F}_0$, i.e., if $\mathcal{F}_{\text{fill}}$ lies on the common tangent to all of the morphologies found from $W = W^*$ to $W = 1 - \phi$, they coexist with each other in a thermodynamic sense. If, however, $\mathcal{F}_{\text{fill}} < \mathcal{F}_0$, the dense filling of the interstitial space is the most favorable configuration. In this case, it shares a common tangent with capillary bridges at a volume below W^* (dashed line in Fig. 6.23), but not with any of the larger clusters. Hence the capillary bridges will then directly decay into homogeneously filled regions which only contain grains and liquid.

By setting $\mathcal{F}_{\text{fill}} = \mathcal{F}_0$, we can obtain a phase boundary in the space spanned by the packing density, ϕ, the contact angle, θ, and the bridge number, k. This boundary is shown, in a projection on the plane spanned by the former two variables, in Fig. 6.24. The three curves correspond to values of $k \in 5, 6, 7$. Obviously, the dependence on k is not very strong. For random piles of spheres, we see that we are invariably in the regime of open structures, as we have found in experiments. However, these systems dwell close to the phase boundary. In fact, with hydrophobized grains one frequently observes morphologies which rather correspond to the closed clusters, i.e., interstitial space homogeneously filled with liquid. In particular, we see that for a hexagonal crystalline packing (either hcp or fcc) with

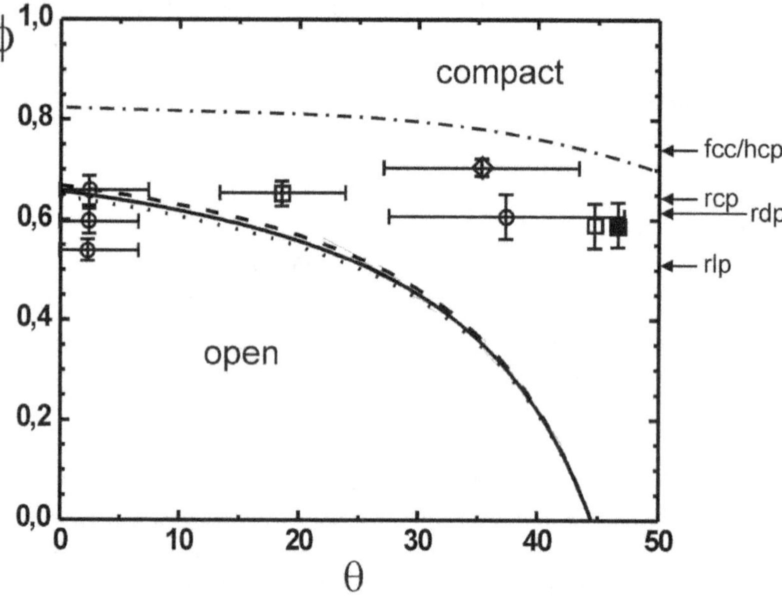

Fig. 6.24   Approximate phase diagram delineating where we should expect to find closed (compact) liquid structures or open structures, respectively. Below the solid line, which corresponds to $k = 6$, only open structures are stable. The dashed and dotted lines are for $k = 7$ and $k = 5$, respectively. Above the dash-dotted line, we should only expect compact morphologies. In the intermediate regime, the morphologies to be found depend upon the history of the system. Open structures prevail in practice because any agitation disrupts existing compact clusters, but there are large barriers for open clusters to decay into compact ones again.

$\phi_{\text{fcc}} \approx 0.74$, we should expect dense liquid clusters even for completely wetting liquids. To the best of the author's knowledge, this rather striking prediction has so far not been investigated.

By calculating the stability of fillings of the interstice between four spherical grains in mutual contact (tetrahedral geometry), one can estimate under which conditions the compact morphology would be the most stable one. This is indicated by the dash-dotted curve in Fig. 6.24. Above this curve, we should only find closed, bulk like liquid filling. Below this curve we should expect a range where the morphology of a cluster is history dependent. As the data show, however, we find almost only open structures in this intermediate range. An obvious explanation is that any stirring or agitation process leads to disruption of clusters, and their reintegration may take very long time. In contrast, there is no natural process which condenses disrupted structures into compact morphologies. Hence we should not be surprised to find primarily open morphologies in the intermediate regime, and only occasionally compact ones.

Another perspective on the liquid cluster morphologies can be gained by resuming the discussion of Fig. 6.15. In Fig. 6.25, similar data are compiled, but in a rescaled way. The surface of the clusters is normalized with respect to the surface of the interstitial space, which is represented by the solid line in Fig. 6.15. For increasing cluster volume, the dots representing the clusters approach a horizontal line well above unity. This indicates the large inner surface, which largely exceeds the surface of the interstitial space. As we add more liquid, larger and larger clusters are encountered. For each value of W, the largest cluster is indicated by an open circle. They all have more or less the same surface-to-volume ratio, corresponding to the horizontal 'asymptote' at $\tilde{S} = 4.0$.

It is straightforward to check that the formation of clusters proceeds in a spatially random, uncorrelated fashion. We can consider the formation of clusters as a bond percolation problem [218]. Capillary bridges are 'glued together' at their perimeters by means of filling the nooks in between them. If this proceeds in a random fashion, percolation theory predicts that the number of clusters of a certain size scales as their size to the power of -2.18 [218]. This is well fulfilled for our samples, as shown in the inset in Fig. 6.25 for two different values of W. In this double-logarithmic plot, the solid line has a slope of -2.18, in reasonable agreement with the data. Hence the growth of clusters is at least consistent with the idea that coalescence of small clusters to larger ones proceeds in a stochastically independent fashion.

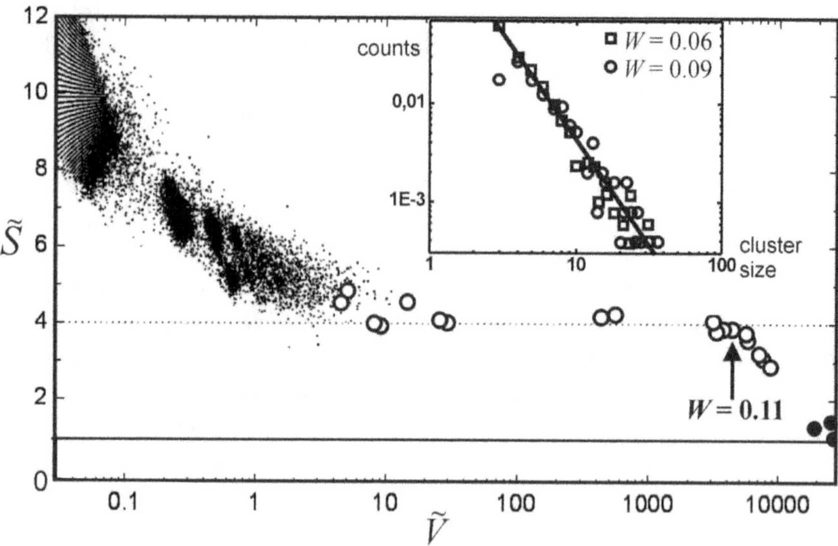

Fig. 6.25　Liquid clusters represented in a similar way as in Fig. 6.15, but with the total cluster surface scaled with respect to what corresponds to the solid line in Fig. 6.15. The small black dots are again for $W = 0.03$. As W is increased, the largest cluster is analyzed and represented by the open circles, respectively. As W comes beyond 0.11, the data leave the dotted line and approach compact filling at large total liquid volumes. Inset: Comparison of the scaling of cluster size with numb er of clusters with the prediction from percolation theory, for two different values of W. Adapted with modifications from [213].

As W exceeds 0.11, however, the behavior changes qualitatively. The data points (open circles) bend off the horizontal line and gradually approach the data obtained for (almost) completely filled samples (full circles). Comparing with Fig. 6.21, we see that this crossover takes place when the liquid percolates within the sample. It is not surprising that percolation plays a decisive role here, as it changes completely the mode by which clusters can grow. When there are still many small clusters, growth can proceed by just connecting small clusters to form larger ones. As soon as most of the liquid is contained in a single cluster, however, this process becomes scarcely possible. Additional liquid must be stored instead within the nooks of the clusters which are already there. These are then inflated until the interstices are completely filled.

Somewhere on the way from percolation of the liquid to complete saturation, the volume fraction of the interstitial air will become smaller than the percolation threshold, and be trapped in occluded bubbles. As Fig. 6.21

shows, this happens around $W \approx 0.28$. As it has no morphological consequences on the liquid structures, it does not show up prominently in Fig. 6.25.

We should finally compare the above findings, which have been invariably obtained with idealized grains such as ballottini, with natural granulates. We do not want to show the full data set here, but present in Fig. 6.26 a comparison of glass beads and sand grains, compiled in the same way as in Fig. 6.15. We clearly see that the behavior of the data is strikingly similar. The only pronounced difference is that in the sand sample, the little 'clouds' representing the different cluster morphologies are not resolved anymore. The overall scaling of the surface of clusters with their volume, however, is very nicely reproduced. Whatever may be hidden behind this commonality, it corroborates the overall impression which recurred already several times in this book: that the properties of wet granular systems tend to depend much less on microscopic details as one might expect.

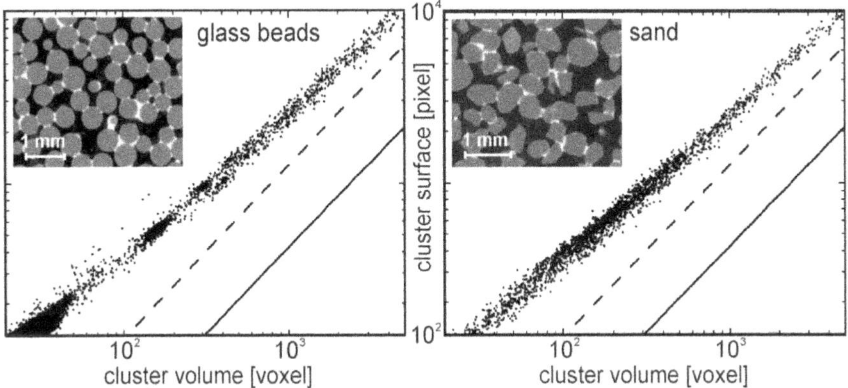

Fig. 6.26 Comparison of cluster statistics for glass beads (left) and sand grains (right). The accordance is striking. Only the individual 'clouds' corresponding to certain cluster morphologies characteristic of piles of spheres of equal size are missing (or blurred) in the right panel. Taken from [213].

6.3 Mechanical properties of wet granular piles

There are several candidate quantities which can be studied to describe the mechanical properties of a granular pile. We have already discussed the dependence of the angle of repose on liquid content in the context of

Fig. 6.8. This is very sensitive to minute amounts of liquid adsorbed to the surfaces of the grains. However, the angle of repose is not a quantity which is useful for studying intensive properties of the granular pile. It is not straightforward to make an connection between the angle of repose and an equation of state, or constitutive equation, of the system under study. In what follows, we will hence try to gain some insight into wet granular matter by means of other quantities, which are more accessible to physical concepts of intensive variables describing the state of the material.

6.3.1   *Tensile strength*

A particularly common quantity describing the mechanical consistency of a granular material is the tensile strength. It determines, e.g., how much of its own weight a wet granular clump sticking to the ceiling would be able to carry, without being torn apart by gravity. There are several ways to measure the tensile strength of cohesive granular matter. Will discuss here a particularly simple and reliable method which can be implemented with very limited effort. The setup is shown in Fig. 6.27 in the top left panel. The granulate is filled into a tube which is mounted horizontally onto a rotation stage. As the tube, one may use a commercial plastic syringe. The piston helps to prevent premature evaporation of liquid, such that the setup can be used as well for granulates wetted by volatile liquids. It is important that the axis of rotation intersects the granular plug in the interior of the tube.

When the rotation stage is set in motion, the centrifugal forces exert a tensile stress on the granular plug, which is highest in the center of rotation (i.e., on the rotation axis). It consists of two contributions. One is from the centrifugal forces exerted on the plug, and scales as the square of the length of that plug, r_p, times the square of the angular rotation frequency. The other stems from the friction of the plug with the walls of the tube, and is proportional to r_p. The speed of rotation is now gradually increased until the plug ruptures above the axis of rotation. This is shown in the top right panel of the figure. The critical angular frequency at which the rupture occurs will be larger the larger the tensile strength of the plug is. If the critical angular frequency at which the plug ruptures is determined for a number of different r_p, the tensile strength can be inferred from the data. The easiest way to achieve this is to plot $r_p \omega^2$ vs. $1/r_p$. According to what was said above, this yields a straight line the slope of which is directly given by the tensile strength of the granular plug, times its mass density.

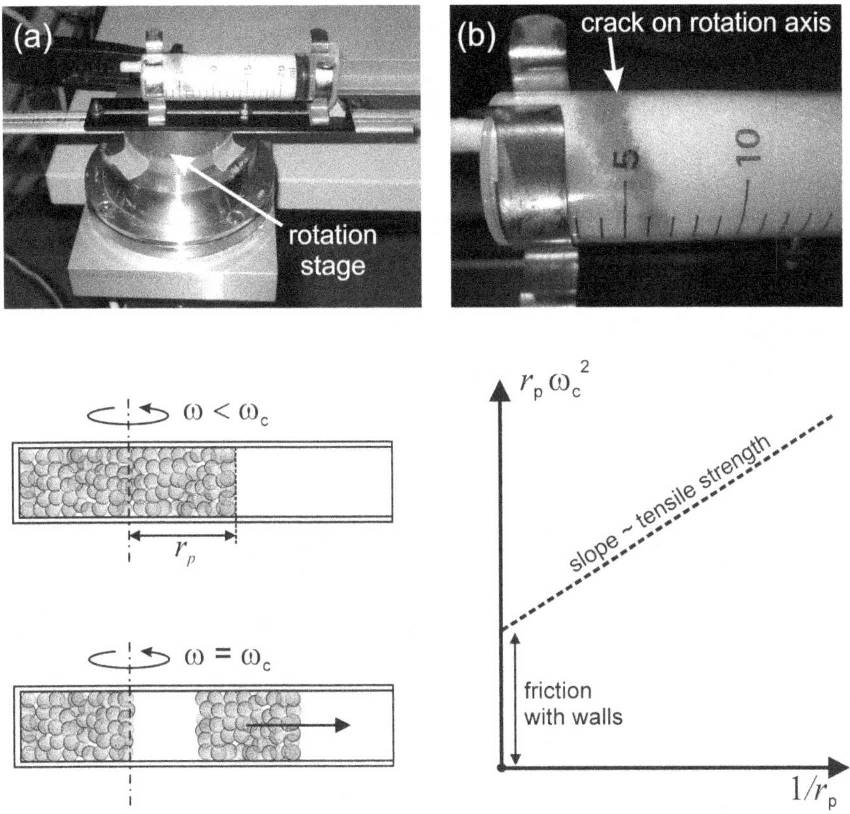

Fig. 6.27 A simple experimental setup for measuring the tensile strength of wet granular matter. The sample is filled in a plastic tube which is then mounted on a rotation stage. The rotation speed at which the granular pile ruptures is a measure of its tensile strength. By varying the length of the granular plug and plotting the critical angular velocity in an appropriate way, the tensile strength of the pile can be quantitatively determined, independently of the friction of the pile with the walls of the tube.

A typical result is shown in Fig. 6.28 for two different samples of spherical grains. First of all, we find that the tensile strength is higher for smaller grains. The most prominent feature, however, is the wide plateau extending from $W \approx 0.01$ to well beyond $W = 0.10$. This is well-known in the engineering sciences, as one can find it in the pertinent literature on granular mechanics for various characteristic quantities [221]. Figure 6.29 shows a figure detail adapted from a well-known text book on wet granular matter [222]. We see that as liquid is added to the granulate, the tensile strength

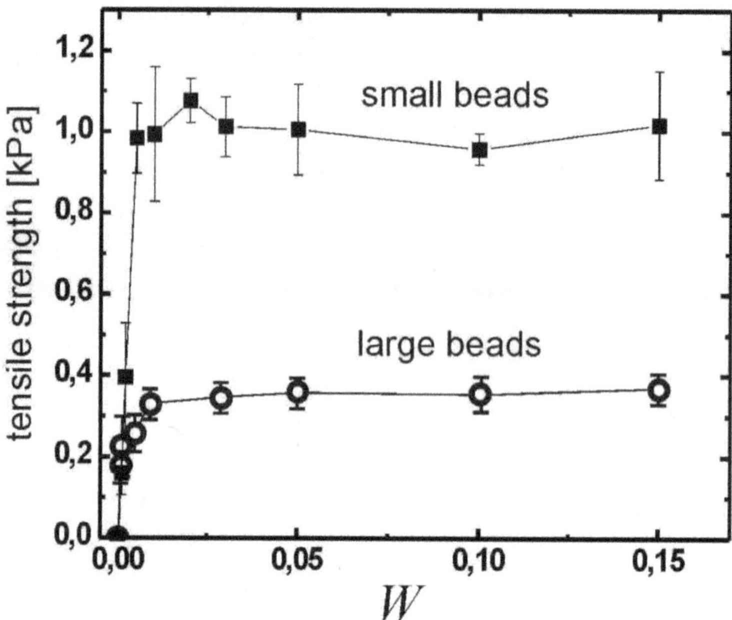

Fig. 6.28   Typical results of measurements of tensile strength as a function of liquid content. One observes a wide plateau, the height of which scales inversely with the size of the grains.

first rises rapidly, but then levels off and remains constant[9] up to a liquid content of about 0.15, very much in line with the results displayed in Fig. 6.28. As more liquid is added, the tensile strength increases strongly, almost up to saturation. This clear separation into two regimes has led to a commonplace assumption which obviously has inspired the inset sketch: that the change in behavior at $W > 0.1$ marks the transition from the pendular to the funicular state. We know already that this is not true, since this transition takes place at much smaller liquid content.

The presence of the plateau, which does not exhibit any feature at the transition from the pendular to the funicular regime calls for an explanation. This is straightforward to obtain from the discussion of Laplace pressure, which lead to Fig. 6.19. First of all, we have seen that the Laplace pressure is virtually constant for $W > W_{cb}$. At the same time, we know from the morphology of the clusters that these appear as individual capillary bridges, 'glued together' only at their outer perimeters. Consequently, the

---

[9]A similar plateau has been reported for the angle of repose [221].

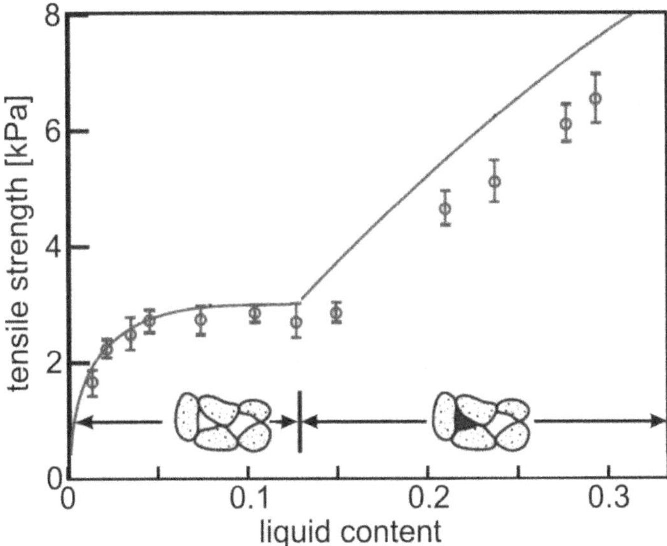

Fig. 6.29 The tensile strength of a wet pile of limestone particles (particle size about 70 μm), as a function of liquid content (adapted with modification from [222]; courtesy of Erin Koos, KIT Karlsruhe).

areas around the points of mutual contact of adjacent grains, on which the (negative) Laplace pressure acts (thus creating the major part of the attractive inter-granular force), do not change very much as more liquid is added. The latter is stored in the nooks between the perimeters of the capillary bridges, with little consequence on the area over which the pressure is effectively exerted. If we assume the attractive force experienced by the grains is mainly given by the Laplace pressure and the wetted area, it should be constant in the whole range of W beyond W^*. For $W_{cb} < W < W^*$, the pile is dominated by individual capillary bridges. We have seen that the force exerted by these bridges is roughly constant as well. Consequently, we expect the attractive forces exerted on the grains, and hence the tensile strength, to be approximately constant for all W larger than W_{cb}, in agreement with the observation.

Yet our argument is not complete so far. As obvious from Fig. 6.16, the perimeter term accounts for a large fraction of the attractive force at the high values of $\beta \approx \pi/6$ pertaining to the funicular state. Hence we must consider the behavior of the forces exerted by the liquid interface onto the grains at the three-phase contact lines somewhat more closely.

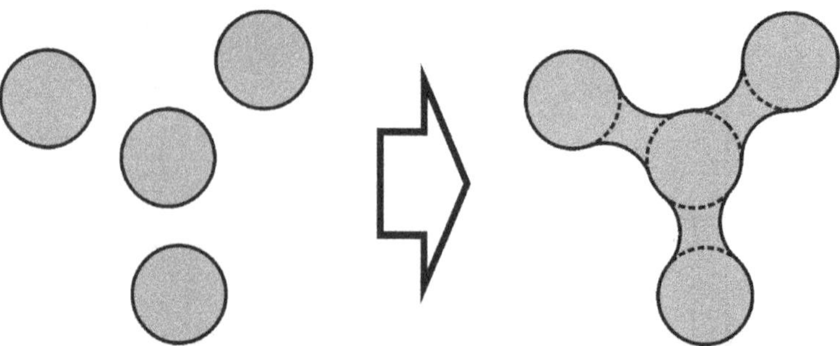

Fig. 6.30 Schematic sketch of the filling of nooks within a liquid cluster, modelled in two dimensions. Left: Wetted regions of grain surfaces, indicated by the grey areas. Right: The total length of the newly created solid curves equals the total length of those which are lost (dashed curves). It is obviously well conceivable that nooks can be filled with liquid without changing the length of the three-phase contact line. This would leave the perimeter term of the inter-granular forces unchanged.

Since the contact angle at which the liquid interface hits the solid is always given by θ, all we have to consider is the length of the contact line. The contribution of the liquid interface to the tensile strength should be proportional to the interfacial tension, γ, times the total length of the contact line. We have already mentioned that the funicular regime is characterized by filling the nooks between the capillary bridges with liquid, thereby forming larger clusters. Figure 6.30 illustrates in a simplified two-dimensional sketch why this is not likely to lead to appreciable changes in the total length of the three-phase contact line. The left graph shows three disjoint wetted areas. When the regions between these are filled, we obtain the graph on the right-hand side. Obviously, the length of the boundaries which have been lost (dashed) equals the length of the newly created boundaries.

Hence the total length of the boundary of the grey region is identical in both graphs, although its area has changed quite a bit. Clearly, this full three-dimensional geometry is a lot more complex, but the simple picture put forward here may well capture the essential physics. If this is the case, we may say that we have developed a reasonably complete understanding why the tensile strength is so remarkably constant over such a large range of wetness.

All contributions to the attractive forces scale in the same manner with the surface tension and the size of the grains, as they are proportional to γ/R. To complete our discussion, it is of interest to check whether this scaling is fulfilled for the tensile strength. Figure 6.31 shows measurements of tensile strength for variable grain size and variable surface tension of the

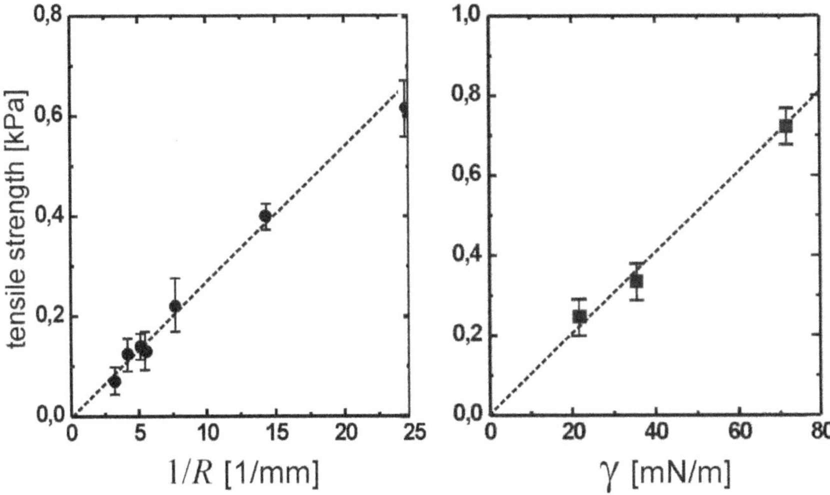

Fig. 6.31 The tensile stress vs. inverse grain radius (left) and vs. the surface tension of the liquid. Both panels demonstrate the expected scaling proportional to γ/R. Data taken from [213].

wetting liquid. Clearly, the data scale proportional to the surface tension and inversely proportional to the grain size, as predicted.

6.3.2 *Yield stress*

So far we have discussed the mechanical properties of wet granular piles in terms of the angle of repose and the tensile strength. The most obvious quantity, however, which comes to mind when one considers the pasty consistency of a wet granulate is the yield stress. A dry granulate has a zero yield stress, as it is not able to stage any noticeable resistance to an applied shear stress. In strong contrast, a wet granulate will retain its shape against, e.g., the stress imparted on it by gravity, which is exploited in the use of expendable molds, or in the construction of sand sculptures.

From a fundamental point of view, it is of interest to investigate whether the yield stress can be traced directly to the extension and rupture of liquid capillary bridges within the pile. We can start by means of a simple estimate of the energy required for these processes. We consider the application of a unity shear to a wet granular sample in the pendular regime. For each capillary bridge, this leads to rupture with a probability p_r which should be of order unity. The number of capillary bridges within the granular pile is given by $3\phi k/8\pi R^3$. With the help of Eq. (4.42), we can then express the contribution due to the elongation and rupture of capillary bridges to the yield stress as

$$\sigma_{cb} = \frac{3\phi k}{8\pi R^3} 4.37 \mathrm{p_r} \gamma R^2 \cos\theta \sqrt{\tilde{V}}. \tag{6.16}$$

With the help of Eq. (6.5) we obtain

$$\sigma_{cb} = 4.37 \frac{\gamma \mathrm{p_r}}{R} \cos\theta \sqrt{\frac{3\phi k W}{8\pi}}. \tag{6.17}$$

In a real granular system, there is friction between the grains, which provides another means of resistance to the applied shear. It scales with the normal forces the capillary bridges exert upon their contact points, multiplied by the friction coefficient, μ. Since granular piles are always mechanically frustrated (cf. Fig. 4.17), we can safely assume that friction fully contributes to the resistance against any site exchange processes among grains. If the pile is under no additional external compressive stress, the tangential forces between grains which are necessary to overcome the friction is directly determined by the attractive capillary forces acting between grains. We then can estimate the contribution of friction to the yield stress using the expression (4.36) for the pressure exerted on the grain contacts. We obtain

$$\sigma_\mu \approx \frac{\gamma \phi k \mu}{2R} \cos\theta. \tag{6.18}$$

Both σ_{cb} and σ_μ represent the energies which must be afforded to achieve a unity shear of the pile. Since energy is additive, so should then be the yield stress contributions. Hence we estimate the yield stress, σ_y, of the granular pile by adding the two contributions and arrive at

$$\sigma_y = \sigma_{cb} + \sigma_\mu \approx \frac{\gamma}{R} \phi k \cos\theta \left(4.37 \mathrm{p_r} \sqrt{\frac{3W}{8\pi\phi k}} + \frac{\mu}{2}\right). \tag{6.19}$$

It is of interest how large the contributions are relative to each other. For $W = W^*$, $k = 6$, and $\phi = 0.6$, we find

$$\sigma_y \propto 0.13 \mathrm{p_r} + \frac{\mu}{2}, \tag{6.20}$$

which shows that both contributions are in fact in the same order of magnitude. It would thus be very valuable to have an experimental method which allows to distinguish between both contributions.

A simple way to do this at least qualitatively is to vary the external mechanical pressure onto the granular pile. This should then increase the

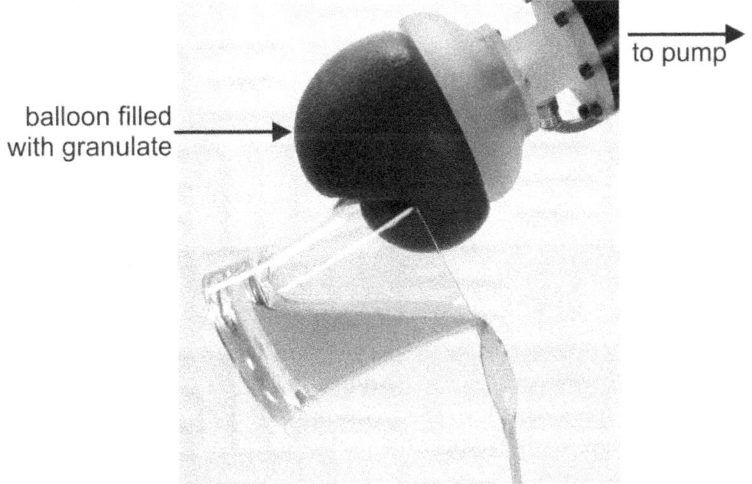

Fig. 6.32   The universal jamming gripper. It consists of a rubber balloon filled with a suitable granulate, and a pump which is used to adjust the gas pressure in the interstices of the granulate [223]. Image courtesy of John Amend, Cornell University.

contribution from σ_μ, but leave σ_{cb} unchanged. That the effect of external pressure onto the yield strength of a granular pile may indeed be substantial is impressively demonstrated by a device called the 'universal jamming gripper' shown in Fig. 6.32. It consists of a rubber balloon filled with a dry granulate (here the granulate was just ground coffee!). As long as the air pressure inside the balloon is the same as outside, the balloon is very flexible and will smoothly nestle any shape it is in contact with. As soon as the pressure inside the balloon is reduced by means of a pump, the hull of the balloon exerts an external pressure on the granulate, impeding any further shear deformation by virtue of the friction between the grains. The granular pile has thus effectively solidified in the imposed shape. It can thus be used as a gripper, suitable for a vast number of different shapes. In the figure, it is used to handle a glass of orange juice.

Efficient control of deformation as well as external compressive stress is possible by so-called triaxial shear cells, which are routinely used as test devices for granular materials in industrial and geo-scientific context [224]. These are rather expensive apparatuses, which are not necessary for the simple investigations we have to perform. It is therefore of interest to shortly discuss a quite simple device which has been introduced to apply shear to a granular pile in a rather homogeneous way [118, 225, 226]. It

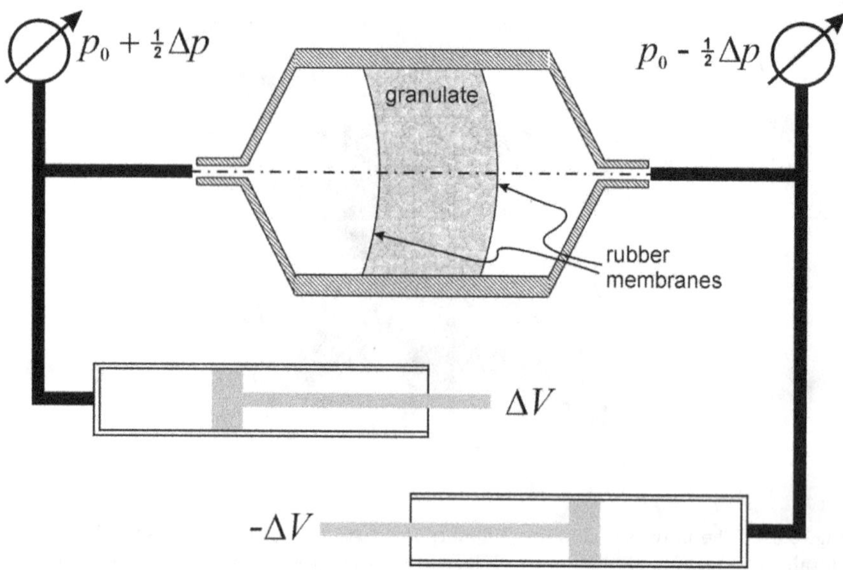

Fig. 6.33 Simple experimental setup for measuring the yield stress of wet granular matter. The solid curves indicate the rubber membranes, which separate the sample from the adjacent water reservoirs. The granulate is sheared with a mean velocity u by changing the content of the two reservoirs. The pressures p_1 and p_2 are continuously recorded.

can be readily set up and implemented for laboratory use. One of its merits is that is quite effectively suppresses the formation of shear bands in the sample.

Figure 6.33 shows a schematic of the setup. The granular material (e.g., spherical glass beads as above) is filled into a cylindrical cell. Each of the flat sides of the cell consists of a thin rubber membrane. Adjacent to each membrane, the cylinder continues as a water reservoir, which is connected to a syringe. The pistons of the two syringes can be moved by computer-controlled stepper motors. When the two pistons are moved at equal speed but in opposite direction, the sample between the membranes is deformed at constant volume and well defined velocity, u. The latter is defined as the rate of change of the reservoir volume, divided by the cross-section of the cylinder. The elastic tension of the membranes ensured a (roughly) parabolic deformation, and thus a flow profile similar to Poiseuille flow within the sample. In this way, avalanching and shear band formation can be kept to a tolerable minimum.

The pressures in both reservoirs are recorded while the sample is sheared

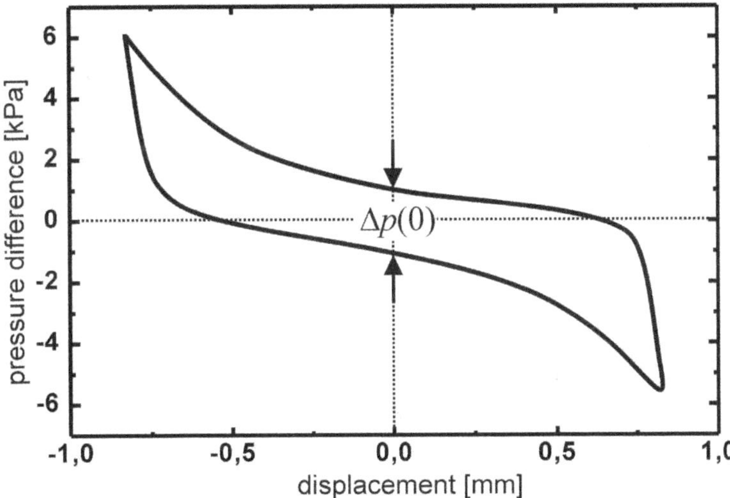

Fig. 6.34 A typical hysteresis loop as obtained with the setup sketched in Fig. 6.33. Data kindly provided by Michael Sipahi, Göttingen.

back and forth with a constant velocity. The corresponding shear rate varies over the sample and is of order u/r, if r is the radius of the sample cell. A plot of the pressure difference, Δp as a function of the displaced volume yields a well developed hysteresis loop, as shown in Fig. 6.34. The slope of the two branches of this loop correspond to the elastic response of the membranes. The opening of the loop at zero strain, $\Delta p(0)$, is a measure of the yield stress. The total strain should be chosen large enough to induce irreversible particle motions [227].

It is advantageous that this simple setup is compact enough to be inserted into an X-ray micro-tomograph. Figure 6.35 shows a typical result. The top row shows tomograms of the whole sample at different stages of shear. The images annotated '0' correspond to the centerline in Fig. 6.34 (zero displacement). In the images annotated with the arrow. The overall deformation of the sample and bend of the bounding rubber membranes is clearly visible. No formation of shear bands is visible. The bottom row shows a closeup sequence of a similar experiment where the total shear has been larger. The two white triangles have been inserted as markers to quantify the applied shear.

It is particularly interesting to observe the transformations in the liquid morphologies. Of the 15 capillary bridges visible in the field of view, four are ruptured and four are being established in the course of the four images.

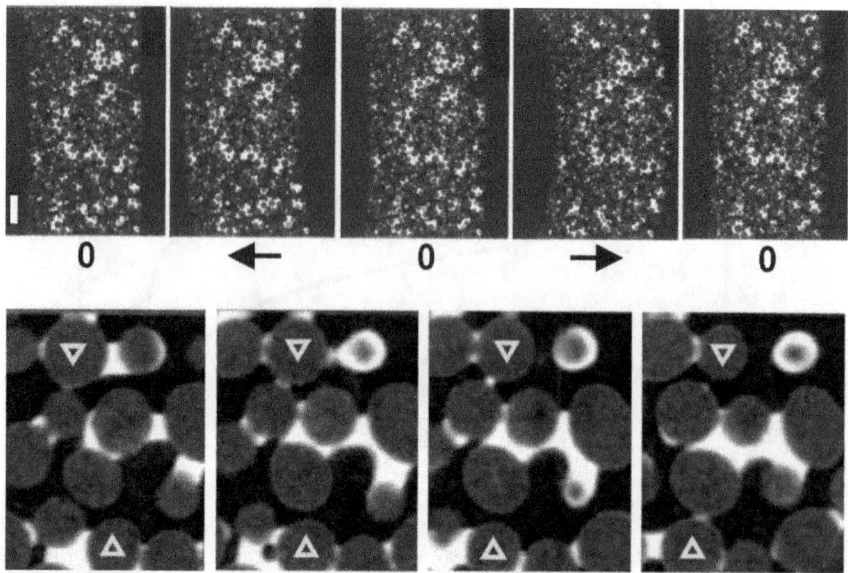

Fig. 6.35 Top row: five different stages of shearing the cell which is sketched schematically in Fig. 6.33. Those annotated '0' have just returned into the un-deformed position again. Those annotated with an arrow have just reached their largest displacement. The bent shape of the membranes is clearly visible (scale bar: 2 mm). Bottom row: closeup of a similar experiment as the one shown in the top row, but at higher maximum shear. The triangles are markers for qualitatively determining the shear. Images courtesy of Ralf Seemann (Saarbrücken), Kamal Singh (ESRF, Grenoble, France), and Mario Scheel (ESRF, Grenoble, France).

Hence the fraction of ruptured bridges is $4/15 = 0.267$. From the positions of the white triangles, we determine the total shear to be 0.38. From this we can estimate the fraction of bridges ruptured at unity shear and obtain $p_r \approx 0.267/0.38 \approx 0.7$.

Let us now turn to the assessment of the contributions through friction, by varying the absolute external pressure, p_0, exerted on the rubber membranes. We should expect a dependence

$$\sigma_\mu \approx \mu \left(p_0 + \frac{\gamma \phi k}{2R} \cos \theta \right), \tag{6.21}$$

hence the contribution from friction should increase linearly with applied pressure. For a dry granulate, only this linear term should be present.

The result of a typical measurement is shown in Fig. 6.36, where data taken for same granulate in the dry (squares) and in the wetted (triangles) state have been included. The contribution from the capillary bridges

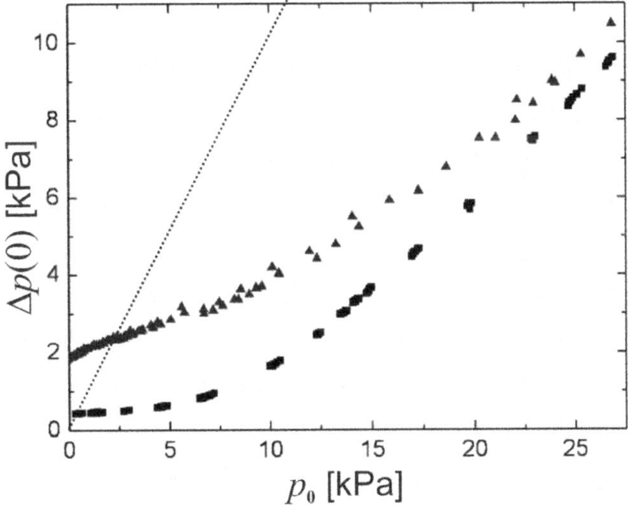

Fig. 6.36 Dependence of the differential pressure Δp on the absolute pressure p in the setup sketched in Fig. 6.33 (data kindly provided by Michael Sipahi, Göttingen).

corresponds to the vertical difference between both curves, extrapolated to zero pressure. One way to isolate the contribution from capillary bridge rupture alone could thus be to extrapolate to $p_0 = -p_{\text{int}}$. Inserting numbers for p_{int}, one sees that this is on the order of one kPa at most, such that the extrapolation can be quite reliably performed [226]. The dotted line indicates equal p_0 and $\Delta p(0)$. Left of this line, the extrapolation seems to be soundly possible. The value of the yield stress, $\Delta p(0)$, extrapolated to vanishing applied pressure is known as the Coulomb cohesion parameter [228] in the engineering literature.

What should surprise us, however, is that the obtained dependencies on p_0 are far more complex than expected. For the dry case, we expected a linear relationship. Instead, the yield stress seems to be unchanged up to an applied pressure of more than 5 kPa, but rises about linearly for larger pressures. That such a more complex behavior is obtained can be qualitatively understood when one contemplates the propagation of forces in a granular pile. Force can be passed on from one grain to the next only if the corresponding contact points are quite accurately positioned oppositely on the grain surface. If this is not the case, the grain will slip away instead. Hence forces organize in granular piles preferentially along paths where contact points are in opposite position on the grains. This leads to the development of so-called force chains.

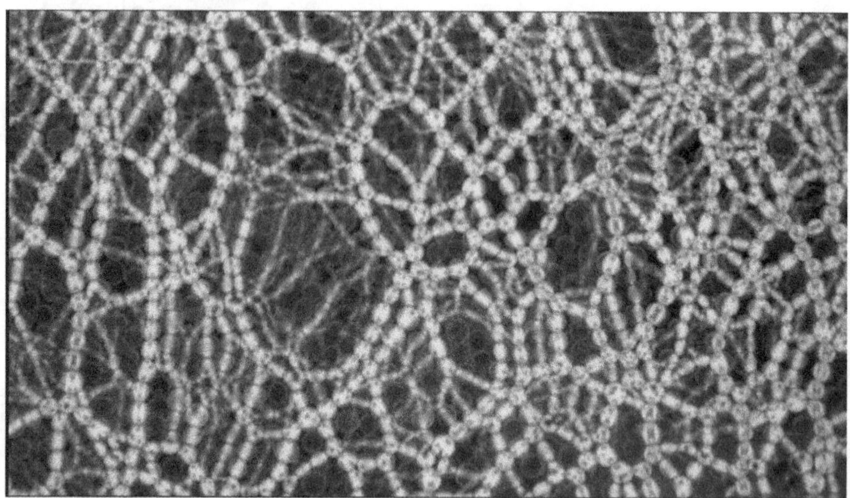

Fig. 6.37   Top: So-called force chains, observed in a two-dimensional pile of plastic discs between crossed polarizers. The 'force chains' are thus very clearly visible, unveiling the structure of force transmission in granular piles (image kindly provided by Jie Zhang and Robert P. Behringer, Duke University, USA). Bottom: The ruins of the aqueduct of Maintenon (close to Paris). The arches remain until today, reminding us of the strength of arch structures in architecture and in granular systems (adapted from Wikimedia. Copyright (2010) Laifen).

These can be nicely observed in two-dimensional model systems [229]. If one piles circular disks made of a suitable polymer between two glass plates and exerts a pressure on the arrangement of disks from the side, the deformation of the disks can be made visible by strain-induced birefringence. All one has to do is mount the whole setup between crossed linear polarizers. Disks which are force free appear dark in transmission, but regions where deformation induces birefringence will light up. A beautiful example of this experiment is shown in the top panel of Fig. 6.37, where a vertical pressure has been applied. Clearly, the forces organize into chain-like structures, an effect which is commonly called *arching*.

The robustness of such arch structures is exploited in architecture since ages, as it is impressively illustrated by the ruins of the aqueduct of Maintenon (a few kilometers southwest of Paris). It was supposed to deliver the water from the river Eure to Versailles, in order to feed the elaborated trick fountains next to the king's palace, but it was never finished. While almost all of the superstructure has been eroded away, the arches still remain until today; and so does the problem to come up with a reliable theory of arching in granular piles [230]. There are meanwhile quite elaborate concepts for describing arching in two dimensions (e.g., the force network ensemble [231]), but the analog in three dimensions is still largely elusive.

Let us therefore return to the measurement of the yield stress, and not discuss any further the possible ways to accurately separate σ_{cb} from σ_μ. Experimental results for σ_y are displayed in Fig. 6.38 for the full range of W up to complete saturation. As the above formula for σ_y suggests, we do not again find a plateau as nice as for the tensile strength. The dashed line is a superposition of a constant (1.4 kPa) and a square root behavior ($1.9\sqrt{W}$), inspired by Eq. (6.19). When the liquid content increases further, the rupture distance increases as well, such that capillary bridges may withstand larger shear without being destroyed. Hence we expect p_r, and alongside σ_y, to decrease at large W. This is in fact what is observed.

However, this simple picture cannot contain the whole truth. The formulas derived above are only valid for the pendular regime. As soon as the funicular regime is entered at $W \approx 0.026$, the bridge angle β becomes constant, and it is not clear how (and if at all) the rupture energy, E_{cb}, should vary as W is increased further. The more liquid is stored in the liquid clusters, the more appropriate is it to treat the physics of the interstitial liquid under isobaric conditions, not isochoric as we have done when treating individual capillary bridges.

Nevertheless, it remains interesting that our data can be fitted so well

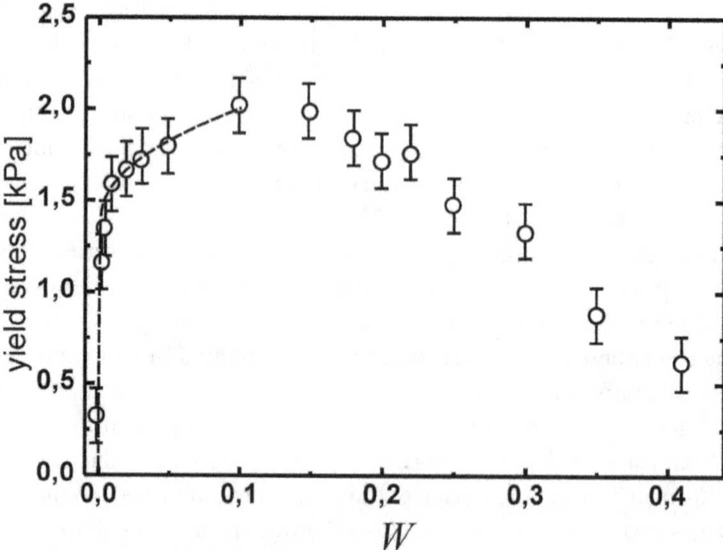

Fig. 6.38 Dependence of the yield stress, as determined with the setup discussed above, on the liquid content of the sample (data taken from [225]).

with the above expressions. In Chapter 5, we have already seen that the scaling of rupture energy with liquid content was just as expected form single capillary bridges, but extended far into the funicular regime (cf. Fig. 5.26). Such commonalities are of great interest, since they may justify the use of simple two-body force models in the simulation of granular systems even at large liquid content. It is thus of particular importance to unravel their (so far elusive) physical origin.

Let us now turn to the dependence of Δp on the shear rate. As Fig. 6.39 shows, Δp decreases significantly as u is increased in the wet, and only in the wet case. This is in qualitative accordance with results obtained before by Tegzes in rotating drum experiments [41]. No clear dependence of Δp on u was found in the dry case. This is not surprising, since the applied absolute pressures were kept close to zero. For the fully saturated case, a small increase of Δp with u for the higher velocities (see Fig. 6.39) was observed, which may be due to the viscosity of the surrounding water.

A certain dependence on shear rate is indeed expected on the basis of Eq. (6.4). The time a capillary bridge has for formation is inversely proportional to the shear rate, such that at higher shear rate, a smaller bridge volume is expected. In particular, from Eq. (6.4) we may expect a

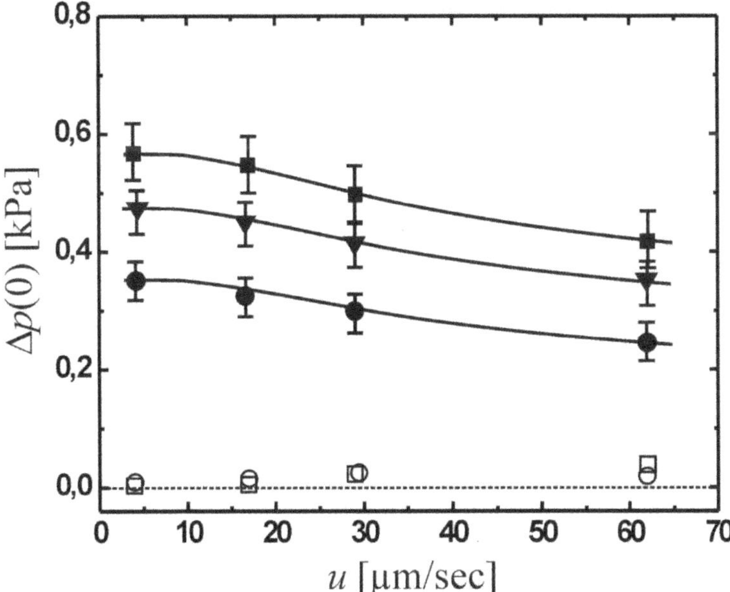

Fig. 6.39 Dependence of the differential pressure Δp on the shearing velocity u, for an absolute pressure of $p = 0.76$ kPa. Open circles: dry sample. Open squares: full saturation. Full triangles: $W = 0.004$; Full squares: $W = 0.014$; Full circles: $W = 0.27$. Data from [118].

dependence of the form

$$\Delta p = \Delta_1 + \Delta_2(1 - e^{-u_0/u}), \qquad (6.22)$$

where u_0 should be roughly the sample radius divided by t_0. With $t_0 \approx$ five minutes, we would expect $u_0 \approx 30$ μm/s. The solid curves in Fig. 6.39 represent Eq. (6.22), with $\Delta_1 \approx \Delta_2$ and $u_0 = 43$ μm/s. This is quite reasonable agreement.

A major conclusion from these data is that the formation dynamics of the liquid bridges may account for shear thinning effects. From our above discussion on the capillary bridge network, we know that the relative change of the coordination number, k, is similar to the relative change in packing density. Since dilatancy changes the density only by a few percent, a dramatic effect on k would not be expected. The strong change observed in Δp (up to 20%) with increasing shear rate can thus not be explained by dilatancy. This is in contrast to a wide spread opinion that dilation effects are generally the main reason for shear thinning. This is of particular interest since shear thinning is a major requirement for shear banding and avalanching to occur.[10] Forthcoming studies will have to unravel which are the dominant effects contributing to shear thinning in wet granular matter. This may have substantial consequences for the understanding of unwanted phenomena like mud flows and land slides, as well as for possible mitigation strategies.

6.4 Phase transitions in wet granular piles

Our considerations have thus taken us back to the Introduction, Fig. 1.5, where we had posed the question whether phenomena like soil liquefaction might be understood in the conceptual framework of phase transitions. In Chapter 5, we have seen that this approach may in cases be quite successfully applied to wet granular gases, and we will now investigate whether the same holds for the dense wet granulates we are dealing with in the present chapter. There is already a large body of literature on the avalanching behavior of wet granular matter, chiefly from geophysics and the engineering sciences. The most frequently used device here is the rotating drum as introduced in Chapter 2. For our approach, however, this setting is not well suited, since the granulate is agitated in a very inhomogeneous way, such that it is difficult to distill any intensive properties of the system from the data. In order to subject the granulate to a homogeneous agitation, we resort to the method of vertical agitation we have introduced and used in Chapter 5. Even in view of settings like the one shown in Fig. 1.5, this is not a purely academic way of posing the problem, since soil liquefaction is well known to occur preferentially in connection with earthquakes. In fact, the little town of La Conchita, where the photo in Fig. 1.5 was taken, is located directly on the San Andreas transform fault in California.

6.4.1 *Melting by vibration*

What we will do in this section is to take a closer look at the horizontal phase boundary shown in Fig. 5.21. At sufficiently strong agitation, the grains start to move in an erratic fashion, very much reminiscent of the molecular motion in a liquid, while the density of the granulate remains close to dense packing [34]. The onset of this motion corresponds to the horizontal phase boundary shown in Fig. 5.24 close to the left axis.

---

[10]It may well be that an avalanche model put forward by Dahmen *et al.* [232] for granular systems without reference to moisture applies particularly well to sheared *wet* granular matter.

In the dry case, it is easy to derive conditions for this type of fluidization to occur. If the vertical position of the container is given by $A\cos\omega t$, where $\omega = 2\pi f_{sh}$ is the (angular) frequency of the oscillation, its maximum acceleration is $A\omega^2$. If this is less than the acceleration of gravity, it is clear that the granulate will remain at rest with respect to the container, and no fluidization takes place. If, however $A\omega^2$ exceeds g, there will be during each oscillation period a short phase where the grains are levitated. Hence they start to move with respect to each other, and fluidization will occur. If we define $\Gamma := A\omega^2/g$ as before (cf. Chapter 5), we can state that $\Gamma > 1$ is a necessary condition for fluidization. Experimentally, one finds that fluidization of dry granulates occurs at $\Gamma_c^{\mathrm{dry}} \approx 1.2$ [34, 233]. It manifests itself by the onset of an irregular, relative motion of the grains, usually accompanied by some convective flow within the container.

We had seen that the critical acceleration at which fluidization sets in shifts considerably if a wetting liquid is being added to the pile. Since this shift is independent of liquid content as the inter-granular capillary force and acceleration is proportional to force, we have called this fluidization transition *force driven*. Let us now try whether we can on the basis of the known capillary forces predict the shift in Γ_c quantitatively. If we consider a single grain, demanding the capillary force and the force of acceleration to be equal, $F_{cb} = m\Gamma_c g$, leads to

$$\Gamma_c = \frac{2\pi R\gamma}{\frac{4}{3}\pi R^3 g\varrho_s\phi} = \frac{3\gamma}{2gR^2\varrho_s\phi}. \tag{6.23}$$

If we put in numbers, we readily see that Γ_c should be a few hundred for sub-millimeter grains. This is clearly at variance with experiment, where Γ_c is almost invariably found to be below 10. It meets, however, well with the experience that a single wet grain (to which Eq. (6.23) should apply) is much harder to shake off the surface it clings to than a whole clump of granulate.

This points to the question whether we have so far been considering the appropriate mass scale. The pile is agitated from its outer boundaries, which coincide with the container walls. If we accelerate the container, the forces which are consequently exerted on the pile are proportional to the *total mass* of the sample. If this is denoted by M, we can write to total acceleration force in the levitating phase of each oscillation cycle as

$$F_{\mathrm{acc}} = (\Gamma - 1)Mg. \tag{6.24}$$

The capillary forces acting between the container walls and the sample are assumed to be similar to the force acting between two beads. The total capillary force is then given by the capillary bridge force times the number of grains touching the container walls. If we perform experiments in a cylindrical container (vertical axis) with radius r, which we fill with granulate to a height H, we have

$$F_{\text{cap}} = F_{cb}\phi_2\pi r(2H\mu + r), \tag{6.25}$$

where ϕ_2 is the two-dimensional density of the bridges formed between the container wall and the glass beads. The first term in Eq. (6.25) contains the friction coefficient, μ, since the vertical container walls can act on the sample only via tangential forces. For hexagonal close packing, $\phi_2 = (2\sqrt{3}R^2)^{-1}$. For random dense packing, which we assume, ϕ_2 is smaller roughly by a factor of $(\phi_{\text{rdp}}/\phi_{\text{fcc}})^{2/3}$, such that $\phi_2 \approx 0.26/R^2$. When evaluating F_{cb}, we must be careful that this concerns bridges between grains and a flat wall. According to Derjaguin's approximation, Eq. (4.29), this results in doubling the force.

Relative motion between the glass pile and the container walls will take place if $F_{\text{acc}} \geq F_{\text{cap}}$. For the condition for the onset of fluidization we then obtain

$$\Gamma_c \approx 1 + \frac{3.27\gamma\cos\theta}{Rg\phi\varrho_s}\left(\frac{2\mu}{r} + \frac{1}{H}\right). \tag{6.26}$$

Inserting numbers again, we find that Γ_c is now predicted to be on the order of 10 or less for a typical sample size of a few cubic centimeters, depending furthermore on μ. It is thus predicted that the extra acceleration required for fluidization, for ideal spheres and high frequency, scales as γ/R, as encountered as well, e.g., for the Laplace pressure and the tensile strength. It is furthermore predicted to be independent of W, as already seen in Fig. 5.22, and it should have an additive term which scales inversely proportional to the filling height, H.

Some experiments have been performed to check these predictions [126, 213]. Cylindrical glass containers with an inner diameter of 26 mm were filled with different granulates with variable grain size, to variable heights H up to about 30 mm. Controlled amounts of wetting liquid were added and the lid closed, in order to prevent evaporation of liquid. The sample was mounted onto an electromagnetic shaker similar to the one described earlier (Fig. 5.17). It was capable of applying amplitudes between 1 μm and 1 mm, at frequencies between $\omega = 20$ and 320 Hz. At fixed agitation frequency, the amplitude was increased until a relative motion of the particles was visible through the glass wall. As expected, the critical acceleration was found not to depend significantly on the agitation frequency (cf. Fig. 5.22).

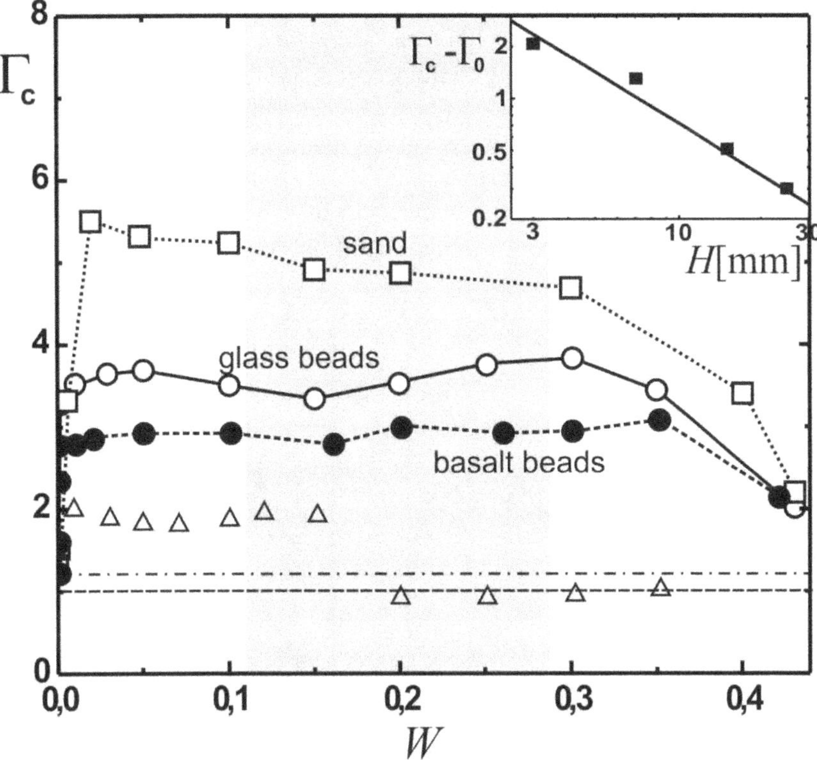

Fig. 6.40 The critical acceleration for fluidization, measured for different systems. Up to a liquid content of about $W \approx 0.15$, a well developed plateau is invariably observed. Sometimes it extends significantly farther. Inset: the scaling with the height of the pile is consistent with the theoretical model, which predicts scaling of the excess amplitude with $1/H$ [213].

Results for three different types of grains are presented in the main panel of Fig. 6.40, for the full range from $W = 0$ to saturation, $W = 1 - \phi$. The behavior is qualitatively the same for sand, glass beads, and basalt beads. As some liquid is added to the dry sample, the critical acceleration rapidly jumps to an elevated value. As more liquid is added, we find a wide plateau, similar to the one encountered already in Fig. 5.22. The grey shaded area indicates the bicontinuous regime, as identified in Fig. 6.21.

The open triangles show data which have been obtained with larger glass beads. Here we find that the plateau does not extend that far, but instead breaks off abruptly somewhere within the bicontinuous regime. It then and continues at values around $\Gamma \approx 1$ [225]. A possible explanation is

that if a large liquid cluster is formed which contains the majority of the
liquid, any acceleration imparted on the sample can give rise to substantial
liquid flow, and hence possibly to fluidization through completely different
mechanisms as the one discussed above. Since such effects are found to be
hard to control and reproduce, we will not dwell on them any further here.
We just mention that the plateau was invariably found to extend at least to
about $W = 0.15$, similar to the plateaus in tensile strength (cf. Figs. 6.28
and 6.29) and Laplace pressure (Fig. 6.19).

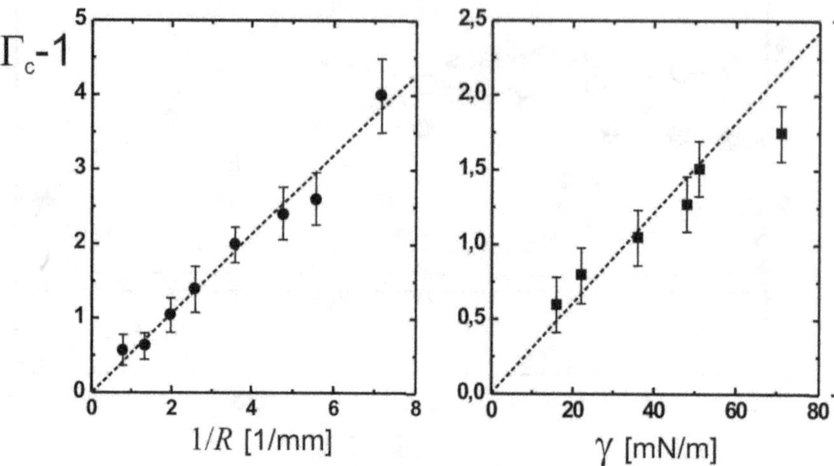

Fig. 6.41   Test of the scaling of $(\Gamma_c - 1)$ with γ and R. Clearly, the predicted scaling
proportional to γ/R is fulfilled. Data from [213].

Just as we did for the tensile strength, we check whether $\Gamma_c - 1$ scales
as γ/R as predicted by Eq. (6.26). This is shown in Fig. 6.41, where results
with glass beads of different size and with a number of different wetting
liquids are compiled. Clearly, the scaling is as predicted as far as both γ
and R are concerned. Finally we should check whether the dependence on
the filling height is as predicted. If we denote by Γ_0 the sum of the first
two terms in Eq. (6.26), we expect that $\Gamma_c - \Gamma_0$ should scale as the inverse
of H. That this is indeed the case is shown in the inset in Fig. 6.40. In
this double-logarithmic plot, the slope of the straight line is minus one, in
accordance with the prediction [225].

These results suggest that fluidization of wet granular matter by vertical
agitation, at least for this model system, can be quite well understood by the
interaction of the pile with the walls of the container. If this is the case, the

fluidization process must sensitively depend upon the capillary interaction of the glass beads with the walls of the container. This hypothesis was tested by hydrophobizing the container walls [225]. They were exposed to a solution of octadecyl trichlorosilane ($C_{18}H_{37}SiCl_3$), which binds covalently to the glass surface. As a result, a monomolecular, close packed layer of chemically bound octadecane forms on the glass, turning it, chemically speaking, into a waxy surface. This results in contact angles (with water) well above 100 degrees. As discussed above (Eq. (4.26)), there should be no attractive force to the wall in this case. In fact, the effect was dramatic: fluidization could not at all be achieved with the apparatus, which was capable of applying accelerations up to $\Gamma \approx 20$.

6.4.2 *Quantum mud*

We have now arrived at a rather thorough understanding of the melting transition we first discussed in Chapter 5, in the context of phase transitions far from thermal equilibrium. We found that it exists in many wet granular systems, and it is of interest to see how far we can go in the variation of parameters. In Section 4.3.3, we have seen that the concept of hysteretic formation and rupture of capillary bridges pertains down to the nanometer scale. We may thus try to find out how small we can make the grains and still find genuinely 'granular' physics, such as the vibration-induced melting transition just discussed.

If the grain diameter comes of the order of micrometer, the physics will change almost certainly, since van der Waals forces between the grains can then not anymore be neglected (cf. Fig. 1.4). A glance at Fig. 1.4, however, reveals that below the typical size of sand grains there should still be several orders of magnitude of room for 'wet granular physics'. All we have to do is look for a liquid with a very small surface tension, in order to reduce the capillary forces accordingly. A suitable paring is liquid helium as a liquid and grains with about 10 μm diameter. The latter are safely large enough for not baking together irreversibly by van der Waals forces alone. Liquid helium is known to wet all solid surface completely, with the alkali metals being the only known exceptions.

In order to set up an experiment on fluidization by vertical acceleration of a granular pile wetted with liquid helium, one needs a suitable cryostat which is capable of accommodating such device. If one keeps the shaker itself at laboratory temperature, the experiment can actually be set up in a rather simple, standard bath cryostat. In this experiment [147], the shaker was mounted above the cryostat, and upside down. Its moving stage was firmly connected to a thin-walled stainless steel tube, which was tightened with respect to the cryostat walls with a flexible stainless steel bellow, and reached down all the way into the cold region of the device. There, at its bottom end, the sample cell was mounted to it, which had a capillary for insertion of helium, and an optical window on its top. Through an optical access into the stainless steel tube above the cryostat, equipped with a tilted mirror inside the tube and a glass window separating the cold region from laboratory air, one could look down into the sample cell. This optical path was used to illuminate the cell with an expanded laser beam, the intensity of which was modulated at the frequency at which the shaker was operated. In this way, stroboscopic images of the granulate in the cold sample cell could be obtained, which immediately showed whether the granulate was fluidized or still.

As the stainless steel tube and sample cell were completely tightened with respect to the cryostat bath, they could be immersed in liquid helium. The temperature of the cell, which was made from oxygen-free copper, could therefore be well controlled through the vapor pressure of the helium bath. At fixed temperature, well-controlled amounts of helium gas were condensed into the sample cell by means of a gauge volume. As the sample volume was known, one could quantitatively predict at which amount of helium the bulk liquid phase would start to condense in the cell.

Figure 6.42 shows results for the critical acceleration for fluidization of a pile of polystyrene beads. As long as only gaseous helium is in the cell, the critical acceleration is identical to what one obtains with the dry sample. This is at values of the scaled amount of helium which are below unity. As soon as helium starts to condense (vertical dashed line), the critical acceleration jumps to a higher value, just as observed with the other systems discussed before. Furthermore, we find the well-known plateau as we increase the amount of liquid.

An amusing side aspect is revealed by the difference between the open and the closed circles, which have been taken at different temperatures. Liquid helium has the unique property to become superfluid at a temperature of 2.17 K. Hence the open circles correspond to the normal fluid, while the closed circles correspond to the superfluid phase. At first glance, one might expect that the superfluid, in analogy to a superconductor, is a fluid with zero viscosity, and hence one can switch off viscosity effects here by just varying the temperature. In view of the discussion in Section 5.1, this could in fact be of great interest.

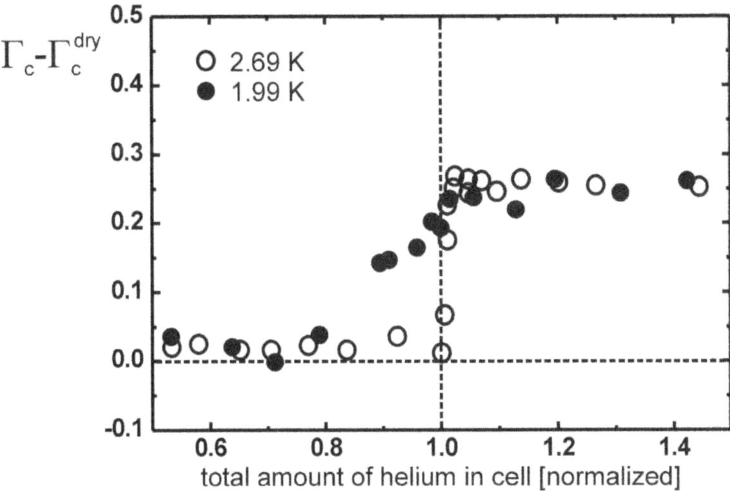

Fig. 6.42 Critical acceleration for mechanically fluidizing a granulate consisting of 10 μm diameter polystyrene spheres wetted by liquid helium. If the helium is in the normal fluid state (open circles), the same behavior is observed as for liquids with larger surface tension. One observes a clear jump when the pile is being wetted, and a wide plateau expressing insensitivity to the exact amount of wetting liquid. If the helium is superfluid (closed circles), one observes a clear deviation towards lower liquid content. This is due to the osmotic pressure of quantum excitations generated by the agitation of the pile. Redrawn after [147].

However, a superfluid is *not* just a fluid with vanishing viscosity [1]. It behaves as such only in certain very particular situations. In a dynamic setting like the one here, a different perspective is appropriate. The superfluid phase may be viewed as an ideal fluid, void of any viscous effects, in which there are quantum excitations propagating, like phonons in a crystal lattice. As each of these excitations has a certain energy, there number is determined by thermal distribution functions. From a statistical physics point of view, the quantum excitations in the inviscid superfluid background phase are analogous to ions dissolved in a solvent. Their density then gives rise to an osmotic pressure. If one heats a superfluid at some place, one creates many quantum excitations there, which by virtue of the osmotic pressure pull more superfluid phase to this region. This is a very strong effect, and is commonly called the fountain effect, alluding to a very impressive fountain-like setup which can be used for demonstration [1].

In a granulate wetted by superfluid helium, quantum excitations will be generated whenever two grains collide with each other at their points of contact. As a consequence, the increased osmotic pressure will attract more liquid to these point, such that the system appears wetter than it actually is. The effect is strong enough to wet the contact points even when there is not enough helium in the cell to form any appreciable amount of condensate. This shows up as the elevated black circles reaching out into the regime left of the vertical dashed line. In summary, we can observe a genuine quantum effect in granular matter, albeit so far of admittedly limited practical relevance. The characteristic phenomenology of wet granular matter can be observed down to a grain size of 10 microns, in accordance with the consideration of energy scales in the introduction.

6.4.3 *Surface melting*

When one has to rely on visual inspection of a granular sample from above, such as in the liquid-helium setup just discussed, one may be betrayed in the accurate determination of the point of fluidization by an effect called *surface melting*, which is well known to accompany melting points in regular solid state matter, but is not so well known for granular piles. We should not close this section without a few words on this phenomenon.

In a cohesive granular pile which is vibrated as discussed above, one may imagine that moving around is easier for a grain sitting on the pile surface than for a particle well inside the pile. While this is an obvious idea, it is not so straightforward to test experimentally, since granular piles are too opaque to look inside. Hence it is difficult to keep track of a particle, as to which layer it is in at which time of the experiment. Of course one can use immersion fluids, but this changes the dynamics of energy injection completely.

It is much easier to use one of the simulation codes we have already successfully applied in the previous chapter and observe the phenomenon *in silico*. What one has to do is to clearly assign each grain at any time to a certain layer of the pile. Partial tracks of grains traversed within a layer may then be compiled to yield information about the averaged mean square displacement, assigned to each individual layer. In this way, a plot like the one displayed in Fig. 6.43 is obtained [176]. The diffusion coefficient is in arbitrary units, but the acceleration, Γ, is scaled as before. For $\Gamma < 1$, the in-layer diffusion coefficient is below 10^{-4}. Above threshold, it rises dramatically over three orders of magnitude.

One can see that for most of the layers this threshold is around $\Gamma \approx$ 2, a typical value for a wet granulate. The top layer, however, fluidizes

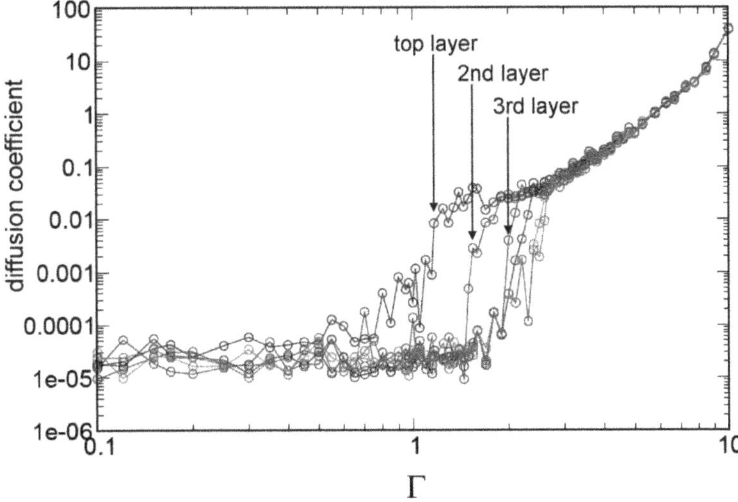

Fig. 6.43 Surface melting, observed by means of molecular-dynamics-type computer simulation of spherical grains, which interact via hysteretic capillary bridges. The diffusivity, which is resolved layer by layer, increases by several orders of magnitude at the melting transition. This transition lies for the first and second layer significantly below the transition for the bulk (data taken from [176]).

immediately above $\Gamma = 1$, shortly after that the second layer fluidizes as well. Obviously, melting of the wet granular pile proceeds as it does for most solid substances: through a molten surface layer, the thickness of which gradually increases as the critical temperature or, in the present case, the critical acceleration for fluidization is approached.

One should note however, that it may not be easy to observe surface melting in any granular medium. In we recall Fig. 5.30, the incomplete restitution present in any real granulate may have an opposite effect. If the surface of the pile is cooler than the interior, the pile may even melt from below. One may anticipate that there can be a complex interplay here of different energy scales and length scales which is still to be unravelled.

6.4.4 *Melting by shear*

Let us again return to the land slide depicted in Fig. 1.5 and our paradigm to see this as a phase transition-like phenomenon. In the previous section, we have focused on agitation by vibration, which may be seen as alluding to the substantial role seismicity is playing in such events. The fact that soil liquefaction is particularly likely after heavy rain fall, when the soil is heavy from the imbibed water, suggests that we should discuss the effects of imposed static force fields as well.

In principle, we have done this already when investigating yield stress. This can as well be seen as a melting transition, which is induced not by a (granular) temperature but by an applied force exceeding a critical value. If we want to embark on conceptual studies, the yield stress is not so well suited for experiments, since we have inevitably friction effects, and thereby arching, which is poorly understood and hard to control (cf. Fig. 6.37). A frictionless wet granulate would be an ideal system, but there is no such thing as a frictionless granulate in nature.

It is therefore useful to resort again to computer simulations, where friction can be completely disregarded. A potential problem here is that the presence of a yield stress leads almost inevitably to shear bands and hence to an inhomogeneous distribution of the shear within the sample. This is disadvantageous if we want to derive constitutive relations for the material. In the experiments presented in Section 6.3.2, we have prevented the formation of shear bands by using rubber membranes as a confinement, and imposed the shear rate on the sample. In simulations, we can exploit the freedom of choosing the applied force field appropriately.

Let us first discuss a two-dimensional system, which is computationally less costly [234]. In the right panel in Fig. 6.44, a snapshot of a simulation of a two-dimensional version of a wet granulate is shown. There are disks of two sizes with a ratio of about 1.5 on order to prevent crystallization. The dark spots indicate capillary bridges. These are modelled as introduced in Section 5.3 (circles in Fig. 5.24) integrating Newton's equations of motion. The system is driven by a force field acting on all 'grains' (disks). The driving force points everywhere in the y direction, and its magnitude varies sinusoidally as indicated in the sketch below the snapshot, $F_d = F_d^0 \cos\left(\frac{2\pi x}{L}\right)$. By choosing this spatial dependence of the force field, we can minimize shear band formation in the sample.

We consider the regime of slowly creeping flow, at the edge of fluidization, of which we extract by proper averaging the y-component, $v_y(x)$. In the left panel of Fig. 6.44, we plot the amplitude of this velocity field as a function of applied force amplitude. There is a clear transition between a moving and a quiescent state, with substantial hysteresis between both states. It is apparent that the solidification comes closer to the critical region than the liquefaction. We therefore concentrate on the former in what follows, in order to extract as much information as possible on the critical nature of the phenomenon involved here.

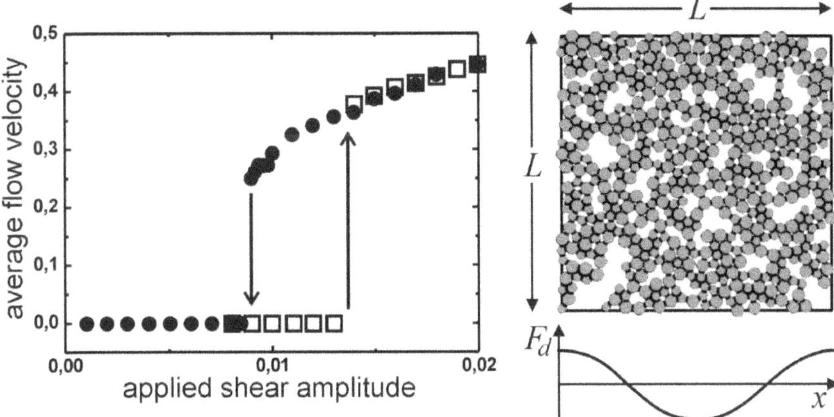

Fig. 6.44   Left: Simulation results for the average flow velocity as a function of the applied shear field amplitude. Right: Sketch of the system under study. The spatial dependence of the applied force field is depicted at the bottom. Redrawn with modifications from [234].

Guided by our discussion of phase transitions far from equilibrium in Section 5.4, we will again adopt the paradigm that in steady state, the injected power must be equal to the dissipated power. The power which is injected by means of the external force acting on the disks is given by

$$P_{\text{inj}} = \int_0^L v_y(x)\ F_d(x)\ n(x)\ dx, \tag{6.27}$$

where $F_d(x)$ is the y-component of the external force and $n(x)$ is the number of particles per unit length dx (i.e. the number of particles in a narrow rectangle of size $L \cdot dx$ aligned parallel to the external field). As pointed out above, the density may be regarded as uniform in the simulation box, such that $n(x) = n$ takes a constant value.

Assuming that the velocity profile is faithfully approximated by its first even harmonic, i.e. $v_y(x) \simeq \Delta v_y\ \cos(2\pi\ x/L)$, we obtain the estimate

$$P_{\text{inj}} \approx \frac{n}{2}\Delta v_y\ F_d^0, \tag{6.28}$$

for the injected power, where n is the density of grains.

In order to calculate the power dissipated by rupturing of capillary bridges, we observe that for each disk the creeping flow enforces a change of neighbors (in the direction of the flow) according to the sher rate, dv_y/dx. If p_r denotes again the average number of capillary bridges which break on a grain upon unity shear, such displacement goes along with an energy dissipation $p_r E_{cb}$. The total dissipated power is thus given by

$$P_{\text{diss}} = \int_0^L n(x) \left| \frac{dv_y}{dx} \right| p_r\, E_{cb}\, dx. \tag{6.29}$$

For every function $v_y(x)$ with period L this integral yields

$$P_{\text{diss}} \approx n \frac{4\, p_r\, E_{cb}\, \Delta v_y}{L}, \tag{6.30}$$

where again we used that the particle density is spatially uniform.

If we now demand the injected power to equal the dissipated power, we obtain the following estimate for the minimum forcing required to maintain the flow

$$LF_d^0 \approx 8\, p_r\, E_{cb}, \tag{6.31}$$

provided that the velocity profile is close to the fundamental harmonic.[11] If we interpret Eq. (6.28) as the product of the shear stress times the shear rate, we immediately find for the critical shear stress

$$\sigma_c \simeq nLF_d^0 \approx 8np_r E_{cb}. \tag{6.32}$$

One would expect that the critical shear stress diverges when approaching random close packing:

$$\sigma_c \propto \left(\frac{\phi_j - \phi}{\phi} \right)^{-\alpha}, \tag{6.33}$$

where α is a positive exponent, and $\phi_j \approx 0.84$ in our bi-disperse system [235]. In this expression the factor $1/(\phi_j - \phi)$ can be interpreted as the smallest area where there is enough *free volume* for two disks to pass, and hence the ratio $\phi/(\phi_j - \phi)$ amounts to the minimum number of disks that must be displaced in order to let two disks pass.

In Fig. 6.45, data are compiled for the critical force amplitude for solidification as a function of ϕ. As anticipated, the force diverges when the close-packing density ϕ_j is approached. The critical exponent is indeed found to be $\alpha \approx 1/2$ [234].

Equation (6.33) still involves an unknown pre-factor which is a function of the rupture energy E_{cb}. Varying the rupture separation at fixed system size L and packing fraction ϕ reveals that this pre-factor scales as $-\ln\sqrt{2E_{cb}}$. Forthcoming work will have to address the question whether

---

[11] Even for the rather extreme opposite case of a plug flow, the estimate (6.31) for F_d^0 is just multiplied by a factor of $\pi/4$ [234].

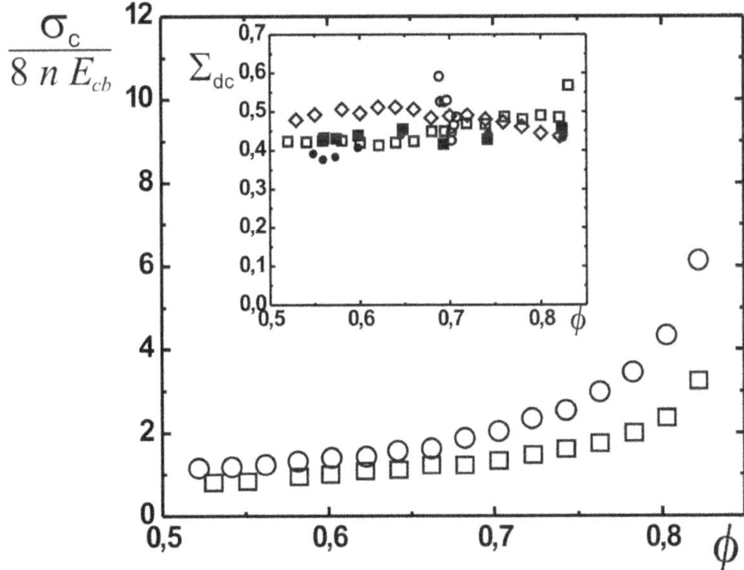

Fig. 6.45 Amplitude of the normalized external force field at solidification, $F_d^0\, L/8E_{cb}$, as a function of the packing fraction, ϕ. The two curves for the solidification refer to rupture energies $E_{cb} = 0.01$ (top) and $E_{cb} = 0.05$ (bottom), respectively. As expected, the forces diverge when approaching random close packing, and a smaller number of capillary bridges break when rearranging a system with a larger E_{cb}. In order to underpin this assertion, the inset shows that Σ_{dc} takes roughly constant values (cf. Eq. (6.35)). The system size is $L = 18$. Redrawn after [234].

this might be attributed to an exponential screening of stresses in the system. Irrespective of the physical origin of this dependence, the scaling of $F_s\, L/8\, E_{cb}$ with E_{cb} allows us to write the critical force at the solidification transition as

$$\sigma_c \approx -B\frac{8\, E_{cb}}{L}\,\ln\sqrt{2E_{cb}}\,\left(\frac{\phi}{\phi_j - \phi}\right)^{1/2},\qquad(6.34)$$

with a numerical pre-factor $B = (0.45\pm0.05)$. The error corresponds to the scattering of the numerical data [234]. Another way to express this result is to define the function

$$\Sigma_{dc} = \frac{-\sigma_c}{8nE_{cb}\ln\sqrt{2e_{cb}}}\left(1 - \frac{\phi_j}{\phi}\right)^{1/2},\qquad(6.35)$$

which should for all data yield the same value of about 0.45. That this is indeed the case is demonstrated in the inset in Fig. 6.45, which contains a compilation of all data obtained so far for this system. Note that the critical exponent of 0.5 with respect to the variation of ϕ close to jamming (ϕ_j) is distinctly smaller than the corresponding exponent for slurries. This is $2.5\phi_j \approx 1.5$ according to Eq. (2.42), and has been reported to be even close to 2 [64, 66, 65, 236].

Schultz *et al.* have performed similar numerical simulations in three-dimensions [237], with the same hysteretic force characteristics. Just as in the two-dimensional simulations, the driving force was a field acting to each grain, and shear bands could not develop. Similar to the results shown in Fig. 6.44, the authors obtained a transition occurring at a finite critical stress. The data were analyzed in terms of the diffusivity of the particles. Diffusion constants referring to the three spatial coordinates were computed separately. They were found to agree quantitatively with each other, which suggests that there was no significant anisotropy in the transport properties of the system. The results are shown in Fig. 6.46 in the left panel, while the right panel shows a rendering of the system.

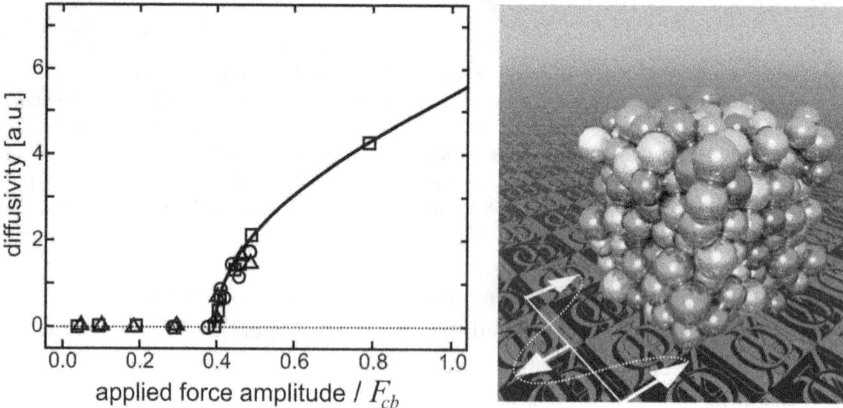

Fig. 6.46   Left: Results of a simulation of yield, similar to the setting in Fig. 6.44, but in three dimensions (see text). Different symbols refer to different directions (x, y, z) with respect to which diffusivity has been evaluated. Right: Rendering of the system under study. The spatial dependence of the applied force field is depicted to the bottom left [237].

A striking difference with respect to Fig. 6.44 is that there is no jump observed in the transport currents, nor any noticeable hysteresis. The solid curve represents a square root behavior, indicating a critical exponent of 1/2. Aside from the missing discrete jump, this is in qualitative accordance with the behavior of the two-dimensional system. Due to the obviously small size of the sample, however, the results must still be taken with care concerning the statistical quality of the data. Whether the absence of the jump and hysteresis (indicative of a second order phase transition) is due to the limited statistics or a consequence of the different dimensionality must be clarified in further studies. These should then also contain some variations in the density, in order to fully compare with the results obtained with the two-dimensional system.

6.5 Conclusions

In the course of this chapter, the following distinctions of wetness regimes of granular piles have been identified:

- The *humidity regime.* This is defined as the range of wetness where no clearly developed capillary bridges can be observed.
- The *pendular regime.* Here we have fully developed capillary (pendular) bridges, each of which is in contact with the surfaces of two neighboring grains.
- The *cluster regime.* This is defined by the presence of many clusters which are larger than an individual capillary bridge.
- The *bicontinuous regime.* This is defined by the presence of at least one large sample-spanning cluster (percolation) which contains a large portion of the wetting liquid (secondary fluid). At the same time, the primary fluid forms a percolated set as well.
- The *bubble regime.* Here the secondary liquid is not anymore percolated, but is contained in individual occluded bubbles.

The humidity regime comprises the asperity and roughness regimes, which are physically hard to distinguish clearly from each other. The cluster regime, the bicontinuous regime, and the bubble regime together correspond to what is classically called the funicular regime. As the most striking facts, we found that closely related quantities such as the matric suction, the tensile strength, and the critical acceleration required for fluidization exhibit wide plateaus, where there is almost no dependence on the liquid content. This goes in parallel with the observation made in the previous chapter that much of the dynamics of a wet granular gas can be described by just using two-body interactions by capillary bridges, even at a wetness where larger liquid clusters should prevail.

In the case of the yield stress, the situation is a little more involved, and a similarly nicely developed plateau is not observed. Nevertheless, it turns out that even in the case of wet granular piles (which are denser than the granular gases of the previous chapter), many phenomena can be understood pretending there are only two-body interactions in the system. Although we have tried to rationalize this striking observation, its deeper provenance is still to be uncovered.

Further reading

The review by Mitarai and Nori about wet granular piles should help clarify many things which may have remained opaque in this chapter [238]. In particular, the authors make a clear connection to notions which are common in the engineering sciences, but not known to most physicists.

For those who have already gained some sovereignty in dealing with different nomenclatures, the comprehensive book by Gudehus [239] will be helpful for appreciating the vast amount of knowledge which has already been accumulated in the engineering sciences about wet granular matter. In particular, we have tried not to trouble the reader with too many technical details. Hence we have chosen not to present the theory of Mohr and Coulomb. This does not bear any new physics beyond what we have presented here, but leads to a more complete and formally elegant picture [228, 239, 35, 240]. It is also worthwhile to enjoy reading the early papers by Rumpf [241, 242], which have already a lot of the physics we have presented here.

One should appreciate a number of papers [141, 243–245] which have greatly contributed to our present understanding of wet granular piles. An important difference to the treatment presented here, however, is that the importance of the hysteretic nature of capillary bridges was not yet appreciated. As a consequence, many physical aspects of slowly deformed granular piles are derived correctly, but aspects owing to hysteretic energy loss are not resolved.

For venturing out into the wide field of geo-technical problems, the books by Dikau [6] and Das [224] may serve as a useful introduction. And, quite the opposite, to dive into the mathematics of packing geometries in a comprehensible way, the review of Torquato and Stillinger [246] is enjoyable to read.

Chapter 7

Special Topics

We have come a long way now. The mixing triangle of grains, liquid, and gas (or primary fluid), which we have introduced at the beginning of this book, has been explored in most of its physical facets, as far as they have been thoroughly investigated until today. It might thus be considered appropriate to terminate our journey through the physics of wet granular matter at this point. However, the author feels that a book on wet granular matter would not be complete without a section dedicated to the physical aspects of sand sculptures, as these represent doubtlessly the most obvious, most spectacular, and paradigmatic examples of this class of material. As such, they are very useful to serve as a model for illustrating wet granular physics. Furthermore, it will be instructive to add a few words on topics which are not strictly covered by the program suggested by the title of this book, but which are sufficiently closely related to wet granular matter to be at least mentioned and shortly discussed.

7.1 The physics of sand castles

7.1.1 *Mechanical stability*

A glance at Fig. 7.1 reminds us that sand sculptures can be created which exhibit slope angles approaching ninety degrees. The technique by which such 'drip towers' are produced shall concern us further below. Here we first want to discuss what determines the stability of sand sculptures, and in what features, slope angles, and feature sizes this may result. To this end, it is illuminating to notice that large sand sculptures almost invariably have (average) slopes much milder than vertical. Figure 7.2 shows a particularly beautiful example. As one can see, the slope angles, in particular in the

Fig. 7.1 'Drip towers' at the beach of Rocky Neck (Connecticut, USA). Photo kindly provided by Jeffrey Hollis.

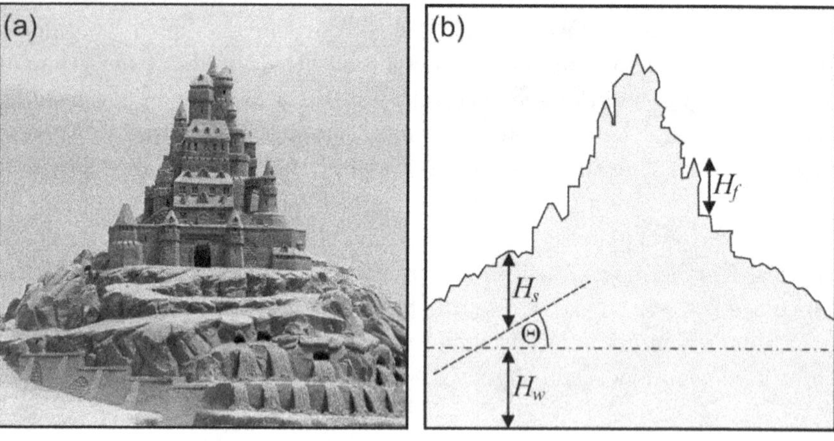

Fig. 7.2 (a) A particularly beautiful example of a sand castle. It consists of a few billions of sand grains, held together by capillary forces between adjacent grain surfaces, mediated by the interstitial water (image taken with permission from www.picture-newsletter.com). (b) A few important quantities inserted into the silhouette of the sand castle of (a). H_w: height up to which the pile is wetted by liquid capillary pressure (matric suction) above the water table. H_s: depth down to which liquid interface forces dominate over gravitational load stress. Θ: slope angle of the pile. H_f: feature height which can be achieved at slope angles strongly exceeding Θ.

lower part of the sculpture, are very much reminiscent of customary slope angles of dry sand piles. Only smaller features terminate as vertical cliffs.

We can easily rationalize this observation by comparing the stresses generated within the pile due to the liquid and due to the mechanical load.

The compressive stress exerted onto the internal structure of the pile, p_{int}, has been estimated in Eq. (4.36), which reads

$$p_{\mathrm{int}} = \frac{\gamma k \phi \cos \theta}{2R}. \tag{7.1}$$

This is the characteristic scale of stresses which can be exerted by the capillary bridges extending between adjacent grains.

It is clear that if there is a large overburden of granular material, there will be as well a considerable mechanical stress,

$$p_{\mathrm{load}} = \varrho_s \phi g H \tag{7.2}$$

if H is the vertical depth below the surface of the pile. The relative size of the two contributions, p_{load} and p_{int}, has been introduced as one of the major dimensionless quantities governing the mechanical properties of a cohesive granulate [247]. It is quantified as the cohesion number,

$$\mathsf{Ch} = \frac{p_{\mathrm{int}}}{p_{\mathrm{load}}}. \tag{7.3}$$

The depth down to which the liquid dominates the stabilization of the pile is then determined by Ch coming of order unity.

A more quantitative estimate can be obtained by resorting to Fig. 6.36. We are interested at which external pressure the influence of the liquid, $\Delta p^{wet} - \Delta p^{dry}$, accounts for only half of the total stability, Δp^{wet}. Data from Fig. 6.36 are replotted accordingly in Fig. 7.3. The ordinete is the ratio $(\Delta p^{wet} - \Delta p^{dry})/\Delta p^{wet}$. The abscissa is the external pressure, p_0, rescaled according to the yield stress extrapolated to vanishing friction effects, $p_0^* = p_0/\Delta p^0$ (Coulomb cohesion parameter, see Section 6.3.2). The curve crosses the value 0.5 at about $p_0^* \approx 6$. Hence we expect that the slopes of a sand pile can be effectively stabilized by a liquid additive down to a depth of about

$$H_s \approx 6 \frac{\gamma k \cos \theta}{2 \varrho_s R g}, \tag{7.4}$$

which is indicated in the right panel of Fig. 7.2. Inserting in numbers shows that for complete wetting ($\theta = 0$) by water and $R = 200~\mu$m, we find that H_s is about 25 cm. At any greater depth, the gravitational stress exerted on the pile by the weight of the overlying granulate exceeds all effects from moisture. The latter can then even have a destabilizing effect if it happens to act as a lubricant for the grains used.

This seems to be at variance with many features of sand sculptures encountered occasionally, which seem to exhibit slope angles well in excess

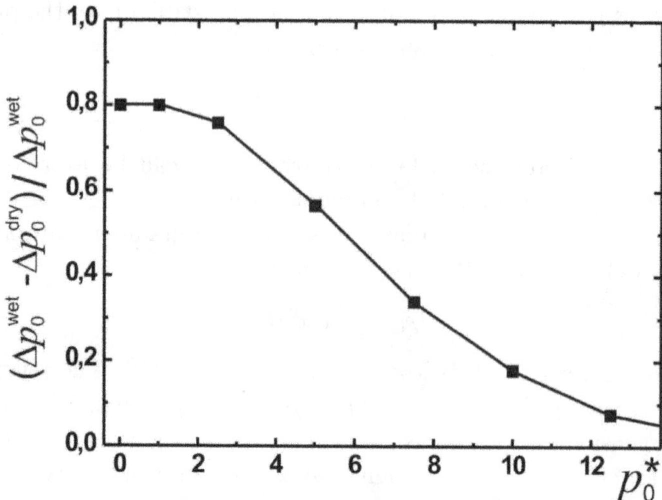

Fig. 7.3 The ratio $(\Delta p^{wet} - \Delta p^{dry})/\Delta p^{wet}$ plotted as a function of external pressure, p_0, normalized with respect to Δp^{wet} at zero p_0. Data from Fig. 6.36.

of what could ever be possible with dry grains alone. The reason is very simple: the sand which is found in natural habitats comes with a large number of additives. On the beaches of the oceans, there are biological remains, aside from the omnipresent salt. As soon as the wet sand is used for sculpturing, part of the liquid evaporates, and some of the solutes and suspended material accumulates at the contact points between adjacent grains, to which the water is attracted by wetting forces. As a consequence, the cohesive strength of the granular pile is significantly increased by simple 'gluing' effects. The importance of such additives becomes evident when the sun dries the sculpture. If it was only the liquid which stabilized it, it would immediately collapse into a structureless heap, resembling a sand dune. This is, however, almost never observed. Aside from a few minor defects, the sand sculptures seem instead like baked together to a material which appears brittle but still similarly resistive as the wet granulate from which it was sculptured.

On many sand sculpture competition festivals, it holds as a rule that no artificial additives may be used. This is why the sand is sometimes transported over some distance to the sites of these festivals, from sources which provide sand with suitable natural additives. Inland sources may contain, e.g., large quantities of clay minerals, which serve similar purposes as the above mentioned additives found in sea water. If, e.g., there are silt particles present with a volume fraction of ϕ_s, they outnumber the larger grains, which are present with volume fraction ϕ_g, by a factor of $(R_g/R_s)^3 \phi_s/\phi_g$, where R_s and R_g are the average radii of the silt particles and grains, respectively. As the physics of each contact between grains is determined by an effective radius dominated by the smaller grain (cf. Eq. (4.29)), by far most of the contacts will be dominated by the small silt particles. These are then dominating the pressure scale, with the radius of the silt particles entering Eq. (7.1), resulting in substantial increase of H_s. Furthermore, it is advantageous to take sand from non-marine sources like rivers or former glacial moraines, where the grains can be much less well rounded than those found at beaches. Although we have seen that the shape of the grains does not influence the mechanical properties of the material as strongly as one might expect at first glance, all of the above mentioned effects contribute to stabilizing the sculptures. A wide grain size distribution may even render some of the results we have obtained with our simple model granulates invalid, such as the insensitivity of the mechanical properties of the amount of liquid. The latter may then become a critical parameter, whose skilful treatment is a matter of the personal experience of the artist.

Playing around with wetted piles of our model granulates, like spherical glass beads, one appreciates quite readily that the estimate we have given above is very realistic. A pile of wet glass spheres shows the same qualitative behavior as a pile of wet sand, but its mechanical stability is much inferior. When the pile reaches a height of several centimeters, substantial problems to stabilize steep slopes become readily apparent [248]. A glass pile exceeding 10 cm in height has already a strong propensity to flow apart and spread over its support.

7.1.2 *Distribution of liquid*

Another question of particular interest is the distribution of liquid in the sand pile. As we have seen in Chapter 6, the matric suction, or (negative) Laplace pressure within the interstitial liquid is approximately given by

$$-p_L = 4.46\frac{\gamma}{R}\cos\theta. \tag{7.5}$$

At the water table, which in Fig. 7.2 is symbolically made to coincide with the lower border of the image, the pressure is zero because the curvature of the free liquid surface vanishes. Below the water table, the granulate is completely saturated, and it is clear that the saturation will monotonically decrease as we go up above the water table. We know that the liquid structure is percolated down to a saturation corresponding to $W = 0.11$, and we know that the matric suction is virtually unchanged down to a saturation corresponding to $W = 0.026$. Throughout the percolated liquid structure, the pressure within the liquid must balance hydrostatic pressure, which is given by

$$p_{\mathrm{hydr}} = \varrho_l H g, \qquad (7.6)$$

if H is the height above the water table. The height which corresponds to the percolation transition, and actually to the preferred pressure of all open liquid structures within the pile, is obtained by setting $p_{\mathrm{hydr}} = p_L$, which leads to

$$H_w = \frac{4.46\gamma\cos\theta}{Rg\varrho_l} = \frac{8.92}{k}\frac{\varrho_s}{\varrho_l}H_s. \qquad (7.7)$$

For fine sand, this can easily reach 30 cm. Above this height, liquid clusters cannot be stable in the interstitial space. As one can see from Fig. 6.19, only individual capillary bridges will be able to balance the elevated matric suction above H_w. Hence there will above H_w be no percolated liquid structure anymore in the pile. These are confined to the space below H_w.

On the grains, however, percolated liquid films can well persist to much higher elevation. As discussed in Section 3.2, this is determined not by the curvature of capillary bridge surfaces, but by the curvatures of the deepest troughs of the roughness in the grains. Corresponding elevations may be well a few meters, up to which, correspondingly, liquid transport through the wetting layers on the grains surfaces may be possible. This may play an important role in may processes taking place in the vadose zone,[1] such as, e.g., the formation of gypsum roses.

7.1.3 *Drip towers*

In Chapter 6, we have seen that wet granular matter, up to a liquid content of about half way to liquid saturation, can be described by only a few key parameters in the classic configurations outlined above. We may now try to explore the range of larger liquid content. We start by returning to the fully saturated state, where the interstitial space of the granulate is completely filled with liquid. This is called a slurry, and we have dealt with its properties already at length in Chapter 2. Since there are no liquid

---

[1]By the vadose zone one denotes the zone above the water table, which is still affected by the presence of the ground water.

interfaces spanning between the grains, it can be seen as just a different type of dry granulate, just that the interstitial fluid has a higher viscosity than the gas. In fact, sand under water does not exhibit any stiffness, and behaves in many aspects as dry sand. This has been exploited in experiments on sand dune formation, where wind was modelled by water flow [50].

It appears tempting to explore the range of large liquid content starting from a slurry and gradually removing liquid (or, equivalently, injecting some air). At first glance, one might then expect to find the same behavior as observed for small liquid content, with the 'capillary bridges' now being formed by air pockets residing at the contact points of adjacent grains. However, if we use the same materials as before, the liquid wets the grains surface, and the air would rather try to accumulate in the pore space between the grains, as small interstitial bubbles (bubble regime, see Fig. 6.11), with the liquid covering at least the majority of the grain surface.

Quite surprisingly, one does not have to go as far as stirring tiny bubbles into the slurry to stiffen it. A much smaller reduction in liquid content is enough, as it has been shown [249]. If a drop of slurry is deposited on an absorptive substrate, such as a consolidated pile of grains, a tiny fraction of the liquid immediately leaves the slurry drop into the substrate, pulling the free liquid interface inward towards the interstitial space between the grains. A schematic representation of the setup is sketched in Fig. 7.4(a). The slurry is deposited drop-wise from a nozzle onto a granular bed consisting of the same grains, but without wetting liquid added. This acts as an efficient blotting medium, draining rapidly away and surplus liquid, and thus leads to an effective drying of the slurry surface.

This can be appreciated qualitatively from Figs. 7.4(b)–7.4(d). Aside from a little tower which has already been built from a few deposited droplets, each panel shows as well the next arriving drop shortly before it hits the tower. At this stage, it still shows a shiny reflection due to the free liquid surface surrounding the grains. After deposition, however, all drops appear matt. The liquid–gas interface has receded towards the interior of the slurry drop, such that grains are sticking out of the liquid surface. Hence the light is diffusely scattered. Clearly, the liquid surface must acquire a negative mean curvature, and correspondingly a negative Laplace pressure, in order to invade the slurry drop by even only one half grain diameter. The remaining interstitial space is still filled with liquid, but there is a considerable external pressure, which stiffens the drop due to the mutual friction between the grains. This leads to a substantial

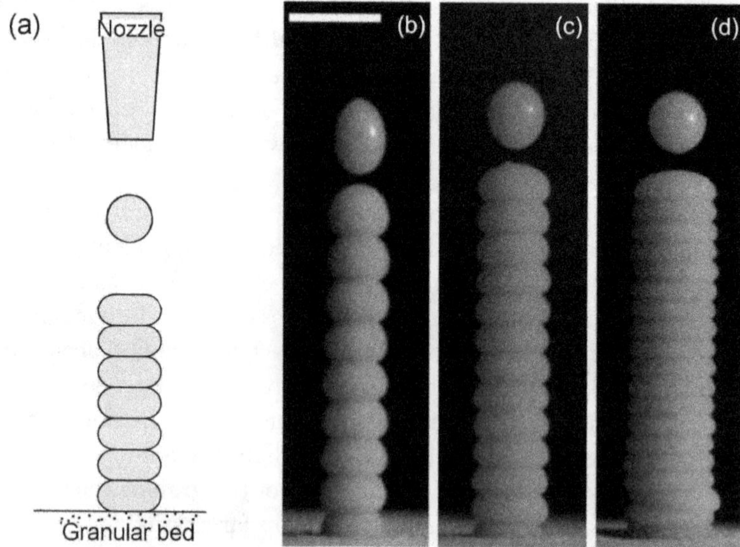

Fig. 7.4 Setup (a) and resultant tower structures (b, c, and d), created by dripping a dense slurry onto a porous support (consisting of the same granulate as suspended in the slurry) [249]. Illumination (far field) is from the left. As it is clearly visible from the shiny reflection on the drops still in free fall, there is a smooth liquid–gas interface constituting the drop surface. As soon as the drop has hit the support, be it the granular bed at the bottom or the already existing tower, the liquid immediately drains away through the porous structure of the pile, leaving behind a surface which exposes the dry grains and thus appears matt. The height of the nozzle above the tip of the tower was (b) 30 mm, (c) 50 mm, and (d) 200 mm. The scale bar is 10 mm (Images b, c, and d reproduced from Fig. 5 of [249]. Copyright (2011) American Physical Society).

increase in yield stress, similar to that discussed in Section 6.3.2. It can be exploited to build so-called 'drip towers', and can be nicely controlled by proper instrumentation [249].

In some sense, the liquid–air interface plays a similar role here as the rubber balloon membrane in Fig. 6.32. As the impressive towers presented in Fig. 7.4 demonstrate, the mechanical stiffness acquired by the slurry is in fact considerable. The accompanying loss of liquid is minute, as it scales only with the surface area of the drops. The liquid content can be roughly expressed by $W = W_{sat}(1 - R/R_{\mathrm{drop}})$, and thus can be extremely close to W_{sat}. In Figs. 7.4(b)–7.4(d), the height from which the drops are deposited is varied. Obviously, the main features do not change, only the width of the resulting tower becomes larger as the height (and thereby the impact velocity) is increased. The particular state in which the wet granulate is encountered in the structures just described is called the *capillary state* in the engineering literature [238], and holds a rich bouquet of phenomena [250, 251].

The technique described above is popular for constructing larger structures as well, like sand castles ('drip castles'). One can construct them by taking a hand full of sand completely saturated with water, and letting the slurry drip down from the palm along the thumb onto the desired place. This is of course less well defined than the setup of Chopin and Kudrolli. A typical result looks like those shown in Fig. 7.1. Some toy stores sell a handy device for this purpose, which is shown in Fig. 7.5. It consists of a conical plastic container which is internally partitioned into a large compartment holding the wet sand, and a small one into which water is being poured. The latter compartment has a small hole at its lower end yielding to the large compartment. There it forms the slurry, which then drains out of the bottom hole. If one takes care to use well-sieved sand, this device allows for the construction of quite homogeneous and well-defined drip-castle structures.

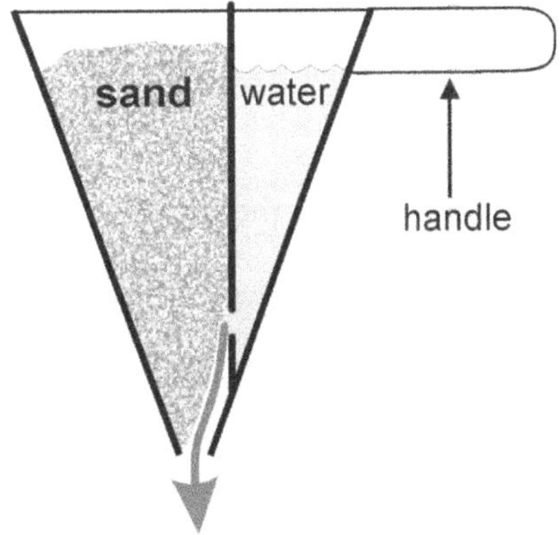

Fig. 7.5 A simple toy device for the construction of drip towers. In its lower part, a slurry is formed from the wet sand and the water which is added in the separate compartment. The slurry then drains out at the bottom (grey arrow). In Germany, the device is sold under the name 'Kleckertüte' ('spill bag').

7.2   Immiscible granular suspensions

7.2.1   *Capillary suspensions*

For a dense granular pile, the liquid content cannot be larger than $W_{sat} = 1 - \phi$. To explore this range nevertheless, we may ask what happens if we reduce instead the solid volume fraction in a slurry to values below random close packing. On the basis of the Krieger-Dougherty relation (cf. Eq. (2.42)), one might expect that this will result in a slurry with rheological properties close to those of the suspending liquid. As already mentioned, it is has been shown that slurries can be very well considered as Newtonian fluids, if the shear rate is not too small.

However, it is well known that for immiscible fluids as the primary and secondary suspending liquids, this may not be the case. If particles are suspended in a non-wetting fluid, and small amounts of an immiscible secondary liquid is added which wets the solid particles, this secondary liquid creates capillary bridges and may cause the slurry to acquire a pasty consistency, very much like the wet sand from which the sculpture shown in Fig. 7.2(a) has been constructed [252]. In a recent paper, it has been shown that this effect goes even well beyond the mere formation of capillary bridges between adjacent particles. It may as well be present when the wetting properties of the liquids are reversed, such that the liquid representing the majority phase does wet the particles, and the secondary liquid does not. In either case, it has become clear that the formation of complex network structures within the slurry plays an important role [253, 254].

The effect is demonstrated impressively in Fig. 7.6 for a suspension of hydrophobized calcium carbonate particles (average radius 0.8 μm) in di-isononylphtalate (DINP, C_6H_4-$(COO$-$C_9H_{19})_2$), which wets the particles well. As it is clearly seen, upon addition of less that 1% of water (which makes a wetting angle of about 140 degrees with the particles under DINP), the fluid-like suspension shown to the upper left is transformed into an elastic gel, after a short time of thorough stirring. The solid volume fraction is only about 11%, which shows that there must be an extended network of particles connected by liquid interfaces, which spans the whole suspension and thus bestows considerable stiffness to it. The yield strength of the stiffened suspension does not depend appreciably upon the solid volume fraction in a wide range [253], from about 11% to almost 30%.

This clearly distinguishes the capillary suspensions discussed here from related customary systems, like wet sand. The ability of the network

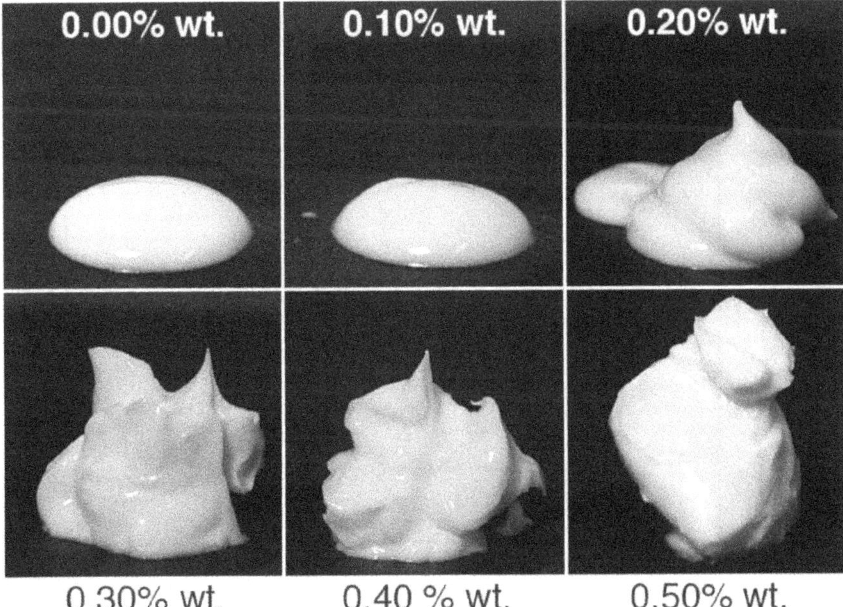

Fig. 7.6   Transition from a fluid-like slurry (micron-size hydrophobized calcium carbonate particles suspended in DINP, solid volume fraction is 0.11) to an elastic gel through addition of a secondary fluid (water) which does not wet the particles. The water thus plays a similar role here as the air in the 'drip tower' example discussed before. The numbers assigned to the images indicate the weight percent of secondary fluid [253]. Note that it remains below 1%! (Image courtesy of Erin Koos, KIT Karlsruhe).

structures to unfold such importance for the mechanical properties certainly lies in the partial density matching provided by the three phases involved. Even if the densities of the liquids and that of the solid particles are not exactly equal, they are much closer to each other than the densities of, e.g., the sand grains, the water, and the interstitial air, as encountered in case of the wet sand example. In the latter system, extended network structures like those present in the capillary suspensions would not be stable, collapsing readily due to gravitational pressure. In micro-gravity conditions, however, it should be possible to create moldable wet granulates with solid volume fractions down to 11%, similar to the capillary suspensions demonstrated on ground.

It should be noted that the difference in the physical mechanisms, or liquid cluster morphologies, for the wetting and non-wetting secondary liquid also show up in macroscopic quantities. Figure 7.7 shows the yield

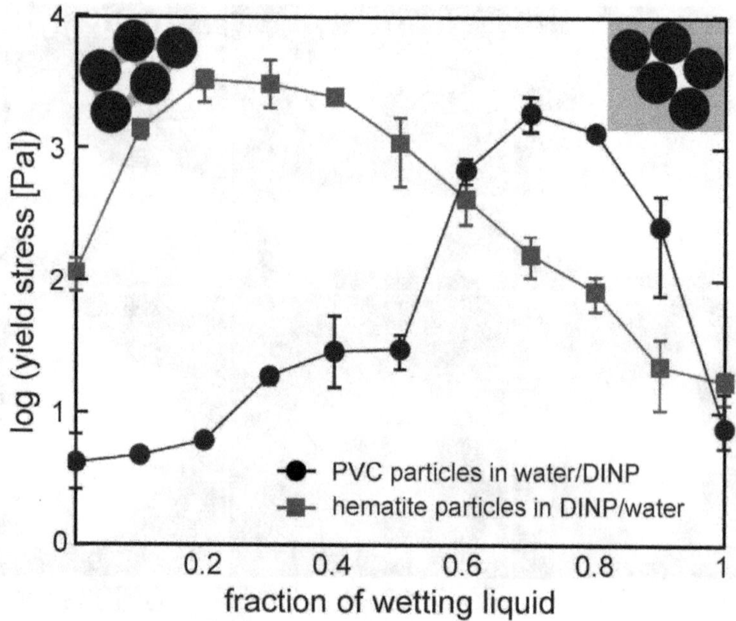

Fig. 7.7 The yield stress of granular suspensions of two different types, one of which strengthens preferentially through capillary bridge networks (squares), the other through the formation of networks of compact clusters (circles) [253]. While the stability of the former is reminiscent to the mechanisms discussed in connection with Fig. 6.19, the latter owes its stiffness rather to a mechanism like the one responsible for the 'drip towers' in Fig. 7.4 (image courtesy of Erin Koos, KIT Karlsruhe).

stress for two different systems as a function of the wetting liquid content. Although in both cases one can clearly see a strong increase, over several orders of magnitude, for an intermediate volume fraction, it is obvious that the details of this dependency are distinctly different for the two cases. This is expected, since the geometry of the liquid interface structure building up in between the solid particles is certainly strongly different, as evidenced by sample numerical simulations of liquid interfaces spanning aggregates of granular particles [253, 214] for different contact angles.

It suggests itself to use mixtures with a critical mixing temperature in order to turn such systems into sensitive thermotropic materials. As long as the temperature is such that the liquids are miscible, the suspension will behave just like a slurry, with a viscosity slightly increased according to, e.g., Eq. (2.42). As soon as the system is brought into the miscibility gap, the fluid segregates into two immiscible phases. One expects that the

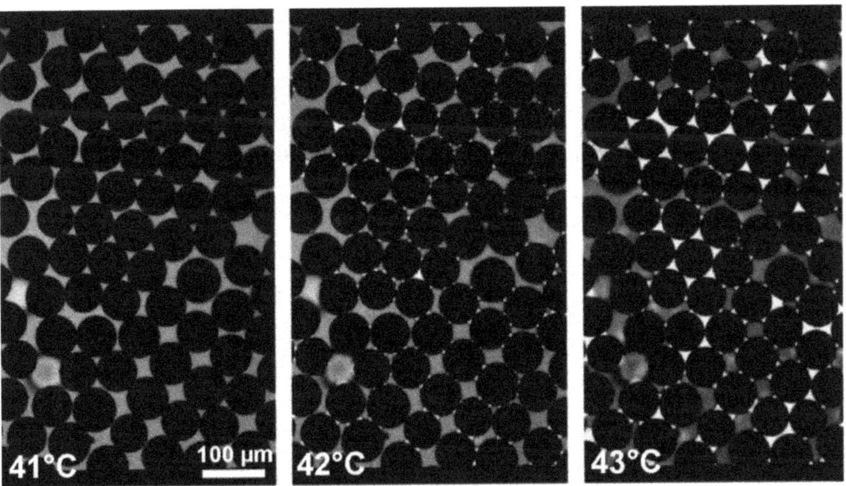

Fig. 7.8 Spherical glass beads (70 microns in diameter) suspended in a mixture of 2,6-lutidine and water. As temperature increases, the mixture decomposes into a primary (lutidine enriched) and a secondary (lutidine depleted) fluid [255]. Center image, capillary bridges have formed at contact points. Right image, larger clusters appear (images kindly provided by Somnath Karmakar and Christoph Goegelein, Göttingen).

phase which better wets the suspended particles will start to form capillary bridges connecting the particles, or the more complex structures described in the work by Koos and Willenbacher mentioned above. If this happens, the suspension should dramatically change its rheological properties.

The tunable formation of capillary bridges out of a fluid in the miscibility gap is demonstrated in Fig. 7.8. Spherical glass beads with a diameter of 70 microns are surrounded in a mixture of 2,6-lutidine ($C_5H_3N(CH_3)_2$) and water in a flat cell with a height of 170 microns. Mixtures of lutidine and water are known to exhibit a lower critical mixing temperature, such that the system decomposed into two phases at higher temperature. At low temperature (left image), the beads are visible as dark disks on a uniform background. As temperature is raised into the (lower) miscibility gap of the lutidine-water system, a lutidine-depleted phase emerges from the mixture which preferentially wets the glass beads. At 42 degrees, the formation of individual capillary bridges at all points of contact between neighboring grains can be clearly observed (center image). At three points, capillary bridges have merged forming trimers. Hence the aqueous (lutidine depleted) phase has a volume fraction close to W^*.

As the temperature is raised further, more aqueous phase is formed, and we see that more capillary bridges coalesce into larger clusters, much in line with what we have discussed in the previous chapter. The process is entirely reversible, such that the rheological properties of the suspension can be switched by only slight temperature variations. It should be noted that temperature is not the only parameter by which miscibility may be tuned. Pressure, for example, allows for particularly fast quenches in and out of the miscibility gap, thus allowing for fast switching of the rheological properties of the suspension.

7.2.2 *Pickering emulsions*

Very closely related to the capillary suspensions just discussed are what has become known as Pickering emulsions. They are named after the author of one of the first papers where they have been described [256, 257]. It was recognized that immiscible liquids could be compatibilized by small solid particles, which played a similar role as surfactants in usual emulsification processes. That solid particles can indeed reduce the interfacial tension between two liquids is illustrated in Fig. 7.9. The only condition is that the contact angle of the liquid interface with the surface of the particles is neither zero nor π. If it has any intermediate value, the particles will try to segregate to the interface, as this enables the formation of the equilibrium contact angle. The reduction in free energy thus incurred leads to an effective reduction of the interfacial tension, which is now considered as an average over the particle–laden interface.

The total interfacial free energy per unit interface can be directly computed from the individual interfacial tensions. In particular, if the grain has equal affinity to both liquid phases, we have $\gamma_{1s} = \gamma_{2s}$. In this case, it is readily verified that the free energy released upon moving the grain from inside one of the two bulk phases towards the interface is given by $\pi R^2 \gamma_{12}$. In other words, the grain in effect 'cuts out' the whole interfacial area corresponding to its cross-section, and sits symmetrically at the interface. Since the maximum two-dimensional packing density of spheres is 0.91, we see that for $\gamma_{1s} = \gamma_{2s}$ the grains act as a very efficient surfactant, reducing the interfacial tension by more than 90%.

Hence we see that solid particles can be quite effective surfactants, although they do not exhibit any amphiphilic structure [258–260], like a hydrophilic and a hydrophobic domain. Such particles can also be used, but they are much more difficult to synthesize. A particular merit of Pickering emulsions is that the particles can be made from very stable materials.

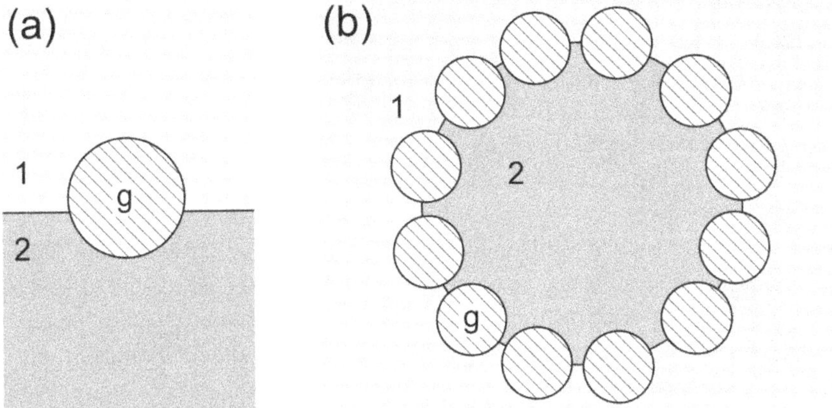

Fig. 7.9   Schematic representation of the inner structure of a Pickering emulsion. (a) A grain at the interface of the two liquids which are to be compatibilized. (b) A droplet of phase 2, stabilized within phase 1. If the grain surface has equal affinity to both liquid phases, the interfacial tension between phases 1 and 2 is reduced according to the two-dimensional packing density of the grains at the interface.

For instance, they can be used as surfactants for metal melts, since many solids have higher melting temperatures than those of most metals. This is very useful, e.g., in the process technology of metal foams and similar formulations.

Surfactants segregate to surfaces and hence can be used in minute quantities. The fact that the granular particles play the role of a surfactant here hints at the place in the mixing triangle, Fig. 1.9, where we can expect to find Pickering emulsions. It is close to the lower boundary, where the system consists almost exclusively of the fluid phases, and the grains play the role of a dilute additive.

7.3   Quicksand and quickclay

Any text on cohesive granular matter would be incomplete without a few words on quicksand and quickclay, although the interactions involved here are not due to wetting forces. While quicksand occurs much more frequently in scary movie scenes than in nature, quickclay is a rather widespread phenomenon and has frequently caused casualties and immense damage to livestock and buildings. The common feature of both quicksand and quickclay is an *irreversible* change in the cohesion forces of a (more or less dense) granular pile completely saturated with an aqueous liquid, which in turn results in an irreversible change of mechanical properties. A soil which has been solid enough to support houses, streets and traffic for years may hence suddenly fluidize and run down even soft slopes over large distances [261]. It is clear that a lot of research is being devoted to the understanding, prediction, and mitigation of these rather frightening events.

The mechanisms behind the quickclay phenomenon are quite reproducible and meanwhile well understood in their main aspects. Quickclay is generated when a clastic sediment (i.e., a granulate) is deposited in the presence of water with a high salinity (usually sea water), which is later leached out by ample supply of freshwater. In an aqueous solution with high ionic content, electrostatic forces have a small range, since charges are effectively screened by the abundant counter-ions. This holds as well for the electrostatic charge on the grains. Almost every mineral surface has some ionic groups which can dissociate, such that the grain surface charges up electrostatically. Although the strength of this charge depends upon parameters such as temperature or pH, it is almost never exactly zero. The density of ions in the surrounding liquid decides on the range over which this charge can be felt by other grains.

Since the grains are usually all of the same material, their charge has the same sign, hence there is an electrostatic repulsion between them. However, as long as there are many ions in the liquid, this force is only very short ranged. At the ionic strength of seawater, for instance, the screening length is only a few nanometers. Hence the repulsion is effectively switched off as long as there are enough ions around, and the grains will sediment into full contact. At contact, other interactions such as, e.g., the van der Waals force, are stronger and lead to the formation of a rather stable granular pile.

It may occur that the interstitial water is then slowly replaced by freshwater. This may happen if a marine sediment is uplifted above sea level by tectonic or isostatic forces, such that rain will gradually replace the seawater still in place. If this proceeds sufficiently slowly, without disturbing the positions of the grains, these stay in contact, still glued together by contact forces which overwhelm the electrostatic repulsion as long as mechanical contact is maintained. However, if the systems is mechanically disturbed, as it may happen in an earthquake, the inter-granular contacts are broken. Once this has happened, the repulsive forces dominate over the attractive forces which are only active at close contact, and the material suddenly fluidizes irreversibly.

There are other variants of this mechanism which do not need the presence of water. For instance, some occurrences of quicksand are thought of as being due to sand grains which have sedimented to a rather low density. This may happen when there is some 'glue' in the system conserving inter-granular contacts once they have been established, possibly some glycopolymers of biological origin. If such a system is disturbed, breaking the contact glue, a similar irreversible fluidization occurs as in the case of the quickclay. This is, however much less frequently encountered than quickclay. While the latter can be considered as a reasonably well understood instance of irreversible changes in a granular system [261, 262], quicksand still bears an aura of legend due to its poor reproducibility.

We should finally note that many other phenomena concerning unsaturated soils are far from understood. Due to the enormous complexity of these systems, textbooks on soil mechanics must in large portions resort to phenomenological descriptions and mere classifications of soils, and of the dynamic phenomena associated with them. A detailed understanding is so far out of reach in many cases. The same is true for other examples of cohesive granular matter which are not directly derived from nature.

The present book has tried to introduce the reader to a physical perspective which focuses on the identification of commonalities, and thereby to the identification of basic underlying mechanisms. While it is rather common to collect as much knowledge as possible on such very complex systems, our approach rather aims at identifying what one does *not* need to know. This is sometimes identical, or at least runs in parallel, to identifying questions to the system which can be answered on the basis of particularly easily collected data, but still lead to the desired insight. We demonstrated this approach for just a few cases and problems, with the emergence of the Great Gamsberg in Namibia as a quite spectacular example from geology. The author believes that there is still a lot to be discovered by treating cohesive granular systems as members of the soft matter family, and looking at them with the eyes of a physicist.

List of Symbols

Symbol	Description
A	Hamaker constant
Bo	Bond number
Ca	capillary number ($\mathsf{Ca} = v\eta/\gamma$)
Cb	capillary bridge number (capillary number of the contact line)
Ch	cohesion number
d_c	vertical distance of the contact lines of a capillary bridge
D	number of dimensions of the system
D_f	fractal dimension
E	elasticity parameter of a wet granulate ($\mathsf{E} = YR/\gamma(1-\nu^2)$)
E_{cb}	energy required to rupture a capillary bridge
E_f	kinetic energy after collision ('final')
E_i	kinetic energy before collision ('initial')
E_{vib}	kinetic energy injected by vibration
E^*	non-dimensional agitation energy ($E^* = E_{\mathrm{vib}}/E_{cb}$)
F_{cb}	force exerted by a capillary bridge
$\tilde{F}_{cb}$	non-dimensional capillary bridge force ($F_{cb}/2\pi R\cos\theta$)
$F^\perp$	normal force
$F_\parallel$	tangential force
g	acceleration due to gravity
g_c	value of the two-point correlation function at particle contact
h	(average) liquid film thickness
H	mean curvature of a surface
$\mathcal{H}$	Hamiltonian
I	inertial number (of a slurry)

Symbol	Description
J	viscous number (of a slurry)
k	average number of capillary bridges in contact with a grain
k_B	Botzmann's constant
k_c	average number of true contacts of a grain with neighbors
Kn	Knudsen number
$\mathcal{L}$	Lagrangian
l_m	mean free path
m	mass of a grain
m_{m}	mass of a molecule
Ma	Mach number
n	number density
n_l	number density of molecules in the liquid
p	probability density
p	pressure
p_f	momentum after collision ('final')
p_i	momentum before collision ('initial')
p_{int}	internal pressure due to capillary forces
$\mathrm{p_r}$	probability to rupture a bridge through unity shear
p_s	saturated vapor pressure
p_L	Laplace pressure
$\mathbf{P}$	pressure tensor
P_{ij}	pressure tensor components
P_{inj}	injected power
P_{diss}	dissipated power
Pd	polydispersity
$\mathbf{q}$	vector of heat flux
r	Wenzel's roughness parameter
R	radius of a grain
Re	Reynolds number
s	separation of grain surfaces
$\tilde{s}$	non-dimensional separation of grain surfaces ($\tilde{s} = s/R$)
$\hat{s}$	normalized non-dimensional separation of grains surfaces ($\hat{s} = \tilde{s}/\sqrt{\tilde{V}}$)
s_c	critical separation at which capillary bridge ruptures
Sk	skewness (i.e., asymmetry) of a distribution function
T	temperature

Symbol	Description
T_g	granular temperature
U_{cb}	capillary bridge pseudo-potential
V	liquid volume of capillary bridge
$\tilde{V}$	non-dimensional volume of a capillary bridge ($\tilde{V} = V/R^3$)
V_i	liquid volume of capillary bridge, immediately after its formation ('initial')
V_f	liquid volume of capillary bridge after long time ('final')
V_g	volume of a grain
w	liquid content ('wetness') of a wet granular gas, or a single grain
w_δ	wetness required to fill the roughness on the grains of a granular gas
W	liquid content of a wet granular pile
W_{cb}	liquid content of a granular pile at which capillary bridges start to form
W^*	liquid content of a granular pile at which liquid clusters start to form
Y	Young's modulus
β	half opening angle of a capillary bridge, as viewed from the grain center
β_0	Lagrange multiplier of energy conservation in statistical physics
γ	liquid surface tension
γ_{sl}	interfacial tension between solid and liquid
γ_{sg}	interfacial tension between solid and gas
Γ	non-dimensional peak acceleration in a vibration setting
Γ_c	(critical non-dimensional) peak acceleration at a phase transition
δ	amplitude of grain surface roughness
δ_{dil}	dilatancy parameter
δ_{ij}	Kronecker's delta
Δ	Laplace operator
$\Delta\mu$	difference from coexistence, expressed in terms of the chemical potential
ε	restitution coefficient
ζ	cooling coefficient
η	viscosity
ϑ	polar angle in spherical coordinates
θ	macroscopic contact angle

Symbol	Description
θ_Y	Young's equilibrium contact angle
θ_a	advancing contact angle
θ_r	receding contact angle
Θ_M	maximum angle in a rotating drum, at which avalanche is likely to set in
Θ_R	angle of repose
μ	friction coefficient; chemical potential
ν	Poisson's ratio
ξ	Gibbs surface excess
$\Pi(h)$	disjoining pressure
ϱ	mass density
ϱ_s	mass density of the (solid) grain material
ϱ_l	mass density of the liquid
ρ	auxiliary radial coordinate
σ_y	yield stress
σ_μ	contribution of friction to the yield stress
σ_{cb}	contribution of capillary bridges to the yield stress
$\sigma_1(z)$	average slope, $\langle|\nabla f|\rangle$, of a function f at height z
$\sigma_2(z)$	average squared slope, $\langle|\nabla f|^2\rangle$, of a function f at height z
φ	azimuthal angle in polar coordinates
ϕ	volume fraction of grains
ϕ_j	maximum volume fraction of random packing ($\phi_j \approx 0.64$ for equally sized spheres)
ϕ_{do}	volume fraction at dilatancy onset
$\Phi(h)$	effective interface potential
ω	angular frequency
∂	partial derivative; boundary of a set
∂_y	partial derivative with respect to the variable y
∇	nabla operator

Bibliography

[1] R. J. Donelly (1991). *Quantized Vortices in Helium II* (Cambridge University Press, Cambridge, UK).

[2] R. P. Feynman (1957). *Rev. Mod. Phys.* **29**, p. 205.

[3] H.-J. Butt, M. Kappl (2010). *Surface and Interfacial Forces* (Wiley-VCH, Weinheim).

[4] J. Grotzinger, T. H. Jordan, F. Press, R. Siever (2007). *Understanding Earth*, 5th edn. (W. H. Freeman and Co., New York).

[5] E. C. Spiker, P. L. Gori (2003). *National Landslide Hazards Mitigation Strategy – A Framework for Loss Reduction* (U.S. Dept. of the Interior, Reston, Virginia).

[6] R. Dikau *et al.* (1996). *Landslide Recognition* (Wiley, New York).

[7] I. Farkas, D. Helbing, T. Vicsek (2002). *Nature* **419**, p. 131.

[8] B. Turnbull (2011). *Phys. Rev. Lett.* **107**, p. 258001.

[9] I. U. Vakarelski *et al.* (2009). *Langmuir* **25**, p. 13311.

[10] R. Tolman (1918). *Phys. Rev.* **11**, p. 261.

[11] E. L. Grossman, T. Zhou, E. Ben-Naim (1997). *Phys. Rev. E* **55**, p. 4200.

[12] N. Brilliantov, T. Pöschel (2004). *Kinetic Theory of Granular Gases* (Oxford Univ. Press, Oxford).

[13] S. Ogawa, A. Uememura, N. Oshima (1980). *J. Appl. Math. Phys.* **31**, p. 483.

[14] N. F. Carnahan, K. E. Starling (1969). *J. Chem. Phys.* **51**, p. 635.

[15] L. D. Landau, E. M. Lifshitz (1986). *Theory of Elasticity* (Butterworth, Oxford, UK).

[16] B. N. J. Persson (2000). *Sliding Friction* (Springer, Heidelberg).

[17] S. F. Foerster, M. Y. Louge, H. Chang, K. Allia (1994). *Phys. Fluids* **6**, p. 1108.

[18] L. Labous, A. D. Rosato, R. N. Dave (1997). *Phys. Rev. E* **56**, p. 5717.

[19] J. M. Lifshitz, H. Kolsky (1964). *J. MeC. Phys. Solids* **12**, p. 35.

[20] C. G. Knight, M. V. Swain, M. M. Chaudhri (1977). *J. Mater. Sci.* **12**, p. 1573.

[21] K. D. Supulver, F. Bridges, D. N. C. Lin (1995). *Icarus* **113**, p. 188.

[22] B. Imre, S. Räbsamen, S. M. Springman (2008). *Computers & Geosciences* **34**, p. 339.

[23] P. K. Haff (1983). *J. Fluid Mech.* **134**, p. 401.

[24] P. Yu, M. Schröter, M. Sperl (2020). *Phys. Rev. Lett.* **124**, p. 208007.

[25] K. Sela, I. Goldhirsch (1996). *Phys. Fluids* **9**, p. 856.

[26] J. Rajchenbach (2000). *Adv. Phys.* **49**, p. 229.

[27] I. Goldhirsch, G. Zanetti (1993). *Phys. Rev. Lett.* **70**, p. 1619.

[28] I. Goldhirsch, M.-L. Tan, G. Zanetti (1993). *J. Sci. Computing* **8**, p. 1.

[29] L. P. Kadanoff (1999). *Rev. Mod. Phys.* **71**, p. 435.

[30] D. van der Meer, K. van der Weele, D. Lohse (2002). *Phys. Rev. Lett.* **88**, p. 174302.

[31] J. Eggers (1999). *Phys. Rev. Lett.* **83**, p. 5322.

[32] K. van der Weele, D. van der Meer, M. Versluis, D. Lohse (2001). *Europhys. Lett.* **53**, p. 328.

[33] R. Mikkelsen, D. van der Meer, K. van der Weele, D. Lohse (2002). *Phys. Rev. Lett.* **89**, p. 214301.

[34] J. Duran (2000). *Sands, Powders, and Grains* (Springer, New York).

[35] H. Hinrichsen, D. E. Wolf (2004). *The Physics of Granular Media* (Wiley-VCH, Weinheim).

[36] P. Chaudhuri *et al.* (2005). *Phys. Rev. Lett.* **95**, p. 248301.

[37] A. Fingerle, S. Herminghaus (2008). *Phys. Rev. E* **77**, p. 011306.

[38] S. R. Nagel (1992). *Rev. Mod. Phys.* **64**, p. 321.

[39] G. Constantini, A. Puglisi, U. Marini Bettolo Marconi (2007). *Phys. Rev. E* **75**, p. 061124.

[40] L. Trujillo, H. J. Herrmann (2003). *Physica A* **330**, p. 19.

[41] P. Tegzes (2002). *Stability, Avalanches, and Flow in Dry and Wet Granular Materials* (PhD thesis, University of Budapest).

[42] P. Tegzes *et al.* (1999). *Phys. Rev. E* **60**, p. 5823.

[43] P. Tegzes, T. Vicsek, P. Schiffer (2002). *Phys. Rev. Lett.* **89**, p. 094301.

[44] K. M. Hill *et al.* (1999). *Proc. Nat. Acad. Sci.* **96**, p. 11701.

[45] G. Seiden, M. Ungarish, S. G. Lipson (2005). *Phys. Rev. E* **72**, p. 021407.

[46] P. Claudin, B. Andreotti (2006). *Earth PLanet. Sci. Lett.* **252**, p. 30.

[47] R. D. Lorenz *et al.* (2010). *Icarus* **205**, p. 719.

[48] B. Andreotti, S. Douady (2002). *Phys. Fluids* **14**, p. 415.

[49] C. Groh, I. Rehberg, C. Kruelle (2009). *New J. Phys.* **11**, p. 023014.

[50] F. Charru, B. Andreotti, P. Claudin (2013). *Annu. Rev. Fluid Mech.* **45**, p. 469.

[51] G. Schneider (2008). *The Roadside Geology of Namibia* (Gebrüder Bornträger, Stuttgart).

[52] R. Wittig (1976). *Geol. Rundsch.* **65**, p. 1019.

[53] E. Horsthemke (1992). *Fazies der Karoosedimente in der Huab-Region, Damaraland, NW-Namibia* (PhD thesis, GAGP, Göttingen).

[54] R. Kuper, S. Kröpelin (2006). *Science* **313**, p. 803.

[55] B. van Bocxlaer, D. Verschuren, G. Schettler, S. Kröpelin (2011). *J. Quat. Sci.* **26**, p. 433.

[56] R. D. Cody (1979). *J. Sedim. Petrol.* **49**, p. 1015.

[57] R. D. Cody, A. M. Cody (1988). *J. Sedim. Petrol.* **58**, p. 247.

[58] Ch. Plet-Lajoux, G. Monnier, G. Pedro (1971). *C. R. Acad. Sc. Paris D* **272**, p. 3017.

[59] P. A. Furley, R. Zouzou (1989). *Scot. Geogr. Mag.* **105**, p. 30.

[60] M. R. Rosen, J. K. Warren (1990). *Sedimentology* **37**, p. 983.

[61] M. Kastner (1970). *Amer. Mineralogist* **55**, p. 2128.

[62] E. F. McBride, H. Honda, A. A. Abdel-Wahab, S. Dworkin, T. A. McGilvery (1992). *Gulf Coast Assoc. Geol. Soc.* **XLII**, p. 543.

[63] I. F. Macdonald, M. S. El-Sayed, K. Mow, F. A. L. Dullien (1979). *Ind. Eng. Chem. Fundam.* **18**, p. 199.

[64] F. Boyer, E. Guazzelli, O. Pouliquen (2011). *Phys. Rev. Lett.* **107**, p. 188301.

[65] M. Trulsson, B. Andreotti, P. Claudin (2012). *Phys. Rev. Lett.* **109**, p. 118305.

[66] C. Bonnoit, T. Darnige, E. Clement, A. Lindner (2010). *J. Rheol.* **54**, p. 65.

[67] I. M. Krieger, T. J. Dougherty (1957). *Trans. Soc. Rheol.* **3**, p. 137.

[68] P. G. Rognon, I. Einav, C. Gay (2011). *J. Fluid Mech.* **689**, p. 75.

[69] H. Schlichting, K. Gersten (2000). *Boundary Layer Theory* (Springer, Berlin).

[70] M. Kittel (2015). Sedimentographie und diagenese der etjo- und twyfelfontein-formation in nw namibia.

[71] R. Siever (1962). *J. Geo.* **70**, p. 127.

[72] B. G. Hoal, L. M. Heaman (1995). *Communs geol. Surv. Namibia* **10**, p. 83.

[73] I. Goldhirsch (2003). *Annu. Rev. Fluid Mech.* **35**, p. 267.

[74] T. Pöschel, D. E. Wolf (eds.) (2012). *Granu. Matter* **14**, pp. 77–301.

[75] N. Brilliantov (ed.) (2011). *Math. Model. Nat. Phenom.* **6**, pp. 1–218.

[76] R. A. Bagnold (1941). *The Physics of Blown Sand and Desert Dunes* (Methuen, London).

[77] R. A. Bagnold (1954). *Proc. Roy. Soc. London A* **225**, p. 49.

[78] M. L. Hunt, R. Zenit, C. S. Campbell, C. E. Brennen (2002). *J. Fluid Mech.* **452**, p. 1.

[79] H. J. Herrmann, S. Luding (1998). *Continuum Mech. Thermodyn.* **10**, p. 189.

[80] H. M. Jaeger, S. R. Nagel (1996). *Rev. Mod. Phys.* **68**, p. 1259.

[81] P.-G. De Gennes (1999). *Rev. Mod. Phys.* **71**, p. S374.

[82] I. S. Aranson, L. S. Tsimring (2006). *Rev. Mod. Phys.* **78**, p. 641.

[83] E. J. Hopfinger (1983). *Annu. Rev. Fluid Mech.* **15**, p. 47.

[84] P. Beghin, G. Brugnot (1983). *Cold Regions Sci. Technol.* **8**, p. 67.

[85] B. Knoller, C. Buckee (2000). *Sedimentology* **47**, p. 62.

[86] J. Etienne, M. Rastello, E. J. Hopfinger (2006). *C. R. Mechanique* **334**, p. 545.

[87] T. Young (1805). *Philos. Trans. Roy. Soc. London* **95**, p. 65.

[88] S. Herminghaus *et al.* (1997). *Annal. Phys.* **6**, p. 425.

[89] C. Rascon, A. Parry (2000). *Nature* **407**, p. 986.

[90] R. Seemann *et al.* (2005). *Proc. Nat. Acad. Sci.* **102**, p. 1848.

[91] R. N. Wenzel (1936). *Ind. and Eng. Chem.* **28**, p. 988.

[92] D. Bonn, D. Ross (2001). *Rep. Prog. Phys.* **64**, p. 1085.

[93] S. Herminghaus (2012). *Eur. Phys. J. E* **35**, p. 43.

[94] R. Dufour. C. Semprebon, S. Herminghaus (2016). *Phys. Rev. E* **93**, p. 032802.

[95] K. Brakke (1996). *Phil. Trans. Roy. Soc. A* **354**, p. 2143.

[96] P.-G. De Gennes, F. Brochard, D. Quere (2004). *Capillarity and Wetting Phenomena: Drops, Bubbles, Pearls, Waves* (Springer, New York).

[97] D. Bonn, J. Eggers, J. Indekeu, J. Meunier, E. Rolley (2009). *Rev. Mod. Phys.* **81**, p. 739.

[98] M. O. Robbins, D. Andelman, J. F. Joanny (1991). *Phys. Rev. A* **43**, p. 4344.

[99] R. Netz, D. Andelman (1997). *Phys. Rev. E* **55**, p. 687.

[100] S. Herminghaus, M. Brinkmann, R. Seemann (2008). *Annu. Rev. Fluid Mech.* **38**, p. 101.

[101] S. Herminghaus (2012). *Phys. Rev. Lett.* **109**, p. 236102.

[102] C. Huh, L. E. Scriven (1970). *J. Coll. Int. Sci.* **35**, p. 85.

[103] L. Bocquet, F. Restagno, E. Charlaix (2004). *Eur. Phys. J. E* **14**, p. 177.

[104] J. L. Barrat, L. Bocquet (1999). *Phys. Rev. Lett.* **82**, p. 4671.

[105] T. D. Blake (2006). *J. Coll. Int. Sci.* **299**, p. 1.

[106] J. G. Petrov, P. G. Petrov (1998). *Langmuir* **14**, p. 2490.

[107] M. N. Popescu, G. Oshanin, S. Dietrich, A.-M. Cazabat (2012). *J. Phys. Condens. Matter* **24**, p. 243102.

[108] L. Bocquet, E. Charlaix, S. Ciliberto, J. Crassous (1998). *Nature* **396**, p. 732.

[109] N. Fraysse, H. Thomé, L. Petit (1999). *Eur. Phys. J. B* **11**, p. 615.

[110] B. J. Ennis, G. Tardos, R. Pfeffer (1991). *Powder Technology* **65**, p. 257.

[111] T. G. Mason, A. J. Levine, D. Ertaş, T. C. Halsey (1999). *Phys. Rev. E* **60**, p. R5044.

[112] W. Losert, J.-C. Geminard, S. Nasuno, J. P. Gollub (2000). *Phys. Rev. E* **61**, p. 4060.

[113] C. Feng, A. B. Yu (2000). *Journal of Colloid and Interface Science* **231**, p. 136.

[114] G. D'Anna (2000). *Phys. Rev. E* **62**, p. 982.

[115] D. Geromichalos, M. M. Kohonen, F. Mugele, S. Herminghaus (2003). *Phys. Rev. Lett.* **90**, p. 168702.

[116] F. M. Orr, L. E. Scriven, A. P. Rivas (1975). *J. Fluid Mech.* **67**, p. 723.

[117] J. Crassous, M. Cicotti, E. Charlaix (2011). *Langmuir* **27**, p. 3468.

[118] M. M. Kohonen *et al.* (2004). *Physica A* **339**, p. 7.

[119] C. D. Willett, M. J. Adams, S. A. Johnson, J. P. K. Seville (2000). *Langmuir* **16**, p. 9396.

[120] P. Moeller, D. Bonn (2007). *Europhys. Lett.*

[121] E. T. Owens, K. E. Daniels (2012). *Soft Matter* **9**, p. 1214.

[122] P. R. Nayak (1973). *Wear* **26**, p. 305.

[123] T. C. Halsey, A. J. Levine (1998). *Phys. Rev. Lett.* **80**, p. 3141.

[124] D. J. Hornbaker *et al.* (1997). *Nature* **387**, p. 765.

[125] F. Restagno, E. Charlaix, H. Gayvallet, C. Ursini (2002). *Phys. Rev. E* **66**, p. 021304.

[126] D. Geromichalos (2004). *Dynamik feuchter Granulate* (PhD thesis, University of Ulm).

[127] E. Riedo, F. Levy, H. Brune (2002). *Phys. Rev. Lett.* **88**, p. 185505.

[128] L. Zitzler, S. Herminghaus, F. Mugele (2002). *Phys. Rev. B* **66**, p. 155436.

[129] H. Choe *et al.* (2005). *Phys. Rev. Lett.* **95**, p. 187801.

[130] R. Szoszkiewicz, E. Riedo (2005). *Phys. Rev. Lett.* **95**, p. 135502.

[131] D. B. Asay, M. B. de Boer, S. H. Kim (2010). *J. Adhesion Sci. Technol.* **24**, p. 2363.

[132] Z. S. Khan, A. Steinberger, R. Seemann, S. Herminghaus (2011). *New J. Phys.* **13**, p. 053041.

[133] D. Langbein (2002). *Capillary Surfaces* (Springer, Berlin).

[134] H.-J. Butt, M. Kappl (2009). *Adv. Coll. Int. Sci.* **146**, p. 48.

[135] H.-J. Butt, M. Makowski, M. Kappl, A. Ptak (2011). *KONA Powder and Particle Journal* **29**, p. 53.

[136] M. Farshchi-Tabrizi *et al.* (2006). *Langmuir* **22**, p. 2171.

[137] D. Huycke (2010). *The Metamorphic Ornament: Re-Thinking Granulation. Een onderzouek naar de transformatiemogelijkheden van granulatie naar sculpturaal zilverwerk* (PhD thesis, K. U. Leuven/U. Hasselt).

[138] S. M. Iveson, J. D. Litster, K. Hapgood, B. J. Ennis (2001). *Powder Technology* **117**, p. 3.

[139] M. Tokuyama, K. Kawasaki (1984). *Phys. Lett. A* **100**, p. 337.

[140] L. D. Landau, E. M. Lifshitz (1987). *Fluid Mechanics* (Pergamon Press, Oxford, UK).

[141] G. Lian, C. Thornton, M. J. Adams (1998). *Chem. Eng. Sci.* **53**, p. 3381.

[142] S. J. R. Simons, R. J. Fairbrother (2000). *Powder Technol.* **110**, p. 44.

[143] P. Rynhart, R. McKibbin, R. McLachlan, J. R. Jones (2002). *Res. Lett. Inf. Math. Sci.* **3**, p. 199.

[144] Ch.-C. Liao, Sh.-S. Hsiau (2010). *Powder Technol.* **197**, p. 222.

[145] M. Scheel, D. Geromichalos, S. Herminghaus (2004). *J. Phys.: Condens. Matter* **16**, p. S4213.

[146] A. Fingerle, K. Roeller, K. Huang, S. Herminghaus (2008). *New J. Phys.* **10**, p. 053020.

[147] K. Huang, M. Sohaili, M. Schröter, S. Herminghaus (2009). *Phys. Rev. E* **79**, p. 010301(R).

[148] K. Huang, K. Roeller, S. Herminghaus (2010). *Eur. Phys. J. Special Topics* **179**, p. 25.

[149] F. Gollwitzer, I. Rehberg, C. A. Kruelle, K. Huang (2012). *Phys. Rev. E* **86**, p. 011303.

[150] C. M. Donahue, C. M. Hrenya, R. H. Davis (2010). *Phys. Rev. Lett.* **105**, p. 034501.

[151] Ch.-C. Liao, Sh.-S. Hsiau, T.-H. Tsai, Ch.-H. Tai (2010). *Chem. Eng. Sci.* **65**, p. 1109.

[152] S. H. Chou, S. S. Hsiau (2011). *Powder Technol.* **214**, p. 491.

[153] S. F. Shandarin, Y. B. Zeldovich (1989). *Rev. Mod. Phys.* **61**, p. 185.

[154] E. Weinan, Yu. G. Rykov, Ya. G. Sinai (1996). *Commun. Math. Phys.* **177**, p. 349.

[155] Y. Brenier, E. Grenier (1998). *SIAM J. Numer. Anal.* **35**, p. 2317.

[156] E. Ben-Naim, S. Y. Chen, G. D. Doolen, S. Redner (1999). *Phys. Rev. Lett.* **83**, p. 4069.

[157] L. Frachebourg (1999). *Phys. Rev. Lett.* **82**, p. 1502.

[158] G. G. Joseph, R. Zenit, M. L. Hunt, A. M. Rosenwinkel (2001). *J. Fluid Mech.* **433**, p. 329.

[159] P. Gondret, M. Lance, L. Petit (2002). *Phys. Fluids* **14**, p. 643.

[160] N. V. Brilliantov, T. Pöschel, W. T. Kranz, A. Zippelius (2007). *Phys. Rev. Lett.* **98**, p. 128001.

[161] D. E. Knuth (1997). *The Art of Computer Programming* (Addison-Wesley, Reading, Mass).

[162] S. Ulrich *et al.* (2009a). *Phys. Rev. Lett.* **102**, p. 148002.

[163] S. Ulrich *et al.* (2009b). *Phys. Rev. E* **80**, p. 031306.

[164] M.-K. Müller, S. Luding (2011). *Math. Mod. Nat. Phenom.* **6**, p. 118.

[165] M. Muthukumar (1983). *Phys. Rev. Lett.* **50**, p. 839.

[166] H. G. E. Hentschel, I. Procaccia (1983). *Physica D* **8**, p. 435.

[167] X.-Z. Wang, Y. Huang (1992). *Phys. Rev. A* **46**, p. 5038.

[168] P. Meakin (1986). *Phys. Rev. A* **34**, p. 710.

[169] A. Santos, M. López de Haro, S. Bravo Yuste (1995). *J. Chem. Phys.* **103**, p. 4622.

[170] J. J. Erpenbeck, M. Luban (1985). *Phys. Rev. A* **32**, p. 2920.

[171] B. J. Alder, W. G. Hoover, D. A. Young (1968). *J. Chem. Phys.* **49**, p. 3688.

[172] J. A. Zollweg, G. V. Chester (1992). *Phys. Rev. B* **46**, p. 11186(R).

[173] A. Fingerle (2007). *Entropy Production and Phase Transitions Far from Equilibrium, with Emphasis on Wet Granular Matter* (PhD thesis, University of Göttingen).

[174] S. Strauch (2011). *Phase Transitions in Agitated Wet Granular Gases at Variable Liquid Content* (Bachelor Thesis, University of Göttingen).

[175] K. Roeller, S. Herminghaus, A. Hager-Fingerle (2012). *Comp. Phys. Comm.* **183**, p. 251.

[176] K. Roeller, S. Herminghaus (2011). *Europhys. Lett.* **96**, p. 26003.

[177] C. Semprebon, M. Scheel, S. Herminghaus, R. Seemann, M. Brinkmann (2016). *Phys. Rev. E* **94**, p. 012907.

[178] K. Roeller (2010). *Numerical Simulations of Wet Granular Matter* (PhD Thesis, University of Göttingen).

[179] K. Roeller, J. P. D. Clewett, R. M. Bowley, S. Herminghaus, M. R. Swift (2011). *Phys. Rev. Lett.* **107**, p. 048002.

[180] F. Varnik, J. Baschnagel, K. Binder (2000). *J. Chem. Phys.* **113**, p. 4444.

[181] J. P. D. Clewett *et al.* (2012). *Phys. Rev. Lett.* **109**, p. 228002.

[182] J. J. Binney, N. J. Dowrick, A. J. Fisher, M. E. J. Newman (1992). *The Theory of Critical Phenomena* (Clarendon Press, Oxford, UK).

[183] P. Glansdorff, I. Prigogine (1961). *Thermodynamic Theory of Structure, Stability, and Fluctuations* (Wiley, London).

[184] H. Haken (1977). *Synergetics* (Springer, Berlin).

[185] J. Keizer, R. F. Fox (1974). *Proc. Nat. Acad. Sci.* **71**, p. 192.

[186] R. Landauer (1975). *Phys. Rev. A* **12**, p. 636.

[187] L. de Sobrino (1975). *J. Theor. Biol.* **54**, p. 323.

[188] S. Herminghaus, M. G. Mazza (2017). *Soft Matter* **13**, p. 898.

[189] B. Altaner, S. Grosskinsky, S. Herminghaus, L. Katthän, M. Timme, J. Vollmer (2012). *Phys. Rev. E.* **85**, p. 041133.

[190] J. R. Dorfmann, P. Gaspard (1995). *Phys. Rev. E* **51**, p. 28.

[191] C. Dellago, W. G. Hoover, H. A. Posch (2002). *Phys. Rev. E.* **65**, p. 056216.

[192] A. Fingerle, V. Zaburdaev, S. Herminghaus (2005). *Phys. Rev. Lett.* **95**, p. 198001.

[193] J. S. Rowlinson (1965). *Rep. Prog. Phys.* **28**, p. 169.

[194] S. Luding (2001). *Phys. Rev. E* **63**, p. 042201.

[195] M. Cross, H. Greenside (2009). *Pattern Formation and Dynamics in Nonequilibrium Systems* (Cambridge University Press, Cambridge, UK).

[196] G. Mason (1968). *Nature* **217**, p. 733.

[197] J. D. Bernal, J. Mason (1960). *Nature* **188**, p. 910.

[198] G. D. Scott (1960). *Nature* **188**, p. 908.

[199] E. A. Peters, M. Kollmann, Th. Barenbrug, A. P. Philipse (2001). *Phys. Rev. E* **63**, p. 021404.

[200] A. Donev *et al.* (2004). *Science* **303**, p. 990.

[201] G.-J. Gao, J. Blawzdziewicz, C. S. O'Hern (2009). *Phys. Rev. E* **80**, p. 061303.

[202] M. Neudecker, S. Ulrich, S. Herminghaus, M. Schröter (2013). *Phys. Rev. Lett.* **111**, p. 028001.

[203] L. E. Silbert *et al.* (2002). *Phys. Rev. E* **65**, p. 031304.

[204] R. K. McGeary (1961). *J. Am. Ceram. Soc.* **44**, p. 513.

[205] F. M. Schaller, M. Neudecker, M. Saadatfar, G. W. Delaney, G. Schrder-Türk, M. Schröter (2015). *Phys. Rev. Lett.* **114**, p. 158001.

[206] C. Gao, B. Bhushan (1995). *Wear* **190**, p. 60.

[207] N. Maeda, M. M. Kohonen, H. K. Christenson (2001). *J. Phys. Chem. B* **105**, p. 5906.

[208] G. Holler, U. Albrecht, S. Herminghaus, P. Leiderer (1993). *Appl. Phys. Lett.* **62**, p. 2877.

[209] M. M. Kohonen, N. Maeda, H. K. Christenson (1999). *Phys. Rev. Lett.* **82**, p. 4667.

[210] H. Musil, S. Herminghaus, P. Leiderer (1993). *Surface Science Letters* **294**, p. L919.

[211] S. Herminghaus, T. Paatzsch, T. Häcker, P. Leiderer (1995). *Europhys. Lett* **31**, p. 157.

[212] R. Seemann, W. Mönch, S. Herminghaus (2001). *Europhys. Lett.* **55**, p. 698.

[213] M. Scheel (2009). *Experimental Investigations of The Mechanical Properties of Wet Granular Matter* (PhD thesis, University of Göttingen).

[214] M. Scheel *et al.* (2008a). *Nature Materials* **7**, p. 189.

[215] M. Scheel *et al.* (2008b). *J. Phys.: Condens. Matter* **20**, p. 494236.

[216] N. Lu, T.-H. Kim, S. Sture, W. J. Likos (2009). *J. Eng. Mech.* **135**, p. 1410.

[217] N. Lu, J. W. Godt, D. T. Wu (2010). *Water Resources Res.* **46**, p. W05515.

[218] D. Stauffer, A. Aharony (1995). *Perkolationstheorie* (VCH, Weinheim, Germany).

[219] P. Schall, M. van Hecke (2010). *Annu. Rev. Fluid Mech.* **42**, p. 67.

[220] R. Mani, D. Kadau, D. Or, H. J. Herrmann (2012). *Phys. Rev. Lett.* **109**, p. 248001.

[221] S. Nowak, A. Samadani, A. Kudrolli (2005). *Nature Physics* **1**, p. 50.

[222] H. Schubert (1977). *Agglomeration 77* (AIME, Interscience, New York, USA).

[223] E. Brown *et al.* (2010). *Proc. Nat. Acad. Sci.* **107**, p. 18809.

[224] B. M. Das (1993). *Principles of Soil Dynamics* (PWS-KENT Publishing, Boston).

[225] Z. Fournier *et al.* (2005). *J. Phys.: Condens. Matter* **17**, p. S 477.

[226] J. E. Fiscina *et al.* (2012). *Phys. Rev. E* **86**, p. 020103(R).

[227] F. Da Cruz, F. Chevoir, D. Bonn, P. Coussot (2002). *Phys. Rev. E* **66**, p. 051305.

[228] J. K. Mitchell, K. Soga (2005). *Fundamentals of Soil Behavior* (Wiley, New York).

[229] T. S. Majmudar, R. P. Behringer (2005). *Nature* **435**, p. 1079.

[230] M. E. Cates, J. P. Wittmer, J.-P. Bouchaud, P. Claudin (1998). *Phys. Rev. Lett.* **81**, p. 1841.

[231] B. P. Tighe, J. H. Snoojer, Th. J. H. Vlugt, M. van Hecke (2010). *Soft Matter* **6**, p. 2908.

[232] K. Dahmen, Y. Ben-Zion, J. T. Uhl (2011). *Nature Physics* **7**, p. 554.

[233] G. H. Ristow (1997). *Europhys. Lett.* **40**, p. 625.

[234] S. H. Ebrahimnazhad Rahbari, J. Vollmer, S. Herminghaus, M. Brinkmann (2010). *Phys. Rev. E* **82**, p. 061305.

[235] N. Xu, C. S. O'Hern, L. Kondic (2005). *Phys. Rev. Lett.* **94**, p. 016001.

[236] E. Khain (2007). *Phys. Rev. E* **2007**, p. 051310.

[237] M. Schulz, B. M. Schulz, S. Herminghaus (2003). *Phys. Rev. E* **67**, p. 052301.

[238] N. Mitarai, F. Nori (2006). *Adv. Phys.* **55**, p. 1.

[239] G. Gudehus (2011). *Physical Soil Mechanics* (Springer, Berlin).

[240] K. K. Rao, P. R. Nott (2008). *An Introduction to Granular Flow* (Cambridge University Press, Cambridge).

[241] H. Rumpf (1962). *Agglomeration* (AIME, Interscience, New York).

[242] H. Rumpf (1970). *Chem. Ing. Tech.* **42**, p. 538.

[243] V. Richefeu, M. S. El Youssoufi, F. Radjai (2006). *Phys. Rev. E* **73**, p. 051304.

[244] C. Thornton, M. T. Cicomocos, M. J. Adams (2004). *Powder Technol.* **140**, p. 258.

[245] T. Mikami, H. Kamiya, M. Horio (1998). *Chem. Eng. Sci.* **53**, p. 1927.

[246] S. Torquato, F. H. Stillinger (2010). *Rev. Mod. Phys.* **82**, p. 2633.

[247] P. G. Rognon, J.-N. Roux, D. Wolf, M. Naaim, F. Chevoir (2006). *Europhys. Lett.* **74**, p. 644.

[248] F. Soulié, M. S. El Youssoufi, F. Cherblanc, C. Saix (2006). *Eur. Phys. J. E* **21**, p. 349.

[249] J. Chopin, A. Kudrolli (2011). *Phys. Rev. Lett.* **107**, p. 208304.

[250] M. E. Cates, M. D. Haw, C. B. Holmes (2005). *J. Phys.: Cond. Mat.* **17**, p. S2517.

[251] F. Pacheco-Vazquez, F. Moreau, N. Vandewalle, S. Dorbolo (2012). *Phys. Rev. E* **86**, p. 051303.

[252] S. van Kao, L. E. Nielsen (1975). *J. Coll. Int. Sci.* **53**, p. 367.

[253] E. Koos, N. Willenbacher (2011). *Science* **331**, p. 897.

[254] H.-J. Butt (2011). *Science* **331**, p. 868.

[255] C. Gögelein, M. Brinkmann, M. Schröter, S. Herminghaus (2010). *Langmuir* **26**, p. 17184.

[256] S. U. Pickering (1907). *J. Chem. Soc.* **91**, p. 2001.

[257] W. Ramsden (1903). *Proc. Roy. Soc. London* **72**, p. 156.

[258] V. B. Menon, D. T. Wasan (1988). *Colloids and Surfaces* **29**, p. 7.

[259] B. P. Binks, P. D. I. Fletcher (2001). *Langmuir* **17**, p. 4708.

[260] R. Aveyard, B. P. Binks, J. H. Clint (2003). *Adv. Coll. Int. Sci.* **100-102**, p. 503.

[261] K. Rankka *et al.* (2004). *Quick clay in Sweden*, SGI Report No 65 edn. (Swedish Geotechnical Institute, Linköping).

[262] A. Khaldoun *et al.* (2009). *Phys. Rev. Lett.* **103**, p. 188301.

Index

Printed in the USA
CPSIA information can be obtained
at www.ICGtesting.com
CBHW051246031223
2319CB00001B/3